AF710600

WITHDRAWN

MUDFLOW RHEOLOGY AND DYNAMICS

INTERNATIONAL ASSOCIATION FOR HYDRAULIC RESEARCH
ASSOCIATION INTERNATIONALE DE RECHERCHES HYDRAULIQUES

IAHR AIRH

MONOGRAPH SERIES

This monograph was effected under the auspices of the IAHR Section on Fluvial Hydraulics

In the same series:

T. Takahashi, 1991. *Debris flow.*

M. Bouvard, 1992. *Mobile barrages and intakes on sediment transporting rivers.*

I. Nezu & H. Nakagawa, 1993. *Turbulence in open-channel flows.*

W. Rodi, 1993. *Turbulence models and their application in hydraulics – A state-of-the-art review* (3rd edition).

Zhaohui Wan & Zhaoyin Wang, 1994. *Hyperconcentrated flow.*

Mudflow Rheology and Dynamics

PHILIPPE COUSSOT
Formerly: Division Protection contre les Erosions, Cemagref, Grenoble
Presently: Laboratoire des Matériaux et des Structures du Génie Civil, Champs-sur-Marne

A.A. BALKEMA / ROTTERDAM / BROOKFIELD / 1997

QE
599
.A2
C687
1997

Authorization to photocopy items for internal or personal use, or the internal or personal use of specific clients, is granted by A.A. Balkema, Rotterdam, provided that the base fee of US$1.50 per copy, plus US$0.10 per page is paid directly to Copyright Clearance Center, 222 Rosewood Drive, Danvers, MA 01923, USA. For those organizations that have been granted a photocopy license by CCC, a separate system of payment has been arranged. The fee code for users of the Transactional Reporting Service is: 90 5410 693 X/97 US$1.50 + US$0.10.

Published by
A.A. Balkema, P.O. Box 1675, 3000 BR Rotterdam, Netherlands
Fax: +31.10.4135947; E-mail: balkema@balkema.nl; Internet site: http://www.balkema.nl

A.A. Balkema Publishers, Old Post Road, Brookfield, VT 05036-9704, USA
Fax: 802.276.3837; E-mail: info@ashgate.com

ISBN 90 5410 693 X

© 1997 A.A. Balkema, Rotterdam
Printed in the Netherlands

Contents

Main notations

A	Hamaker constant
α_Σ	kinetic energy coefficient (eq. 10.16)
$\boldsymbol{b}$	body force
b_x, b_y, b_z	body force components
B	Bingham number (eq. 7.16)
Ba	Bagnold number (eq. 3.20)
β	momentum coefficient (eq. 10.10)
C	torque
χ	wave number
D	diffusion coefficient for Brownian motion (translation)
or	dimensionless distance from an external part of the deposit
D_0	channel length or granular flow length
$\boldsymbol{D}$	strain rate tensor
D_{II}	second invariant of $\boldsymbol{D}$
d	sphere diameter
or	distance between particles
$d\sigma$	surface element
$d\Omega$	volume element
Δp_g	driving pressure loss
E	mean specific energy in a fluid cross-section
ε	ratio of the mean distance between two spheres to their radius
or	-1 or 1 depending on flow direction
ε_0	liquid permittivity
F	drag force due to flow around a particle
Fr	Froude number (eq. 7.12)
ϕ	solid volume fraction (concentration)
ϕ_a	clay volume fraction within the clay-water system
ϕ_c	critical (percolation) solid fraction
ϕ_i	critical sedimentation volume fraction
ϕ_m	maximum solid packing fraction
ϕ_S	sand volume fraction
ϕ_s	critical fracture volume fraction
Γ	dimensionless shear rate (eq. 3.12)
G'	elastic (storage) modulus
G''	viscous (loss) modulus

G^*		wave rigidity modulus
G		gravity number (eq. 7.12)
H_b		Herschel-Bulkley number (eq. 7.12)
$\boldsymbol{g}$		gravity force
γ		shear rate
γ^*		shear strain
H		dimensionless flow depth (=G)
He		Hedstrom number (eq. 8.1)
η		apparent viscosity
h		flow depth
h_0		plug height (asymptotic fluid depth) (eq. 8.16)
h_c		critical flow depth
h_n		normal (uniform) flow depth
i		channel slope
I		particulate inertia number (eq. 3.19)
j		imaginary unit
j		slope of the equivalent channel for which a uniform flow with h and Q is obtained
K		parameter of the Herschel-Bulkley model
k		Boltzmann constant (1.38066 10^{-23} J/K)
κ		Debye length (eq. 1.6)
L		channel width
	or	half flow width
	or	tube length
	or	characteristic length
$\mathcal{L}$		free surface width
λ		structural parameter
m		$= 1/n$
	or	particle mass
μ		viscosity
μ_0		interstitial (Newtonian) fluid viscosity
μ_B		Bingham plastic viscosity
$\boldsymbol{n}$		outer unit normal vector
n		parameter of the Herschel-Bulkley model
u, v, w		velocity components
P		wetted perimeter
Pe		Peclet number (general) (eq. 3.13)
Pe_h		hydrodynamic Peclet number (eq. 3.14)
Pe_c		colloidal Peclet number (eq. 3.15)
p		pressure
p_0		atmospheric pressure
Q		flow rate (discharge) through a cross-section
Q_0		dimensionless flow rate
θ		absolute temperature
q		flow rate by unit width
Re		Reynolds number (general) (eq. 3.16)
Re_p		particle Reynolds number (eq. 1.25)

Re_h	hydrodynamic Reynolds number (eq. 3.17)
Re_c	colloidal Reynolds number (eq. 3.18)
Re_b	"Herschel-Bulkley" Reynolds number (eq. 7.15)
R	spherical particle radius or semi-cylindrical channel radius
R_H	hydraulic radius
r	distance from central axis
ρ	suspension or fluid density
ρ_s	solid particle density
ρ_w	water density
ρ_f	interstitial fluid density
Σ	fluid surface
$\boldsymbol{\Sigma}$	local stress tensor
S	wetted area for a channelled flow
or	unit area
σ_{xx}, σ_{xy},...	components of the stress tensor ($\boldsymbol{\Sigma}$)
$\boldsymbol{T}$	extra-stress (deviatoric) tensor
T_{II}	second invariant of $\boldsymbol{T}$
T	dimensionless shear stress (τ/τ_c)
τ	shear stress
t	time
τ_c	yield stress
τ_p	wall shear stress magnitude
τ_{xx}, τ_{xy},...	components of the extra-stress tensor ($\boldsymbol{T}$)
τ_{rr}, $\tau_{r\theta}$,...	components of the extra-stress tensor in cylindrical coordinates
U	mean fluid velocity through a channel cross section
or	dimensionless mean velocity along the x direction
$\boldsymbol{u}$	local fluid velocity
u_r,u_θ,u_z	components of the velocity in cylindrical coordinates
V	relative particle velocity through a fluid
or	characteristic velocity
V_s	free surface velocity
W	dimensionless mean velocity along the z direction
Ω	fluid volume
or	rotation velocity
X,Z	dimensionless coordinates respectively along the x and z directions
y	height from channel bottom
ψ_0	potential of particle surface
y_0	thickness of the sheared region
y_G	depth of the gravity center of a wetted area
ζ	shear rate function

Muddy debris flow cutting across the fields at St-Paul de Varces (Isère, France).
[C. Deymier - P. Mériaux, 1984]

Preface

This book is intended to present current knowledge in the field of mudflows. This includes both rheological aspects (the constitutive equation of flowing mud) and flow characteristics in open channels. The type of material considered is mud at large, or, in rather more physical terms, a suspension containing a large fraction of colloidal particles, in general clay, which gives it a more or less unctuous aspect and in general controls the rheological behaviour of the whole material. It is a matter of fact that, although the exact corresponding limit of grain size distribution is not yet known, suspensions containing a very large proportion of non-colloidal solid grains behave differently from those containing a sufficient proportion of colloidal particles. Typical examples of such materials (see also Chapter 6) are mud suspensions which flow in mountain streams after heavy rain (these are called mudflows and can be considered as a subclass of debris flows), drilling muds composed of water-bentonite mixtures with oil and grain, or red muds, the waste materials from the bauxite refining industry. Under some conditions, flows of similar materials such as sewage sludge, coal slurry or fresh concrete might be described using the tools developed here for mudflows.

The topic of this book was originally motivated by practical concerns, namely an attempt to develop scientifically founded countermeasures for debris flows in the French Alps. The material in debris flows may vary a great deal at least in terms of clay type, grain size distribution or solid fraction. Here we shall only focus on the rheological and flow properties of muddy materials as defined above. This book differs from earlier IAHR monographs by Takahashi (*Debris flows*) and Wan and Wang (*Hyperconcentrated flows*) because it is mainly concerned with homogeneous, concentrated muddy mixtures and because it does not deal with natural events at all. Moreover, we are above all concerned with free surface flows with negligible thermal exchanges. An introduction to confined flows and heat transfers with clay-water mixtures considered as viscoplastic fluids has been provided by Bird et al. (1982. *Rev. Chem. Eng.*, 1, 1-70). Finally I wish to emphasize that this book deals with a rather interdisciplinary topic. Although shortcomings are apparent in current knowledge of this topic and our aim is not to provide a complete review of all sub-topics, the book seeks to show the connections between mudflow rheology and dynamics and several scientific fields. An additional aim of the book is to provide an overview of the problems that have yet to be solved in this field.

This book is separated in two parts. The first one has a fundamental character: it aims to provide all the basic elements for appreciating the constitutive equation of clay-water or mud suspensions. We start this part with a review of the physical properties of these materials (Chapter 1). Then we present the basic tools for describing flows from a general point of view (Chapter 2). We continue with an analysis of the rheological behaviour of suspensions at large, as a function of their components (Chapter 3). Shear rheometry with suspensions remains a far more difficult task than it might *a priori* appear and we review the problems that have been highlighted in literature on this subject (Chapter 4). Then we focus on rheological properties of clay-water mixtures (Chapter 5) and end this part with a review of the behaviour of other types of mud suspensions or slurries (Chapter 6). The second part concerns free surface flow problems for mud suspensions considered as yield stress fluids. This part has a more applied character, since we always try to provide straightforward, mathematical expressions of the solution to flow problems, even if, in some cases, the price to pay is that these expressions are approximate. Free surface flows whose solutions have so far only be expressed numerically are thus not considered here. The basic equations for the 3D treatment of these flows are introduced first (Chapter 7). We start the developments with the simplest problem: uniform flows (Chapter 8). We examine the question of the stability of these flows in the next section (Chapter 9). Gradually varied flows and slow transient flows are then examined respectively in Chapters 10 and 12. The force and impact that mudflows may exert on obstacles is analyzed in Chapter 11.

I could not end this introduction without expressing my gratitude to Maurice Meunier, who offered essential human and working support throughout the years I devoted to mudflows. Ten years ago, at the instigation of Raymond Mura, Maurice Meunier initiated a vast research program on debris flows. In particular he promoted the construction of a special, wide, "torrential platform" (property of LHF (Laboratoire d'Hydraulique de France, Grenoble)) and the construction of a unique large-scale rheometer device to test debris flow material in situ. He also persuaded me to do a Ph.D. at the Laboratory of Rheology, which proved to be a great experience. For the parts of this work which might be considered as significant advances it must not be forgotten that I have tried to do my best to abide by the scientific approach demonstrated by Jean-Michel Piau (currently head of this laboratory). During these years I have always been usefully encouraged by Rémi Pochat (Head of Department at Cemagref). Jean-Pierre Feuvrier and Raymond Mura as successive heads of the "Division Protection contre les Erosions" at Cemagref have also helped me enormously in my work. All the people from this Division and the service "Restauration des Terrains de montagne" were present when I needed help from them. Didier Richard presently in charge of the "Erosion-hydraulique torrentielle" research group at Cemagref and Michel Frémond, currently Head of LMSGC, made possible the practical achievement of the book. My achievements at Cemagref were also made possible by the work and ideas of several students: Pierre Cholin, Jérome Jost, Hans Krage, Emmanuel Bossan, Christophe Ancey, Stéphane Boyer, Bruno Pratx, Sébastien Proust. Special thanks to Christophe Ancey who revised the first version of the manuscript and has always been a useful source of energy within Cemagref. I would also like to thank H.J. Overbeek, IAHR Secretary General, and Pr. Di Silvio (then Pr. Scheuerlein) as Chairman of the IAHR Fluvial Hydraulic Section, who supported the publication of this book as an IAHR monograph, Pr. Yalin, who kindly directed the edition, Pr. Seminara and Pr. Takahashi who had the hard task of reviewing the book and contributed to making a great many improvements.

CHAPTER 1

Physical properties of mud suspensions

Phenomenological studies of fluid behaviour via direct measurements may be sufficient for liquids whose components are simple and fixed. For more complex fluids such as mud suspensions it is also useful to try to relate their rheological properties as closely as possible to their microstructure and physical parameters. This will help promote an overall understanding of possible types of behaviour as a function of physical characteristics which should facilitate rapid, relevant, rheological studies of mud suspensions and the modification or prediction of the behaviour of various similar systems. This "rheophysical" characterization of complex fluids is the basis of a solid rheological study.

In the following, we shall start by examining the basic components of natural mud suspensions and the main physical parameters used to characterize them. The main forms of interaction which may occur between these components, inside a mud suspension, are then reviewed. finally we shall focus on two particular phenomena which may may play an important role in mud suspensions but are generally not included within the scope a rheological study, namely Brownian motion and sedimentation.

1.1 COMPONENTS AND PHYSICAL PARAMETERS

When studying debris flow or mudflow dynamics, we are essentially interested in natural water-debris mixtures consisting of water and grains of various sizes. As will be seen later, this type of material is often basically a suspension of coarse grains in a colloidal dispersion (clay-water mixture). Many natural or industrial materials are also basically suspensions of coarse grains in (colloidal) dispersions (see Chapter 6), though they may involve other particle types. As a consequence, although we focus on suspensions of clay (as colloidal particles) and coarser sediments (as grains), the following review provides some fundamental elements and a general methodology for the rheophysical study of these similar materials.

1.1.1 *Generalities*

A natural mud suspension consists mainly of water and grains whose size ranges from about 0.1 μm to about 10 m. Other components such as air, plant or animal debris, etc., whose concentration is in general negligibly small, will not be taken into account in the following. From a structural point of view, solid grains can be separated into two main different groups which fortunately correspond approximately to two distinct size ranges: clay, up to 2 μm; grains, from 2 μm. The structure of clay particles is radically different from that of other grains. Moreover, due to their small size, water-clay particle interaction is quite different from grain-water interaction. In particular, unlike most other grains, for sufficiently slow flows (Cf Chapter 3) clay particles are fundamentally Brownian, colloidal particles. However, concerning the colloidal aspect the 2 μm limit for this separation into two classes is very approximate and arbitrary. In reality, it is likely that a fraction of small grains whose size is smaller than say 100 μm could be considered either as colloidal or non-colloidal particles depending on flow conditions and suspension characteristics. In some of our discussions it will nevertheless be helpful to neglect this problem and consider only two strictly different classes.

1.1.2 *Components*

1.1.2.1 *Water*

Water density (ρ_w) is very close to 1000 kg/m^3. It consists basically of polar H_2O molecules whose size is approximately 1 Å. Natural water also contains mineral and organic impurities and disolved gases. Even for mountain waters, which are among the purest, this implies that various types of ions and molecules may be found in suspension, depending on its history. A fraction of water adsorbed on solid particles (clay and grains) does not evaporate at temperatures lower than about 300 °C and exhibits high viscosity. The remaining water ("free water") evaporates completely around 105°C. In the following analyses, because of its particular properties, we shall consider adsorbed water as a part of clay particles, while free water will be considered as the surrounding liquid. Far from colloidal particles, free water exhibits a viscosity close to 10^{-3} Pa.s at 20 °C. The problem is more complex for the free water embedded within double-layers (see 1.2.2.2) since additional ions interacting with solid particles are dispersed within it (see below). As an initial approximation, we shall only retain that the viscosity of the free water decreases the further it is from colloidal particles.

1.1.2.2 *Clay particles*

Clay particles originate in the physical or mechanical disintegration of rocks followed by a chemical transformation. Each of them consists of several hundred mono-layers stacked on each other. For a given clay type, these mono-layers are formed by a fixed combination of aluminium-oxygen octahedral ($Al(OH)_3$) sheets and silicon-oxygen tetrahedral (SiO_2) sheets. Various clay types exist: kaolinites, illites, smectites (montmorillonites), chlorites, sepiolites, attapulgites, etc., associated with different combinations [for more details see for example Grim (1953), Caillère et

Hénin (1963) or Caillère et al. (1989)]. Within the lattices of a mono-layer, some aluminium or silicon atoms are replaced by atoms of lower positive valence, which results in an excess of negative charge. This charge may be compensated by the adsorption of cations, such as Ca^{2+}, Li^{+}, K^{+}, Fe^{2+}, on the layers, which are too large to be accommodated inside the lattice. These "exchangeable" cations play an important role in the microstructure and behaviour of the clay-water system. For most common clays, each particle has the form of a leaf (fiber or platelet) whose length is large compared to its thickness. The density of clay particles is about 2650 kg/m^3.

1.1.2.3 *Grains*

The grain group includes various particle types consisting of inert mineral and rock fragments which may be roughly distinguished by their size. Depending on the field of research, slightly different grain size classifications have been proposed in the literature but this problem has no impact on the following analysis. Thus we will arbitrarily use the following simple classification: below 2 µm, clay; from 2 µm to 20 µm, silt; from 20 µm to 200 µm, fine sand; from 0.2 mm to 2 mm, coarse sand; from 2 mm to 20 mm, gravel; from 2 cm to 20 cm, pebbles; above 200 mm, boulders (see Fig. 1.1). These grains are formed by the physical or mechanical disintegration of rocks but keep the structure of their parent rock. Their density is generally greater than 2650 kg/m^3 and smaller than 2750 kg/m^3. Depending on the geological history of the material a large variety of grain types in terms of internal structure, surface roughness, hardness, and geometrical shape, may be found. Grains have a shape which ranges from platelet to sphere. In practice, a set of grains is first characterized by its grain size distribution (see Section 1.1.3.2). Various other geological characteristics may be listed but this is beyond the scope of this book. Anyway with regards to the complexity and diversity of these characteristics, it would certainly be a hard and arduous task to systematically relate them to the rheological characteristics of associated suspensions. Moreover, in general, the basic rheological effect (viscosity increase) of the presence of grains in a suspension simply arises from the rigid volumes they occupy and which may move in relation to others (see 3.1.2). Except under some conditions (for example when the solid fraction is very close to the maximum packing fraction) their roughness, hardness, geometrical shape, etc. play a negligible role in suspension behaviour.

Clay	Silt	Fine sand	Coarse sand	Gravel	Pebbles	Boulders

2 µm 20 µm 200 µm 2 mm 2 cm 0.2 m

Increasing particle diameter →

Figure 1.1: Grain size classification. [This system is the one used by the International Society of Soil Science.]

1.1.3 *Physical parameters*

Here we list the main physical parameters which appear to have a significant effect on the behaviour of mud suspensions.

1.1.3.1 *Concentration*

Since we are dealing with the mixture of a liquid with solid particles, it is of prime importance to have an estimate of the ratio of the solid amount to the liquid amount. Various parameters are then used depending on the research field (Meunier 1991). In our case, the best approach is to consider the solid concentration, i.e. the ratio of the amount of solids to the total amount of material, in particular because it may be used directly to deduce the proportion of water. Here, instead of solid weight concentration, we recommend the use of solid volume concentration, also referred to as the solid fraction in the following, ϕ, which is the ratio of solid volume to total volume (water + solid):

$$\phi = \frac{V_s}{V_s + V_w} \tag{1.1}$$

where V_w and V_s are, respectively, the volumes of water and solid in the sample. The use of this specific parameter is motivated by the valuable observation that, in general, the behaviour of a suspension is not directly related to the density, and thus to the weight, of its elements, whereas it is strongly related to the solid volume fraction that these elements occupy. Furthermore, the viscosity of a suspension tends to infinity when the solid volume concentration tends to a critical value (see 3.1.2) which mainly depends on grain size distribution. This critical concentration, which we shall refere to as ϕ_m, corresponds to the geometrical packing limit. At this stage, all the liquid pores are so small that no solid particles can be put in any of them.*

In practice, the solid volume fraction may be determined by weighing a given amount of mixture before and after drying at 105°C until complete evaporation. This procedure makes it possible to evaporate free water only and thus is consistent with our definition of the solid phase (see Section 1.1.2.1). If we assume that all solid particles have the same density ρ_s, ϕ is computed by:

$$\phi = \frac{\dfrac{M_d}{\rho_s}}{\dfrac{M_d}{\rho_s} + \dfrac{M_w - M_d}{\rho_w}} \tag{1.2}$$

where M_w and M_d are, respectively, the initial (wet) and final (dry) sample masses. Considering the range of density values for grains and clay (see 1.1.2) we think it is

---

* From this definition it also results that ϕ_m depends on particle configuration (see Chapter 3).

preferable, for the sake of simplicity, to use as a first approximation the single value 2700 kg.m^{-3} for ρ_s. The mean density of the mixture is then:

$$\rho=\phi\rho_s + (1-\phi)\rho_w \quad (1.3)$$

1.1.3.2 *Grain size distribution*

For a granular mass, a grain size distribution is associated to a series of diameter ranges which covers exactly (without overlaping) the whole diameter range. The grain size distribution is given in terms of the numbers (or volumes or weights) of grains in each diameter range of this series. For solid grains of uniform density, weight percentage is equivalent to volume percentage. In general, the grain size distribution is expressed as the weight percentages of grains whose diameter is less than a given series of diameter. If necessary, the weight fraction for each diameter range may easily be deduced using this sort of scheme.

The grain size distribution may be determined by different methods, the most common of which are sedimentation testing, dry sieving, and wet sieving (see McIntyre and Loveday (1974)). For a given material the most appropriate technique depends on the characteristics of the material. The dry sieving method involves agitating or vibrating grains in successive sieves of decreasing mesh size. At the end of the test the grains which belong to each diameter range remain in the sieve whose mesh size is the lower boundary of this range. This method is appropriate only for sufficiently coarse granular suspensions because, within a natural dry suspension, forces of attraction may prevent a proper separation of small particles through the mesh simply by vibration.

As a consequence, wet sieving is recommended in most cases since then the solid phase is brushed and agitated in water so that grains move away from one another and pass more easily through the meshes. However, it is this still hardly possible to sieve particles with a diameter smaller than approximately 20 μm. Additionally, with this method, some particles types which look solid within the suspension may be disintegrated during sieving, which may lead to a final grain size distribution finer than the real initial one.

The sedimentation method is generally used for particles whose diameter is smaller than 40 μm but it is not possible to determine particle diameters smaller than approximately 1 μm. Here, the size of a particle is determined as a function of its stationary fall velocity in water. A small amount of material is dispersed in water (in general at a concentration of 20 g per litre). The particles in the uniform suspension obtained after stirring are then allowed to settle. From the amount of material remaining in the suspension at a depth, h, after a settling time, t, the fraction of solid particles with a settling velocity less than h/t is determined. In practice the solid concentration around a given depth is determined approximately (by drawing off a suspension sample) at different times. Formulae for the settling velocity given in Section 1.3 can then be used to relate the obtained concentration to particle diameter. In these calculations, it is assumed that solid particles instantaneously reach their steady velocity independently of the presence of other particles and as far as possible it is necessary to take into account the non-spherical nature of natural grains.

For boulders neither sieving nor sedimentation tests are appropriate and in general it is necessary to carry out a direct measurement of grain size in the field.

1.1.3.3 *Ion concentration*

Ion type and concentration play an important role in double-layer potential (see 1.2.2.2) and thus in suspension behaviour (see 3.1.5). These properties are mainly described by pH, electrolyte concentration and ion types in solution. The pH can be measured with a pH-meter. Since the electrolyte concentration and ion type are not easily measured directly, the suspension may first be purified to control these parameters during subsequent tests (Caillère et al. 1989, Zhao et al. 1991, Rand et al. 1980). Unlike purified clay-water mixtures, natural mud suspensions are often cation-saturated and their pH is close to 7 because of the various transformations they undergo at least after flow initiation in nature. Nevertheless further systematic field studies might determine mudflow pH and electrolyte changes during formation and flow processes and their effect on behaviour.

1.2 MAIN INTERACTIONS

In the following, we shall review the main forms of interaction between the various components of a mud suspension. We shall simplify the problem by only considering interaction between elements of the main different groups as distinguished above (water, clay, grains) without taking into account the specificities of the different elements of each group. The order chosen for this review is based on the idea that the suspension may be made by adding progressively coarser solid particles, in other words clay then grains to an initial quantity of water. We shall then describe each successive form of interaction, within this framework.

1.2.1 *Interaction within water*

Within water, molecules are closely packed (in comparison with a gas) and each of their positions corresponds to a balance between short-distance repulsive forces and attractive Van der Waals forces (see paragraph 1.2.2.1). Water molecules interact through momentum transfers during relative motions of close molecules. At a large scale compared to molecular scale, a large proportion of these interactions give rise to a single, albeit complex, type of interaction, referred to as "hydrodynamic" type. This means that the volume of water is so large that it can be considered as a homogeneous viscous liquid and a relative particle motion yields a force which may be computed from liquid viscosity and particle velocity. In the suspension, when the ratio of the water molecule diameter to any grain diameter is extremely small, energy dissipation within water can be considered as hydrodynamic, since the size of pores between particles is much larger than the diameter of the water molecule. However, due to the existence of double-layers (see below), interaction between small (clay) particles via water is in general more complex.

1.2.2 *Interaction between clay particles immersed in water*

Clay particles belong to the class of colloidal particles whose diameter ranges roughly from 1 nm to 10 μm (Schowalter 1989). When colloidal particles are suspended in a liquid, the obtained mixture is often referred to as a dispersion (Mewis and Spaull 1976). An important feature of colloidal dispersions is the large area-to-volume ratio of the solid matrix. Colloidal systems exhibit specific mechanical, electrical or optical properties. In the following, we are mainly concerned with the mechanical effects. The main forms of interaction between clay particles immersed in water will be considered: van der Waals attraction, Born repulsion, double-layer interaction. Other colloidal forces, which are due to solvation, hydration or particle structure and appear at small separating distances, are generally negligible within natural mud suspensions but should be taken into account for studying mud suspensions containing polymer additives or organic matter, for example.

1.2.2.1 *Van der Waals attraction*

The electric charges of the various components of a water molecule are not distributed symmetrically so that the centre of gravity of positive charges does not fall at the centre of gravity of negative charges. This molecule constitutes a permanent or transient dipole which may be attracted by other differently orientated molecules. This force is referred to as the (London-) Van der Waals attraction and is proportional to d^{-7}, where d is the mean distance between two molecules. The corresponding potential energy is thus proportional to d^{-6}.

When one considers not individual atoms or molecules but much larger solid objects, interaction can be found in thermodynamic fluctuations in the electromagnetic field in and around the particles. It thus depends on the shape, structure and components of the particles. The resulting potential energy decays much less rapidly than the energy between individual molecules or atoms. As a first approximation, it may computed assuming that attractions between all interparticle molecule pairs can be added up. As an example, the energy from the Van der Waals attraction per unit surface between two identical spheres is given as (Hamaker 1937):

$$V_a = -\frac{A}{6}\left[\frac{2}{x^2-4} + \frac{2}{x^2} + \ln\left(\frac{x^2-4}{x^2}\right)\right] \qquad (1.4)$$

where x is the ratio of the distance between particle centres to particle radius, and A the Hamaker constant, which, in fact, depends on sphere characteristics (structure, material type, etc.) and separating medium. This type of calculation is based on the assumption that electromagnetic waves propagate at infinite velocity. Thus expression (1.4) is only valid for separating distances less than about 0.1 μm. In other cases, the retardation may be taken into account and the obtained potential is lower than the one given by (1.4) (see Zeichner and Schowalter (1977) or Schowalter (1984)).

For most materials in a vacuum the Hamaker constant ranges from about 10^{-20} to 10^{-19} J. However the presence of a liquid dispersion medium rather than a vacuum (or air), between the particles, notably reduces the energy from the Van der Waals interaction. In (1.4) A must then be replaced by an effective Hamaker constant.

1.2.2.2 *Electrical double-layers*

At the surface of the mono-layers of clay particles, adsorbed exchangeable cations (see 1.1.2.2), also referred to as counter-ions, may slightly diffuse in water, while remaining attracted by, and thus close to, the particle surface. In parallel, some ions of opposite charge (co-ions) which should be repelled from the surface, will tend to get closer to the surface in order to compensate for the diffusion of counter-ions. This leads to the formation of an electric double-layer made up of the charged surface and a neutralising excess of counter-ions over co-ions distributed in a diffuse manner. The quantitative treatment of the electric double-layer presents a partly unsolved problem. The simple theory due to Gouy (1910) and Chapman (1913), and then Stern (1924) (see Shaw (1980)) provided guidance for subsequent deeper but more speculative theories in practical situations. For a more complete view of the problem the reader is referred to textbooks related to colloid science (for example Kruyt (1952), Shaw (1980) or Hunter (1987)). Here, we shall only present a simple, approximate, theoretical formula, based on Gouy's theory, for the potential energy between two identical spheres with interacting double-layers (Verwey and Overbeek 1948):

$$V_r = 2\pi R\varepsilon_0 \psi_0^2 \ln[1 + \exp(-\kappa R(x\text{-}2))] \qquad (1.5)$$

with:

$$\kappa = \sqrt{\frac{2ne^2z^2}{\varepsilon_0 k\theta}} \qquad (1.6)$$

where k is the Boltzmann constant, θ the absolute temperature, n the bulk concentration of ions, e the unit charge, z the valence of ions, ε_0 the dielectric constant of the liquid, and ψ_0 the potential along each surface (here assumed to be constant). κ, also referred to as the Debye length, is such that κ^{-1} represents a characteristic length (thickness) of the double layer. Equation (1.5) is valid for thin double-layers ($R\kappa > 10$), low surface potentials ($ez\psi_0/k\theta < 2$), and symmetrical electrolytes.

1.2.2.3 *Consequences for colloid stability*

In the vast majority of cases, Van der Waals forces are attractive. When they are more intense than all other forces between particles, the particles can be stuck to each other but, at very short distances, the Born repulsion between the electronic clouds of surface atoms and molecules prevents a closer mutual approach of particles. In this situation, particles aggregate more or less irreversibly because the energy needed to separate particles is generally larger than the viscous energy within the flow. Other forces may induce an additional long range repulsion. When its intensity is sufficiently strong this creates a potential barrier which can hardly be overcome by thermal agitation. The suspension is considered as "stable" if these repulsive forces are sufficient to maintain particles far from each other. On the contrary, without these repulsive forces, the suspension would flocculate, with aggregates forming until they attain a size of the order of a millimeter. Since the van der Waals force is rather weak at long distances the probability of one particle capturing another is much smaller for a dilute suspension than for a concentrated one.

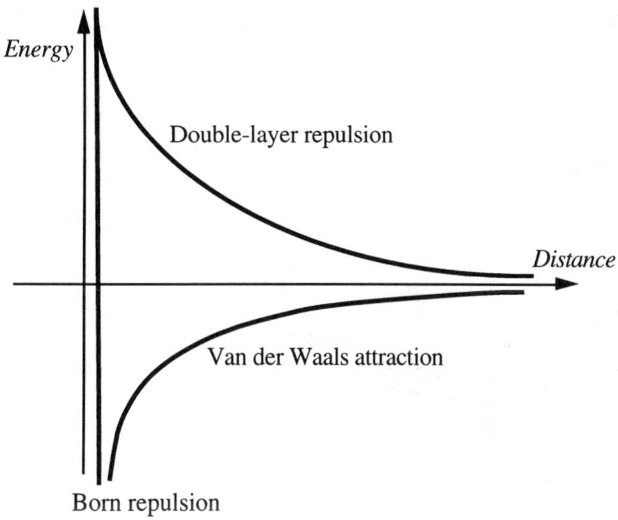

Figure 1.2: Qualitative aspects of potential associated to basic interactions as a function of distance between interacting particles.

To understand the variation of the resulting mutual interaction between clay particles within water, it may be helpful to consider the ideal case of interaction between two particles as a function of the distance that separates them (d). One must keep in mind that this only provides basic qualitative ideas on the forms of interaction within clay suspension. The reality is more complex: a clay suspension consists of a large number of clay particles, of various shapes and sizes, piled in an apparently haphazard manner, all taking part in the mutual interaction between two neighbouring particles.

The total potential energy between two particles is obtained from the sum of the forces quoted above: repulsive force at very short-distances (Born repulsion), attractive force at short distances, repulsive force at large distances. The qualitative aspect of potential energy associated with each of these forces is shown in Figure 1.2. By adding them one obtains a curve (see Fig. 1.3) which gives the form of the total potential energy as a function of the relative distance between spheres. In the absence of sufficiently large repulsive forces, the Van der Waals attraction is dominant at long distances and there exists only one equilibrium distance corresponding to the balance of Born repulsion with van der Waals attraction (curve (1)). The particles can therefore be closely aggregated at this distance and this situation is referred to as flocculation. When repulsive forces at long distances are sufficiently large, there exists a maximum for total potential energy which acts as a barrier, that, in general, prevents closer aggregation (curve (2)). In this case, colloidal dispersion is referred to as being stable. There may also exist a second equilibrium distance, much larger than the first one, for which the van der Waals attraction balances the double-layer repulsion (curve (3)). When particles are aggregated at this distance we are dealing with weak flocculation, since particles may easily (when subjected to hydrodynamic forces, for example) be moved apart from each other and later reaggregated.

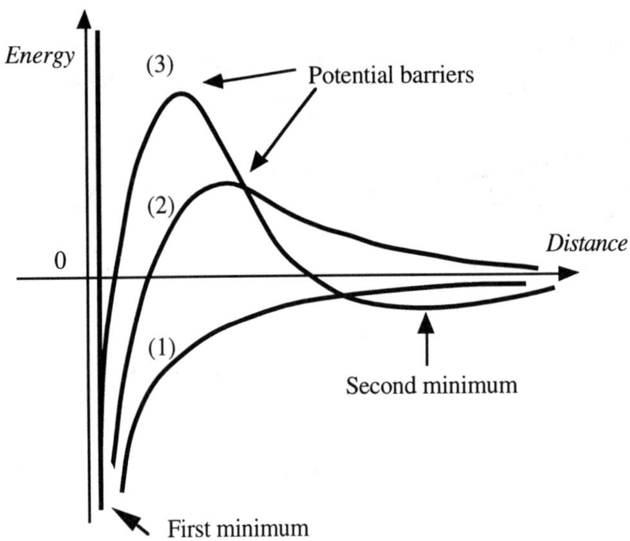

Figure 1.3: Three possible aspects of total potential energy between two particles submitted to the basic interactions considered in Figure 1.2.

As already mentioned, in practice, the flocculation or stability of a colloidal dispersion also depends on hydrodynamic forces. This can be expressed in terms of the relative values of two dimensionless numbers (Zeichner and Schowalter 1977): the repulsion number:

$$N_r = \frac{\varepsilon_0 \psi_0{}^2 R}{A} \tag{1.7}$$

and the flow number:

$$N_f = \frac{6\pi\mu_0 R^3 \gamma}{A} \tag{1.8}$$

where R is the particle radius, μ_0 the viscosity of the suspending medium and γ the shear rate (see Chapter 2). N_r expresses the ratio of the double-layer repulsion to the Van der Waals attraction. N_f expresses the ratio of the viscous drag due to flow to the Van der Waals attraction. For a large enough value of N_r and a small enough value of N_f the dispersion flocculates in a second equilibrium. Natural mud suspensions are generally in this situation. For small enough values of N_r and N_f the dispersion flocculates in a first equilibrium. For large enough values of N_f dispersion is stable.

Many factors may affect the state of the dispersion. An increase in electrolyte concentration increases n which consequently decreases V_r. Then the repulsive force decreases. A pH change also influences V_r since other ions can take the place of counter-ions. There is a specific pH value (neutral stability) for which the double-layer charge falls to zero. The type of exchangeable cations also plays a significant role on account of their specific valence (Wang et al. 1994). Though some studies have focused on these aspects and their consequences on rheological behaviour, we are a long way from fully understanding their possibilities.

1.2.2.4 *Consequences for the structure of clay-water suspensions*

For clay particles, various models for charge distribution and interaction type depending on pH have been presented in literature for different clay types. In general, for high pH values the double-layer has a negative charge whereas when pH drops below 7, this double-layer becomes positive. However, in general neutral stability is not obtained exactly for a pH of 7. Since the octohedral sheet of a clay particle is disrupted at its edge, it has been argued that the edges should possess a positive double-layer in an acid solution and a negative one in a basic solution. Under some conditions, the double-layers of the edges and faces of clay particles might be of opposite signs, which could finally give rise to edge-to-face aggregation via electrostatic attraction between particles. Thus the dependence of edge and face charges on pH may be used to control the flocculation state and rheology of the suspension. In the literature, essentially simple, ideal interaction types have been assumed and used to explain macroscopic behaviour changes with pH: electrostatic edge-to-face attraction, electrostatic face repulsion, edge-to-edge or face-to-face Van der Waals attraction (see Van Olphen (1977)). However the exact nature of interactions in specific cases remains a subject of dispute between researchers (Zhao et al 1991, Melton and Rand 1977, Chang et al. 1993, Van Olphen 1956, Street 1956, Nicol and Hunter 1970, Rand et al 1980, Granquist 1959, Chen et al 1990, M'Ewen and Pratt 1957, M'Ewen and Mould 1957).

Recent studies tend to demonstrate that the form of the interactions could be much more complex (Van Damme et al. 1991). The experimental results of Van Damme et al. (1985), Touret et al. (1990) and Morvan et al. (1994) have shown that no edge-to-face link can be observed in montmorillonite-water mixtures of which the clay particle structure would probably be closer to that of a more or less self-similar millefeuille. In other cases authors have concluded that the particles are associated with one another in tactoids - piles of two to four individual platelets separated by a few layers of water (Fripiat et al. 1982, Rosta and Von Gunten 1990). Using small-angle neutron scattering Ramsay and Lindner (1993) detected short-range correlation and alignment between the particles, in a structure which is isotropic over long ranges. Finally Pignon et al. (1996), using light scattering measurements, have concluded that the structure of laponite-water suspensions was fractal. Because of the diversity of clay types and the complexity of the problem, there is not yet a general theory available for all clays.

Since water-clay suspensions are formed of a large number of small particles able to aggregate, fractal structures may often be encountered in flow experiments. Let us simply mention some examples of such phenomena recently observed. During sedimentation, if repulsive forces are screened by the presence of an electrolyte, falling clay particles may aggregate and form clusters which grow according to the diffusion limited aggregation theory (Couch and Hinch 1991, Gouyet 1992). At a critical size, however, this aggregation may be limited by the surrounding fluid flow. The final precipitate obtained by particle deposition on the bottom can also have some fractal features (Couch and Hinch 1991, Meakin et Jullien 1991). Fractal structures may also be observed when a fluid (water or air) is injected into a mud suspension from a single source (Van Damme 1989, Lemaire et al. 1993). A fractal structure is obtained when the fluid invasion is sufficiently slow because particles have time to rearrange. At larger fluid invasion velocities, fracture lines develop within the suspension. More generally Van Damme (1995) has shown that the hydric properties (adsorption capacity, hydraulic conductivity, pore space, etc.) of soils and clays can adequately be described by scaling laws. These properties of the microstructure of clay-water systems could in particular suggest explanations to the similarity of the rheological behaviour of some clay-water systems (see Chapter 5).

1.2.3 *Interactions between grains within a clay-water suspension*

For grains of sufficient size, under usual conditions, forces induced by the colloidal effects quoted above are generally negligible compared to other forces acting on the particles. The corresponding exact critical particle size ranges in general from 1 μm to 100 μm and, as usual in such cases, there is a diameter range within which the nature and intensity of grain interactions are transitional and complex. In practice the transition range depends on flow characteristics (see Chapter 3). In the following, we will only consider particles for which colloidal effects are strictly negligible under usual conditions. We first consider the simple problem of the interactions between spheres immersed in a simple fluid. Then we consider the case of grains in a clay-water mixture.

1.2.3.1 *Interactions between two spheres*

1.2.3.1.1 Hydrodynamic interaction.

Let us consider solid spheres suspended in a perfectly homogeneous fluid. When the spheres are far from each other they interact hydrodynamically. That means that their relative motion due to suspension flow induces a flow of the interstitial fluid which dissipates viscous energy. As long as the particles remain far from each other, the energy dissipated in that way is slightly larger than the energy which would be dissipated if the particles were replaced by fluid, and the corresponding suspension viscosity is slightly larger than the fluid viscosity (see Chapter 3). When the particles are closer to each other the energy dissipated by their relative motion increases. As two spheres come closer the relative force due to fluid flow between the spheres noticeably increases. In particular, when the ratio (ε) of the distance between the spheres to their radius is much smaller than 1 and the interstitial flow remains laminar (low Reynolds number) the force may be estimated from some basic approximate formulae derived from lubrication theory. Let us consider two neighbouring particles in relative motion immersed in a Newtonian fluid whose viscosity is μ_0. The relative particle velocity may be written as the sum of two components: V_N the normal component, along the axis going through the two particle centres; and V_T the tangential component, perpendicular to the centre axis. At leading order the normal force induced by V_N is (Cf Cox (1974)):

$$f_1 = - \mu_0 \frac{1}{\varepsilon} \left(\frac{3\pi R}{2}\right) V_N \tag{1.9}$$

The tangential force that originates in the tangential component of the velocity is (Cox 1974):

$$f_2 = \mu_0 \left| \ln\varepsilon \right| \pi R \, V_T \tag{1.10}$$

and the normal force due to the tangential component of the relative velocity is (Cameron 1976):

$$f_3 = \mu_0 \sqrt{\frac{1}{\varepsilon} \left(\frac{6\pi R}{5}\right)} \, V_T \tag{1.11}$$

For uniform spheres suspended in a fluid, the mean value for ε may be estimated using the following equation:

$$\varepsilon = 2\left(\left(\frac{\phi_m}{\phi}\right)^{1/3} - 1\right) \tag{1.12}$$

assuming the maximum packing concentration (ϕ_m) is obtained by increasing the particle radius to its maximum possible value, while keeping the same configuration. In the case of crystal configurations it is easy to demonstrate that equation (1.12) is exact (see Bagnold (1954)). For other configurations the expression (1.12) only approximates to the mean value for ε.

1.2.3.1.2 Direct contact

When the separating distance falls to zero, the two grains are "in direct contact". In a simplified view of this problem we can consider that two ideal spheres in vacuum interact through friction for long periods of contact or through collision for short periods of contact. It is worth noting that, even if from a mechanical point of view, friction and collision are well defined, they are much more difficult to separate in practice: it is likely that micro-collisions occur during the friction process while some energy is dissipated via friction during rapid collisions. This results from the complex phenomena which occur at a microscopic scale: solid surfaces enter in contact via multiple discrete contact spots; deformations occur in the region of contact spots; these deformations may be either elastic, plastic, viscoelastic or viscoplastic, depending on surface topography, material properties and normal and shear forces (Bhushan 1996, Tabor 1981).

For pure friction the particles slip in relation to each other and the relative tangential force is proportional to the normal force, as predicted by the Coulomb's law. For a collision without spin between two spheres (1 and 2), sphere motions are basically predicted from the conservation of the total momentum and the conservation of the total kinetic energy along with the condition:

$$\mathbf{k}.(\mathbf{p}'_1-\mathbf{p}'_2)=-r\mathbf{k}.(\mathbf{p}_1-\mathbf{p}_2) \tag{1.13}$$

where $\mathbf{p}_{i=1,2}$ is the momentum of the particle i before the collision, $\mathbf{p}'_{i=1,2}$ is the momentum of the particle i after the collision, **k** the vector joining the particle centres and r the coefficient of restitution. For a purely elastic collision we have $r=1$. However, when the (generally extremely small) deformations due to the collision are not perfectly elastic, a certain amount of energy is dissipated. This process is taken into account thanks to the use of a coefficient of restitution smaller than 1. The case of collision between rotating spheres or oblique collision is more complicated. Then, in practice one often introduces both normal and tangential coefficients of restitution (see the review of Ancey et al. (1996)) corresponding to the transmission of relative forces respectively parallel and perpendicular to the line joining the particle centers.

For spheres immersed in a liquid, processes of interaction by direct contact are different from those occurring in a gas or in a vacuum, as soon as viscous dissipations due to interstitial fluid flow originating from the relative motion of particles are significant. In many cases the dissipation of a considerable amount of energy counteracts direct contacts.

When ε tends towards zero, the theoretical expressions for the three basic hydrodynamic forces (see 1.2.3.1.1) tend towards infinity so that as two particles come closer to each other the resistance to their relative motion becomes infinite. This should in principle preclude direct contact between the particles. In fact, when the interstitial fluid layer becomes extremely thin, various complex phenomena may play a role: since the particles are not perfectly smooth, extremely localized direct contacts can take place (Smart and Leighton 1988); with a gas as the interstitial fluid, when the separating layer is of a size comparable to the mean free path of gas molecules, the formulae above for relative forces are no longer valid (Sangani et al. 1996); with a liquid as the interstitial fluid, slight particle deformations or a slight fluid compression may occur just before contact [Cf Davis et al. 1986, Kytomaa and Schmid 1992, Lian et al. 1996, van Vliet 1996].

In any case the large forces developed when two particles approach each other should in general preclude frictional and collisional processes within suspensions of solid particles in viscous fluids except at solid fractions close to the maximum packing fraction. In that case, particles can enter into direct contact because the normal forces exerted on them as a result of the presence of other particles are relatively large. Then particles are pushed towards each other and friction may occur at low shear rates whereas collisions may occur at high shear rates.

1.2.3.2 *Interaction between grains in a clay-water mixture*
In practice, in a granular suspension, when considering collisions, in addition to the processes described above, it is necessary to take into account the particle form and roughness, the role of the interstitial fluid and possible multiple contacts (three or more particles entering in contact at the same instant). A comprehensive theoretical treatment embracing all these aspects is still lacking.

In the case of grains immersed in a clay-water suspension the problem is even more complex since here the interstitial fluid (clay + water) contains solid particles whose size may be of the same order as that of the distance between grains. In order to compute a *separating distance* between these grains, we shall consider the distance between their external shells (including their roughness). When these grains are separated by a layer of water-clay suspension which is thick compared to the clay particle diameter, they interact "hydrodynamically". Relative motion by the two particles induces macroscopic shear in the water-clay layer. Despite some works on lubricating flows of Bingham fluids (Tichy 1991), the force corresponding to relative motion between two close particles in a viscoplastic (yield stress) fluid such as most clay-water mixtures has not been established so far. When the thickness of the separating water-clay layer is roughly equal to the diameter of a clay particle, this layer can no longer be considered as a continuous medium. The exact nature of interaction under these conditions has not been determined so far. Moreover this type of interaction may be complicated by the motion of the water-clay mixture inside and outside the holes forming the rough surface of the grain. When the grains effectively come into direct contact the nature of interaction should be more or less similar to what is described in the previous Section (friction or collision) but the presence of clay particles which may be jammed in the grain holes might also complicate the processes.

To sum up, we have seen that the classification and description of the interactions, which is not particularly clear for the ideal case of two spheres within a perfect Newtonian fluid, is even more complex for two grains in a clay-water suspension. For all these systems the separation into hydrodynamic, collisional and frictional interactions remains the only, albeit approximate, approach that can be suggested at this stage.

1.2.4 *Interaction within a mud suspension*

In order to have a comprehensive view of the possible interactions between particles within the bulk, a planar fictitious scheme of the interior of a mud suspension (see Fig. 1.4) is helpful. Depending on the scale and area of observation one might distinguish different local predominant interactions as described above: water molecule interaction, clay particle interaction, friction or collision between grains, clay-water flow, etc. Nevertheless, even on the basis of a clear view of the distribution of these types of interaction, it is not *a priori* possible to distinguish the forms of interaction which would play a major role in the flow of a mud suspension. Indeed the macroscopic stress tensor depends on stress transfer through the material (Batchelor 1967) but the relative importance of transmitted stresses via the various interactions cannot be determined *a priori*. As far as possible, it is necessary to compare the energy dissipated through each type of interaction as a function of suspension and macroscopic flow characteristics (see Section 3).

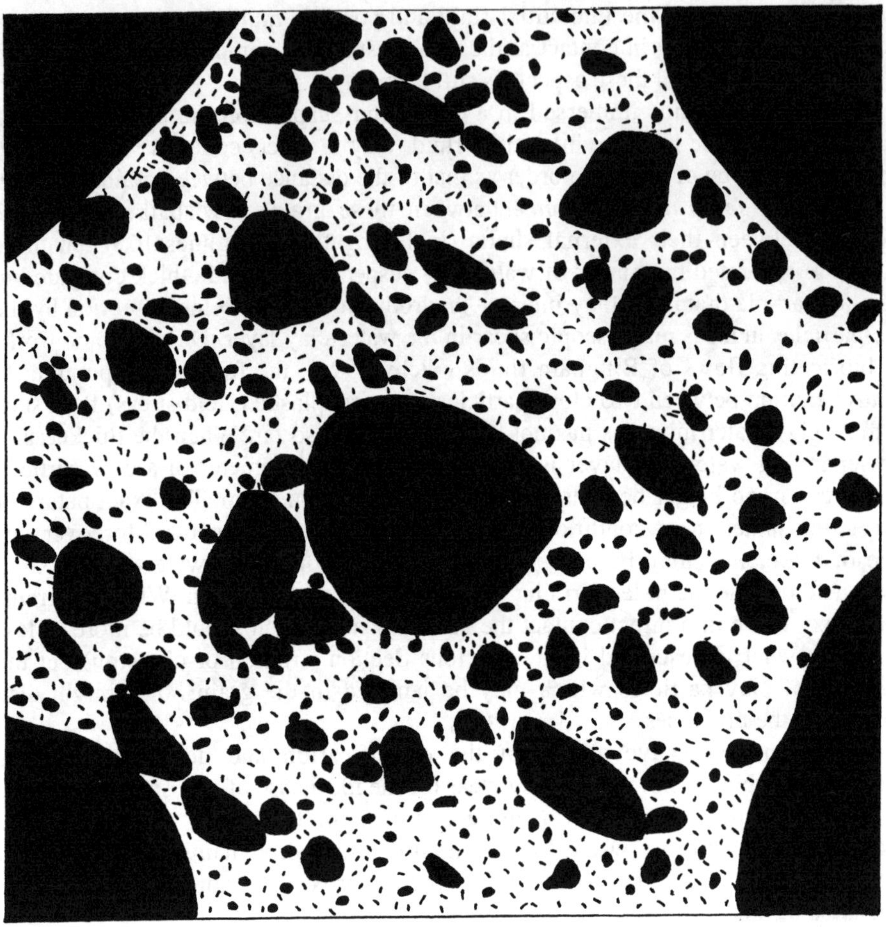

Figure 1.4: Imaginary planar view of the internal structure of a mud suspension. Clay particles (short lines), grains (points of various areas) have been dispersed at random on the plane (water in white) with no account for any model for particle interaction. The approximate length of a side of the picture is 0.1 mm.

1.3 PARTICULAR PROPERTIES

1.3.1 *Brownian motions*

Let us consider one particle suspended in a liquid whose basic elements (molecules) are small compared to this particle. Due to thermal agitation, at each instant, molecules collide with this particle. If the latter is large enough, collisions are numerous and induce an isotropic pressure. On the contrary for a small enough (colloidal) particle, the distribution and intensity of collisions fluctuate from one region to another at the particle surface. These fluctuating forces give rise to rapid, disordered, rotational and translational motions of the particles. Then, in the absence of other forces, colloidal particles diffuse across the liquid.

The diffusion due to Brownian motion can be described by Fick's law:

$$\frac{dm}{dt} = -DS\frac{d\phi}{dx} \tag{1.14}$$

where D is the diffusion coefficient and dm/dt the mass of colloidal substance diffusing in the x direction by unit time across the area S. From (1.14) it can be deduced that the average displacement during a short period t in the direction x is Δx (see for example the simplified demonstration of Shaw (1980) or the generalized approach of Russel (1981)) with:

$$\Delta x = \sqrt{2Dt} \tag{1.15}$$

and $$D = \frac{k\theta}{f} \tag{1.16}$$

where f is the frictional coefficient. For simple liquids f is the ratio of the force exerted on a particle moving through the liquid to the particle velocity. For an isolated sphere (of radius R) in an infinite volume (dilute suspension) of a Newtonian liquid, f is deduced from Stokes' law (1.19) so that we have:

$$f = 6\pi\mu_0 R \tag{1.17}$$

When the suspension is not dilute, hydrodynamic interaction due to the presence of other particles (which also diffuse) is not negligible and thus f is slightly larger than the value given by (1.17) (see Russel (1981)). As far as we know Brownian diffusion through non-Newtonian liquids has not been studied. However, as a first approximation, f may be again estimated as the ratio of the steady force (F) exerted on a sphere moving through the fluid (see 1.3.2) to particle velocity.

1.3.2 *Sedimentation*

1.3.2.1 *Sedimentation velocity*

Let us consider the problem of a solid particle of volume Ω falling at a constant velocity through a static homogeneous fluid. The particle velocity can be computed by writing the balance of the resistance to motion (F) exerted by the surrounding fluid with the force of gravity minus the force of buoyancy. Though it has so far only been systematically demonstrated for Newtonian fluids or for simple flows of Non-Newtonian fluids (see Chapter 8), we shall assume that the pressure distribution around the particle is hydrostatic, that is, it varies linearly with depth. The buoyancy force is equal to the weight of the displaced fluid (the fluid volume removed for immersing the particle) and the force balance can be written:

$$(\rho_s-\rho_f)g\Omega = F \tag{1.18}$$

where ρ_f is the density of the interstitial fluid and g the gravity. Solving (1.18) makes it possible to deduce the steady sedimentation velocity. The key problem remains how to determine F, whose value depends on the rheological parameters of the surrounding fluid and on the velocity and shape of the particle.

For a single sphere slowly falling through an infinite volume of a static Newtonian fluid, the resistance due to fluid flow around the sphere is given by the Stokes' law:

$$F = F_0 = 6\pi\mu_0 RV \tag{1.19}$$

It results from (1.18)-(1.19) that the stationary settling velocity of the sphere is:

$$V = \frac{2(\rho_s-\rho_f)gR^2}{9\mu_0} \tag{1.20}$$

Equation (1.19), and consequently (1.20), must be modified as soon as the fluid volume around the sphere cannot be considered as infinite. In particular, for a non-dilute (ϕ>1%) suspension of identical settling spheres, F is noticeably larger than F_0. This phenomenon can be represented by a hindered settling function $s(\phi)$ such that:

$$F = \frac{F_0}{s} \tag{1.21}$$

For colloidal suspensions, s will also depend on the nature of the interparticle forces. For example predominant attractive forces will increase sedimentation velocity whereas predominant repulsive forces will decrease sedimentation velocity (see Wan and Wang (1994) for a review of various possible settling phenomena within colloidal suspensions).

However even for non-colloidal systems the exact expression for s is not yet exactly known. From experiments and theory it appears that s may vary from an expression of the type:

$$1 - a\phi^{1/3} \tag{1.22}$$

to an expression of the type:

$$1 - b\phi \tag{1.23}$$

where a and b are respectively about 1 and 5. (1.22) is obtained for a more or less ordered particle configuration within the fluid whereas (1.23) is obtained for a random configuration. For example, for a random dispersion the theory of Batchelor (1972) gives b=6.55 and for a simple cubic array of particles, Hasimoto (1959) and Sangani and Acrivos (1982) found a=1.76. Nevertheless, in practice the most commonly used empirical correlation is the one attributed to Richardson and Zaki (1954) and used for ϕ up to ϕ_m:

$$(1 - \phi)^n \tag{1.24}$$

in which, according to Garside and Al-Dibouni (1977), a value of n=5.1 most accurately represents the data. Theoretical explanations of these formulae are not yet entirely satisfactory. In their complete review of this problem Davis and Acrivos (1985) concluded that there could exist in sedimenting suspensions a microscale "structure" whose origin is at present unknown. For moderately concentrated suspensions of spheres in a Newtonian fluid, the sedimentation velocity has also been estimated (Beenakker and Mazur 1985, Noetinger 1989) taking into account local and global motion (in particular, the formation of a dense front propagating downstream) which depends on the size and form of the container.

In contrast to monodisperse suspensions, in polydisperse suspensions, particles of different sizes move in relation to one another as a result of gravity, since sedimentation velocity is more or less proportional to particle surface (see (1.20)). Sedimentation processes are consequently quite complex. For a more complete review of the effects associated with sedimentation of polydisperse suspensions the reader is referred to Davis and Acrivos (1985).

When sedimenting in a Newtonian fluid an isolated non-spherical particle keeps its initial orientation when its velocity is sufficiently small, that is, if the "particle Reynolds number" defined as:

$$Re_p = \frac{\rho_f V R}{\mu_0} \tag{1.25}$$

is much smaller than 1. However, in that case, the force exerted by the fluid depends on particle orientation and form (see for example Munson et al. (1994)). When Re_p is not much smaller than 1, the particle will eventually adopt an orientation that is independent of its initial orientation. This subject is reviewed in more detail by Leal (1980) for Newtonian fluids or slightly non-Newtonian fluids, but it has never been studied in the case of yield stress fluids. From a more general point of view, for moderate or rapid particle motions (Re_p much larger than 1), the expression of the force exerted by the flow over a particle is a much more complex function of velocity, particle radius and particle roughness (see for example Munson et al. (1994)).

For a large particle falling in a fine, concentrated, mud suspension we are in general dealing with the problem of sedimentation in a non-Newtonian (yield stress) fluid (see Chapter 5 and 6). When F is known it is possible to estimate the steady fall velocity of the particle from (1.18). In Chapter 11, which concerns more generally the force exerted by a mud suspension on an obstacle, we discuss the form of F as a function of rheological parameters and particle velocity. Basically the expressions obtained for viscoplastic fluids take the form of a sum of a constant proportional to yield stress and a velocity-dependent term. *As a consequence, for yield stress fluids, a particle denser than the ambient fluid may stay at rest if the fluid yield stress is sufficiently large.*

After determining the equilibrium sedimentation velocity, one can compute the ratio of the characteristic time of sedimentation (L/V), where L is a characteristic length of the suspension sample, to test or flow duration. When this ratio is much larger than 1, sedimentation can be neglected.

1.3.2.2 *Sedimentation equilibrium*

For very dilute clay-water systems, particle interactions may be negligible. In that case it is worth noting that diffusion due to Brownian motion may tend to counteract sedimentation. For the sake of simplicity we shall assume here that these particles are spherical. For other particle shapes the qualitative form of the obtained concentration distribution (eq. (1.29)) would be identical.

For a given particle the mean velocity induced by translational Brownian diffusion in a particular direction (x) may be computed taking into account that the mass flux can be written as:

$$\frac{\mathrm{d}m}{\mathrm{d}t} = \phi S \frac{\mathrm{d}x}{\mathrm{d}t} \qquad (1.26)$$

From (1.14) and (1.26) we obtain:

$$\frac{\mathrm{d}x}{\mathrm{d}t} = V = -\frac{D}{\phi}\frac{\mathrm{d}\phi}{\mathrm{d}x} \qquad (1.27)$$

Sedimentation balances Brownian diffusion when the mean velocity induced by the latter (eq. (1.27)) counterbalances the steady velocity of the former (eq. (1.22)), which, taking into account (1.16) and (1.17), gives:

$$\frac{\mathrm{d}\phi}{\phi} = -\frac{4(\rho_s-\rho_f)g\pi R^3}{3k\theta}\mathrm{d}x \qquad (1.28)$$

(1.28) may be integrated to obtain:

$$\phi = \phi_0 \exp\left(-\frac{4(\rho_s-\rho_f)g\pi R^3 x}{3k\theta}\right) \qquad (1.29)$$

where ϕ_0 is the particle concentration for x=0 (which corresponds to the bottom level). (1.29) is the particle density distribution arising from the balance between Brownian effects and sedimentation for an ideal dilute dispersion.

1.4 REFERENCES

Ancey, C., Evesque, P., and Coussot, P. 1996. Motion of a single bead on a bead row: theoretical investigations. *J. Phys. I (France)*, 6, 725-751.

Bagnold, R.A. 1954. Experiments on a gravity free dispersion of large solid spheres in a Newtonian fluid under shear. *Proc. R. Soc. London*, A225, 49-63.

Batchelor, G.K. 1967. *Introduction to Fluid Dynamics*. Cambridge University Press, Cambridge.

Batchelor, G.K. 1972. Sedimentation in a dilute dispersion of spheres. *J. Fluid Mech.*, 52, 245-268.

Beenakker, CW.J. and Mazur, P., 1985. Many-sphere hydrodynamic interactions. IV. Wall-effects inside a spherical container. *Physica*, A 131, 311-328.

Bhushan, B. 1996. Contact mechanics of rough surfaces in tribology: single asperity contact. *Appl. Mech. Rev.*, 49, 275-299.

Caillère, S., and Hénin, S. 1963. *Minéralogie des Argiles*. Masson, Paris.

Caillère, S., Hénin, S., et Rautureau, M. 1989. *Les argiles*. Septima, Paris.

Cameron, A. 1976. *Basic Lubrication Theory*. John Wiley & Sons, New York.

Chang, S.H., Ryan, M.E., Gupta, R.K. 1993. The effect of pH, ionic strength, and temperature on the rheology and stability of aqueous clay suspensions. *Rheol. Acta*, 32, 263-269.

Chen, J.S., Cushman, J.H., and Low, P.F. 1990. Rheological behavior of Na-montmorillonite suspensions at low electrolyte concentration. *Clays Clay Minerals*, 38, 57-62.

Couch, M., and Hinch, E.J. 1991. Sedimentation, aggregation and compaction. in *Physics of Granular Media*, D. Bideau and J. Dodds (Ed.), Nova Science Publishers, Commack.

Cox, R.G. 1974. The steady motion of a particle of arbitrary shape at small Reynolds number. *J. Fluid Mech.*, 23, 625-643.

Davis, R.H., and Acrivos, A. 1985. Sedimentation of noncolloidal particles at low Reynolds numbers. *Ann. Rev. Fluid Mech.*, 17, 91-118.

Davis, R.H., Serayssol, J.M., and Hinch, E.J. 1996. The elastohydrodynamic collision of two spheres. *J. Fluid Mech.*, 163, 479-497.

Fripiat, J., Cases, J., François, M., and Letellier, M. 1982. Thermodynamic and mycrodynamic behavior of water in clay suspensions. *J. Colloid Interface Sci.*, 89, 378.

Garside, J., and Al-Dibouni, M.R. 1977. Velocity-voidage relationship for fluidization and sedimentation in solid-liquid systems. *Ind. Eng. Chem. Process Des. Dev.*, 16, 206-214.

Gouyet, J.F. 1992. *Physique et structures fractales*. Masson, Paris.

Granquist, W.T. 1959. Flow properties of dilute montmorillonite dispersions. *Proc. Nat. Conf. Clays Clay Minerals*, Sixth Conf., NRC publication, 207-219.

Grim, R.E. 1953. *Clay Mineralogy*. McGraw Hill, New-York.

Hamaker, H.C. 1937. The London-van der Waals attraction between spherical particles. *Physica*, 4, 1058-1072.

Hasimoto, H. 1959. On the periodic fundamental solutions of the Stokes equations and their application to viscous flow past a cubic array of spheres. *J. Fluid Mech.*, 5, 317-328.

Hunter, R.J. 1987. *Foundations of colloid science*. Clarendon Press, Oxford.

Kruyt, H.R. 1952. *Colloid Science*. Elsevier, Amsterdam.

Kytomaa, H.K., and Schmid, P.J. 1992. On the collision of rigid spheres in a weakly compressible fluid. *Phys. Fluids*, A4, 2683-2689.

Leal, L.G. 1980. Particle motions in a viscous fluid. *Ann. Rev. Fluid Mech.*, 12, 435-476.

Lemaire, E., Daccord, G., Levitz, P., and Van Damme, H. 1993. From viscous fingering to viscoelastic fracturing in colloidal fluids. *Phys. Rev. Lett.*, 67, 2009-2012.

Lian, G., Adams, M.J., and Thornton, C. 1996. Elastohydrodynamic collisions of solid spheres. *J. Fluid Mech.*, 311, 141-152.

McIntyre, D.S., and Loveday, J. 1974. Particle size analysis. In *Methods for analysis of irrigated soils*. J. Loveday (Ed.), Tech Comm., 54, Comm. Bur. Soils, Comm. Agric. Bur., Farham Royal, England, 88-99.

M'Ewen, M.B., and Mould, D.L. 1957. The gelation of montmorillonite. Part II: The nature of interparticle forces in sols of Wyoming bentonite. *Trans. Faraday Soc.*, 53, 548-564.

M'Ewen, M.B., and Pratt, M.I. 1957. The gelation of montmorillonite. Part I: The formation of a structural framework in sols of Wyoming bentonite. *Trans. Faraday Soc.*, 53, 535-547.

Meakin, P., and Jullien, R. 1991. Simulations of particle deposition and segregation. in *Physics of Granular Media*, D. Bideau and J. Dodds (Ed.), Nova Science Publishers, Commack.

Melton, I.E., and Rand, B. 1977. Particle interactions in aqueous kaolinite suspensions - Part I. *J. Colloid Interface Sci.*, 60, 308-336.

Meunier, M. 1991. *Eléments d'hydraulique torrentielle*. Cemagref Editions, Paris.

Mewis, J., and Spaull, A.J.B. 1976. Rheology of concentrated suspensions. *Adv. Colloid Interface Sci.*, 6, 173-200.

Morvan M., Espinat D., Lambard J. et Zemb T. 1994. *Coll. Surf. A: Physicochem. Engin. Aspects*, 82, 193-203.

Munson, B.R., Young, D.F., and Okiishi, T.H. 1994. *Fundamentals of fluid mechanics*. John Wiley & Sons, New York.

Nicol, S.K., and Hunter, R.J. 1970. Some rheological and electrokinetic properties of suspensions. *Austr. J. Chem.*, 23, 2177-2186.

Noetinger, B. 1989. A two fluid model for sedimentation phenomena. *Physica* A 157, 1139-1179.

Pignon, F., Piau, J.M., and Magnin, A. 1996. Structure and pertinent length scale of a discotic clay gel. *Phys. Rev. Lett.*, 25, 4857-4860.

Ramsay, J.D.F., and Lindner, P. 1993. Small-angle neutron scattering investigations of the structure of thixotropic dispersions of smectite clay colloids. *J. Chem. Soc. Faraday Trans.*, 89, 4207-4214.

Rand, B., Pekenc, E., Goodwin, J.W., and Smith, R.W. 1980. Investigation into the existence of edge-face coagulated structures in Na-montmorillonite suspensions. *J. Chem. Soc. Faraday Trans.*, 76, 225-235.

Richardson, J.F., and Zaki, W.N. 1954. Sedimentation and fluidization: Part I. *Trans. Inst. Chem. Eng.*, 32, 35-53.

Rosta, L., and Von Gunten, H.R. 1990. Light scattering characterization of laponite sols. *J. Colloid Interface Sci.*, 134, 397-406.

Russel, W.B. 1981. Brownian motion of small particles suspended in liquids. *Ann. Rev. Fluid Mech.*, 13, 425-455.

Sangani, A.S., and Acrivos, A. 1982. Slow flow through a periodic array of spheres. *Int. J. Multiphase Flow*, 8, 343-360.

Sangani, A.S., Mo, G., Tsao, H.K., and Koch, D.L. 1996. Simple shear flows of dense gas-solid suspensions at finite Stokes numbers. *J. Fluid Mech.*, 313, 309-341.

Schowalter, W.R. 1984. Stability and coagulation of colloids in shear fields. *Ann. Rev. Fluid Mech.*, 16, 245-261.

Schowalter, W.R. 1989. Quelques aspects rhéologiques des fluides colloïdaux. *La Houille Blanche*, 1, 32-42.

Shaw, D.J. 1980. *Introduction to Colloid and Surface Chemistry*. Butterworths, London.

Smart, J., and Leighton, D.T. 1989. Measurement of the hydrodynamic surface roughness of noncolloidal spheres. Phys. Fluids, A1, 52-60.

Street, N. 1956. The rheology of kaolinite suspensions. *Austr. J. Chem.*, 9, 467-479.

Tabor, D. 1981. Friction - The present state of our understanding. *J. Lubrication Technol.*, 103, 169-179.

Tichy, J.A. 1991. Hydrodynamic lubrication theory for the Bingham plastic flow model. *J. Rheol.*, 35, 477-496.

Touret O., Pons C.H., Tessier D. et Tardy Y. 1990. Etude de la répartition de l'eau dans des argiles saturées Mg^{2+} aux fortes teneurs en eau. *Clay Minerals*, 25, 217-233.

Van Damme H., Levitz P., Fripiat J.J., Alcover J.F., Gatineau L. et Bergaya F. 1985. Clay Minerals: A molecular approach to their fractal microstructure. in *Physics of Finely Divided Matter*, N. Boccara et M. Daoud (Ed.), Springer-Verlag, Berlin, 24-30

Van Damme, H. 1995. Scale invariance and hydric behaviour of soils and clays. *C.R. Acad. Sci. Paris*, 320, IIa, 665-681.

Van Damme, H. 1989. Flow and interfacial instabilities in Newtonian and colloidal fluids", in *The Fractal Approach to Heterogeneous Chemistry*, D. Anvir (Ed.), Wiley, Chichester, 199-226 (1989).

Van Damme, H., Lemaire, E., et Levitz, P. 1991. La physique de la boue. *Pour la Science*, 167, 82-90.

Van Olphen, H. 1956. Forces between suspended bentonite particles. *Proc. Nat. Conf. Clays Clay Minerals*, Fourth Conf., NRC publication, 204-224.

Van Olphen, H. 1977. *Introduction to Clay Colloid Chemistry*. Wiley & Sons, New-York.

van Vliet, J.H. 1996. The elastohydrodynamic lubrication force in the squeeze motion of Hertzian deformable bodies. *J. Phys.*, 6, 97-104.

Verwey, E.J.W., and Overbeek, J.Th.G. 1948. *Theory of the stability of Lyophobic colloids*. Elsevier, Amsterdam.

Wan, Z., and Wang, Z. 1994. *Hyperconcentrated flow*, IAHR monograph, Balkema, Rotterdam.

Wang, Z., Larsen, P., and Xiang, W. 1994. Rheological properties of sediment suspensions and their implications. *J. Hydr. Res.*, 32, 495-516.

Zeichner, G.R., and Schowalter, W.R. 1977. Use of trajectory analysis to study stability of colloidal dispersions in flow fields. *AIChE J.*, 23, 243-254.

Zhao, H., Low, P.F., and Bradford, J.M. 1991. Effects of pH and electrolyte concentration on particle interaction in three homoionic sodium soil clay suspensions. *Soil Sci.*, 151, 196-207.

CHAPTER 2

Introduction to rheology and fluid mechanics

In this chapter, we shall present the basic tools for a mathematical approach to mudflow rheology and dynamics. Within a certain range of solid concentrations (see Chapter 6), mudflow or debris flow materials undergo very large, approximately continuous deformations. As a result, the most immediate way of describing these phenomena is to consider them within the frame of fluid mechanics. However the materials under consideration generally offer resistance (viscosity) to shear, and thus to flow, which is much larger than water flow resistance and is *a priori* unknown. Consequently, we must start by studying the mechanical behaviour of this kind of material in greater detail. Indeed, this is the aim of rheology.

In the following pages, we shall start by providing some simple physical explanations of the phenomena under study: fluid, flow, viscosity, forces, etc. Then we shall present the various tools that may be useful for a mathematical treatment of viscous flows under the continuum hypothesis and in the absence of thermal changes: forces and stresses, mass and flow equations, and the constitutive equation. It is not our aim to provide a complete presentation of fluid mechanics and rheology. We shall restrict ourselves to reviewing the basic knowledge required to understand the developments presented in the following chapters. Readers wishing to explore these fields in greater depth should consult various textbooks such as the ones written by Batchelor (1967), Truesdell (1974) or Munson et al. (1994) for fluid mechanics, Coleman et al. (1966), Astarita and Marrucci (1974), Bird et al. (1977), Tanner (1988), Barnes et al. (1989), Ferguson and Kemblowski (1991) or Midoux (1993) for rheology, and Cherimissinoff (1988) for both aspects in conjunction with an overview of the mechanical behaviour of all types of materials.

2.1 GENERALITIES

2.1.1 *The concept of fluid*

A common definition of a fluid is a material which takes the form of the container it occupies. A more sophisticated one is a material such that the relative positions of the elements change significantly when suitabe forces, however small in magnitude, are applied to the material. Within this framework, a more complete definition making it possible to strictly distinguish solid from fluid was given by Noll (1955,

1958): the former has a preferred shape to which a configuration of its elements is associated while the second has not; moreover, fluids may remain undeformed only when subjected to an isotropic pressure. However, though these definitions made possible a clear mathematical classification, many natural or industrial materials such as mud, paint, jellies, polymers, etc., cannot be included in any of these classes (Batchelor 1967). Indeed these materials may exhibit a behaviour resembling at first sight a solid while under other conditions they will look like fluids.

In order to present a rapid but consistent picture of fluid mechanics and rheology which, in addition, will include flows of the above materials, we shall use another definition for fluids here: "materials which may undergo as large strains as possible without their mechanical properties being affected". At the very worst they should recover their initial properties after a sufficiently long rest*. This definition does not exclude the possibility of large deformations under small external forces but this is no longer a necessary condition. Furthermore a fluid has no preferred configuration, but it can stay at rest even if subjected to an anisotropic pressure. Finally this definition makes it possible to consider as fluids those materials (yield stress fluids) which behave as liquids (i.e. they undergo irreversible, continuous, inexorable deformations) under some particular conditions, for example when shear stress exceeds a critical value, but behave as solids otherwise, i.e. remain rigid or return to their initial configuration after stress release. Finally the above definition also makes it possible to refer to thixotropic materials which may restructure at rest (see 2.3.3), as fluids.

2.1.2 *The concept of viscosity*

When a fluid undergoes deformations (or flows) the positions of its various elements (configuration) change. During this process, unless the whole fluid is displaced without being deformed, some of its elements move in relation to each other. Since the deformation must be continuous, at least a few neighbouring elements are subjected to a relative motion. Either in a gas or a liquid, neighbouring particles in mean relative motion develop some forces towards each other because of momentum transfers resulting from their interaction. Thus, at microscopic scale, a certain force is needed to provoke relative motion of fluid elements. As a consequence, at macroscopic scale, a force is needed to deform the fluid: it resists flow. This resistance is associated with the notion of viscosity. More precisely, an external stress applied to a fluid tends to deform it rapidly but the internal frictional forces delay the rate of deformation and cause an equilibrium state in which a constant external stress results in a constant rate of deformation. The ratio of the stress to the rate of deformation, which is constant for simple fluids (Newtonian), is referred to as viscosity. A strict definition of viscosity requires some mathematical tools (see 2.3).

---

* We shall see that this definition cannot include some concentrated granular suspensions which, in the absence of significant Brownian motions of particles at rest, remain in a state depending on the previous flow history (Cf 3.1.3.2).

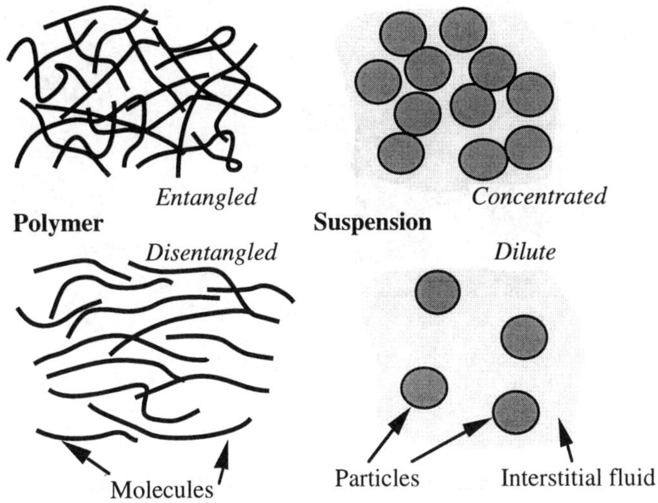

Figure 2.1: Model suspensions: dependence of viscosity on microstructure. Roughly speaking the entangled polymer solution offers more resistance to flow than the disentangled one because the various elements move more easily in relation to each other. For the same reason, the viscosity of the concentrated suspension is greater than the viscosity of the dilute suspension.

This description should help the reader to understand that, if we had sufficient knowledge of microstructural interaction within a given material, a realistic and useful relationship between the physical properties and the mechanical behaviour of the material could be established. To achieve a microstructural approach of this type for rheology we need to select the dominant types of interaction among collisions, friction, entanglements, double-layer attraction, hydrodynamic interaction, etc., between some corresponding basic fluid elements (polymer molecule, grain, clay particle, blood cell, etc.) at an appropriate scale, and study how the distribution and changes of these forms of interaction during flow influence fluid resistance. This has been achieved completely in a very few cases: Eyring's model for simple liquids (Glasstone et al. 1941), the gas kinetic theory (Chapman and Cowling 1952), the Doi-Edwards model for polymers (Doi and Edwards 1986). However, for most other fluids, the proposed models are based on speculative assumptions concerning microstructural processes and have not yet been sufficiently compared with reality. The microstructural approach to fluid behaviour remains the main challenge for rheology in the future, which should provide many improvements in industrial processes. Even if this problem is not completely solved even in the case of mud suspensions, we think it is essential to try to explain mechanical properties in terms of the physics of microstructures. We hope that this will provide a better understanding and a logical scheme for the vast subject of mudflow rheology and should make it possible to extrapolate or use some of these results for other material types.

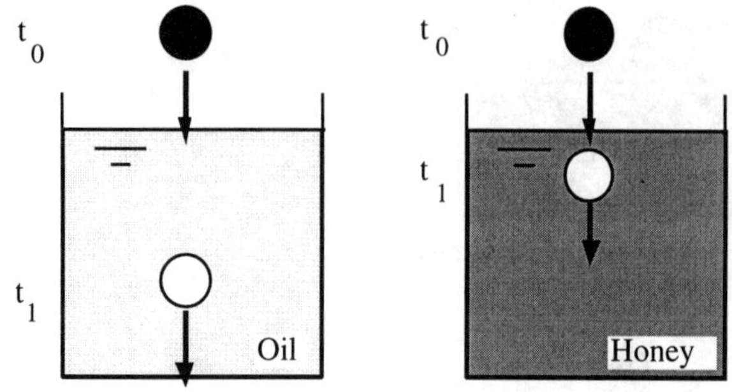

Figure 2.2: Influence of viscosity on the motion of a sphere, subject to gravitational forces, falling through various simple fluids.

It is now easy to understand why the resistance to flow (or viscosity) varies from one fluid to another (see viscosity value examples in Table 2.1): their microstructure (or more precisely their mesostructure*) and thus the interactions between their basic elements are different (Cf Fig. 2.1). For example, consider water in liquid or gaseous state. Under usual conditions, both materials are fluids. In a gas the molecules are far from each other and the viscosity is mainly due to momentum transfers via a relatively small number of random collisions during flow. For liquid water, molecules are comparatively closely packed and the resistance to flow is mainly due to momentum transfers via mean relative motions of molecules close to each other. Without going into greater detail, it is clear that the number and types of interaction and thus the corresponding viscosities will be quite different in these two cases (Cf Table 2.1).

The first macroscopic property of practical interest for civil engineers is that, under the same external conditions, a fluid with a high viscosity flows less rapidly than one with low viscosity. This can be appreciated in various situations: at normal temperatures, air is less viscous than water and will flow more easily when pushed through a pipe; a sphere dropped into oil will reach the bottom of the container faster than if it is dropped into honey (Fig. 2.2); water will flow more easily than honey on an inclined plane; etc. It should be borne in mind that, on each occasion, we tend to compare flows of different materials under "the same external conditions". As will appear more clearly in Section 2.2.6, this is because the tools of basic continuum mechanics (mass and flow equations) are formally valid and similar for all fluids and thus predict the same flow characteristics for fluids of identical viscosity (and density), under the same boundary and initial conditions.

* Throughout the book the term "microstructure" is used though it tends to systematically refer to the structure of material at microscopic scale. It would be more precise to employ the much less-known term "mesostructure" since we are in fact concerned with the material structure at a scale, generally intermediate between the macroscopic and microscopic scales, where most forms of interaction between basic fluid elements take place.

Table 2.1: Viscosity of a number of materials

Material	*Temperature* (°C)	*Viscosity* (Pa.s)
Hydrogen	0	$8.4.10^{-6}$
	100	$10.3.10^{-6}$
Air	-100	$1.16.10^{-5}$
	0	$1.71.10^{-5}$
	100	$2.18.10^{-5}$
	1000	$4.82.10^{-5}$
Water	0	$1.787.10^{-3}$
	100	$0.295.10^{-3}$
Mercury	0	$1.7.10^{-3}$
	50	$1.4.10^{-3}$
Acetone	0	$0.4.10^{-3}$
	50	$0.25.10^{-3}$
Olive oil	20	0.081
Lubricating oil	25	0.3
Glycerol	0	12
	50	1.5
Honey	25	≈ 50
Debris flow	15	10-1000*

* apparent viscosity

Some of the characteristic differences in flows for fluids of different viscosities may also be observed in the example in Figure 2.3 which has clear implications for natural free surface flows. Under the same initial conditions (fluid supply) and boundary conditions (channel geometry), the mean flow rate is higher and the flow depth is smaller for lower fluid viscosity. As a consequence, all other things being equal, in order to avoid overflow we need to devise channels with larger cross-sections for fluids with higher viscosity. Moreover, in the example in Figure 2.3, the final deposits for the first three fluids will be more or less* the same after different durations, if they are relatively simple (Newtonian) fluids under these particular flow conditions. However, for concentrated mud suspensions, the final deposit is generally quite different (see Chapter 12). If the amount of fluid supplied upstream is insufficient (though identical to others) this material will stop finally within the channel. This cannot be explained by a surface tension effect since the surface tension of mud is probably comparable to that of water. [This "yielding behaviour" has already been mentioned in the initial general discussion.] This particular experiment also demonstrates the non-linearity of some types of fluid behaviour. For simple (Newtonian) fluids, flow resistance is characterized by a single parameter to which, under simple boundary conditions, the flow velocity is more or less

* If they are all Newtonian fluids their final state will be mainly dictated by their surface tension.

proportional. With concentrated mud suspensions, motion can exhibit a singularity due to yield stress: the fluid remains static below a critical stress and flows for larger stresses. More generally the viscous properties of a fluid are not related to a single coefficient of whatever size but may vary with shear intensity or, to a greater extent, with the history of motion. In order to take these variations into account we shall speak of the "behaviour" of the fluid, or from a mathematical point of view, the "constitutive equation".

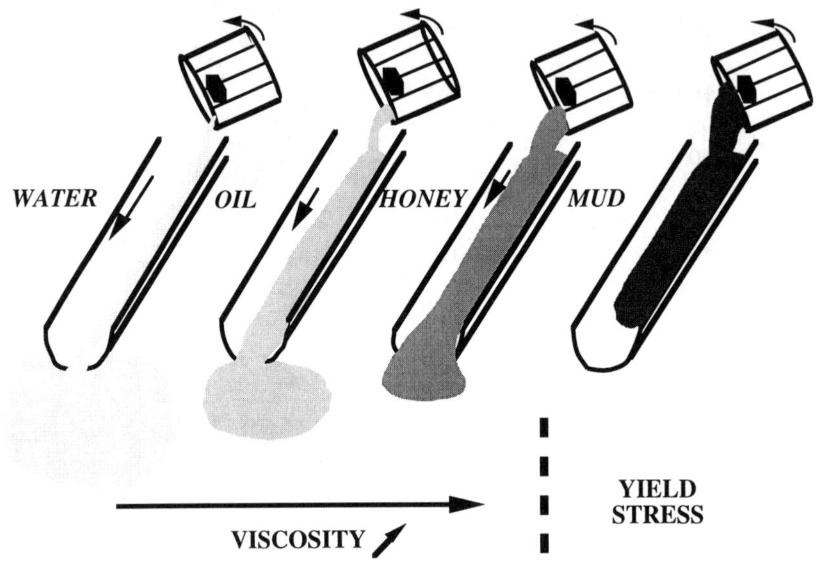

Figure 2.3: Influence of fluid behaviour on the flow characteristics of various fluids. Flow aspect at a given time after emptying a given fluid volume at the top of identical inclined, open channels.

2.2 BASIC FLUID MECHANICS

2.2.1 *The continuum hypothesis*

The basic assumption for a mathematical treatment of fluid dynamics is one of continuum: the material is assumed to be such that its physical characteristics remain continuous in time and space. Theoretically this means that these characteristics, defined by integrals on volumes or surfaces, may be continuously differentiated in time and space. In practice any material is discontinuous at least at atomic scale. Thus the continuum assumption is necessarily associated with a range of scales of observation (in time and space). For the chosen scale, the deviations from the continuity conditions must be negligible.

Under the continuum assumption, velocity variations are not localized along some specific surfaces. More precisely, at our scale of observation, the relative velocity between two close elements must be small compared to any macroscopic characteristic velocity difference. Furthermore the size of fluid elements must be small compared to any macroscopic characteristic length of the flow (in general the smallest distance between boundaries). However these conditions only constitute the conditions necessary for the validity of the continuum assumption and are not always sufficient. In practice, the continuum hypothesis must always be validated empirically by direct observations or measurements under different conditions.

Let us consider the specific case of (natural) mudflows. Here, we are dealing with concentrated mixtures of particles of different sizes and types in water that it would be natural to consider *a priori* as multiphase flows. That means that the motion of each particle might be treated separately taking into account the specific constitutive equation of this particle and its various interactions with the other elements of the mixture. This may be carried out in various ways (Drew 1983, Batchelor 1989). However this is not necessary in the case of mudflows because they can often be considered as single-phase flows at our scale of observation:

- Mudflow density is high and relatively close to the density of solid particles; coarse particle sedimentation may be neglected as a first approximation; thus, on average, the vertical velocity difference between some particles and the rest of the mixture is negligible.
- Mudflow viscosity is fairly high and so is the drag force exerted by the surrounding mixture on a given particle; here, on average, longitudinal velocity differences are negligible.
- Some large boulders, which might locally disrupt the material continuity when their diameter is not much smaller than the channel width, are transported by a mudflow; if they are not too numerous they can often be neglected as a first approximation.
- It generally appears from direct observations (by video) of mudflows that the free surface of the bulk is continuously sheared, which may be extrapolated to the entire flowing mass.

Thus, as a first approximation, a given mudflow may be treated as a single homogeneous, viscous fluid. These statements obviously fail, for example, for flows around boulders of a size comparable to the channel width, excessively liquid mudflows through which coarse grains rapidly settle, or very concentrated mud suspensions which tend to fracture when strained (see Chapter 6).

2.2.2 *Forces*

Dynamics consists in studying the rearrangment (flow) of the various elements of the material under the action of external forces. At this stage forces should be explained and defined. However the concept of force remains relatively mysterious and, in practice, hardly separable from motion since "a force is always seen as an interaction able to induce a motion". We shall not discuss this problem further. For practical purposes, we shall separate forces acting on a fluid element into two classes depending on whether they are proportional to the surface or to the volume of the element (Batchelor 1967). The first class, i.e. surface forces, includes short range forces, which have a direct molecular origin, decrease extremely rapidly with an increase in distance between interacting elements and are negligible unless there is direct contact between these elements. The second class, i.e. body forces, includes long range forces which, under usual conditions, decrease slowly with an increase in distance between interacting elements: electromagnetic, gravity or fictitious (inertial) forces.

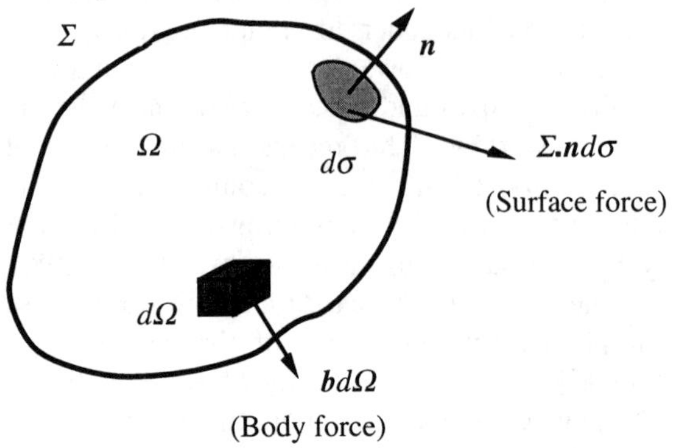

Figure 2.4: Main types of force acting within a fluid volume.

Surface forces give rise to stresses. Consider a volume Ω fixed in space and immersed in the fluid. Its external surface is Σ. One element of its external surface is $d\sigma$ (see Fig. 2.4). It can be shown (Batchelor 1967, Truesdell 1966) that the force acting on this surface may be expressed as follows:

$$(\Sigma.\boldsymbol{n})d\sigma \tag{2.1}$$

where Σ is the symmetrical, local stress tensor and $\boldsymbol{n}$ the outer unit normal vector. The proportionality, previously mentioned, of this force to the fluid element surface appears clearly in (2.1).

Now, take a small fluid element whose volume is $d\Omega$. Body forces exerted on this element are expressed as follows:

$$\boldsymbol{b}d\Omega \tag{2.2}$$

where $\boldsymbol{b}$ is the volume force density.

2.2.3 *Mass conservation (equation of continuity)*

The local fluid density and velocity are ρ and $\boldsymbol{u}$. All the variables depend on time and on the local position whose components in an orthogonal reference frame $(\boldsymbol{i},\boldsymbol{j},\boldsymbol{k})$ are (x,y,z). The equation of continuity expresses that, in the absence of mass sources, any group of fluid elements, though possibly deformed or dilated, keeps the same total mass. Mathematically this may be written as follows:

$$\frac{\partial \rho}{\partial t} + \nabla . \rho \boldsymbol{u} = 0 \tag{2.3}$$

whose developed expression is:

$$\frac{\partial \rho}{\partial t} + \frac{\partial \rho u}{\partial x} + \frac{\partial \rho v}{\partial y} + \frac{\partial \rho w}{\partial z} = 0 \tag{2.4}$$

where (u,v,w) are the components of $\boldsymbol{u}$. The second term of equation (2.3) expresses the mass change due to material flux in space while the first term expresses the local density change with time. The balance between these two types of quantities can be better understood with the help of the integral formulation of the mass conservation principle:

$$\int_{\Omega} \frac{\partial \rho}{\partial t} \mathrm{d}\Omega] = - \int_{\Sigma} (\rho \boldsymbol{u}.\boldsymbol{n}) \mathrm{d}\sigma \tag{2.5}$$

Equation (2.5) simply expresses that a mass change with time within a given volume is necessarily compensated by a mass flux through this volume.

2.2.4 *Conservation of momentum (equation of motion)*

As has already been mentioned, this principle is very closely related to the definition of force: it expresses the fundamental rule of mechanics that a force applied to a particle induces a proportional acceleration. Taking a fluid element we thus have to balance its acceleration by the sum of surface and volume forces acting on this element. This gives:

$$\rho \frac{\mathrm{d}\boldsymbol{u}}{\mathrm{d}t} = \boldsymbol{b} + \nabla . \Sigma \tag{2.6}$$

where $\frac{\mathrm{d}\boldsymbol{u}}{\mathrm{d}t} = \frac{\partial \boldsymbol{u}}{\partial t} + \boldsymbol{u}.\nabla\boldsymbol{u}$ is the material derivative of $\boldsymbol{u}$. The developed expression of (2.6) is:

$$\rho\left(\frac{\partial u}{\partial t} + u\frac{\partial u}{\partial x} + v\frac{\partial u}{\partial y} + w\frac{\partial u}{\partial z}\right) = b_x + \frac{\partial \sigma_{xx}}{\partial x} + \frac{\partial \sigma_{xy}}{\partial y} + \frac{\partial \sigma_{xz}}{\partial z} \tag{2.7a}$$

$$\rho\left(\frac{\partial v}{\partial t} + u\frac{\partial v}{\partial x} + v\frac{\partial v}{\partial y} + w\frac{\partial v}{\partial z}\right) = b_y + \frac{\partial \sigma_{xy}}{\partial x} + \frac{\partial \sigma_{yy}}{\partial y} + \frac{\partial \sigma_{yz}}{\partial z} \tag{2.7b}$$

$$\rho\left(\frac{\partial w}{\partial t} + u\frac{\partial w}{\partial x} + v\frac{\partial w}{\partial y} + w\frac{\partial w}{\partial z}\right) = b_z + \frac{\partial \sigma_{xz}}{\partial x} + \frac{\partial \sigma_{yz}}{\partial y} + \frac{\partial \sigma_{zz}}{\partial z} \tag{2.7c}$$

where $(\sigma_{xx},\sigma_{xy},...)$ are the components of the stress tensor Σ and b_x, b_y and b_z the components of $\boldsymbol{b}$. In the special case where gravity ($\boldsymbol{g}$) is the only external force applied to the fluid, we have $\boldsymbol{b}=\rho\boldsymbol{g}$. The expressions of Equations (2.4) and (2.7) in cylindrical coordinates may be found in Appendix 1.

Motion equation can also be expressed in integral form:

$$\int_{\Omega} \frac{\partial \rho \boldsymbol{u}}{\partial t} \mathrm{d}\Omega + \int_{\Sigma} \rho \boldsymbol{u}(\boldsymbol{u}.\boldsymbol{n})\mathrm{d}\sigma = \int_{\Omega} \boldsymbol{b}\mathrm{d}\Omega + \int_{\Sigma} (\Sigma.\boldsymbol{n})\mathrm{d}\sigma \qquad (2.8)$$

where the right-hand side corresponds to the sum of external forces applied to material volume while the left-hand side expresses the momentum change.

2.2.5 *Boundary and initial conditions*

In order to solve the complete set of equations (2.4) and (2.7) and thus predict flow characteristics from a given time, one needs the set of values of density and velocity at this time. These are the initial conditions.

Boundary conditions correspond to the set of values of variables at the boundary of the considered volume. In fact it is not necessary that the value of each variable be known along the boundary. It is sufficient to have, at each point of the boundary, the value of three orthogonal components among those of the stress and velocity. This expresses mathematically as follows:

For i=1,2,3 $(\Sigma.\boldsymbol{e}_i).\boldsymbol{n}$ is known along Σ_{Ti} and u_i is known along Σ_{ui}

$$\text{with } \sum_{i=1}^{3} (\Sigma_{Ti} \cup \Sigma_{ui}) = \Sigma \text{ and } \Sigma_{Ti} \cap \Sigma_{ui} = \varnothing \qquad (2.9)$$

where $\boldsymbol{e}_{i=1,2,3}$ are three orthogonal directions at a given point of Σ (they may vary from one point to another), u_i is the velocity component in direction $\boldsymbol{e}_i$, and Σ_{Ti} and Σ_{ui} are subparts of Σ.

A typical boundary condition is that of a rigid wall along which the fluid does not slip. There we have $\boldsymbol{u}$=0 along the wall surface. Another typical boundary condition is that of a free surface (interface with air). There the stress at the boundary is such that $\Sigma.\boldsymbol{n} = -p_0\boldsymbol{n}$ along this surface, where p_0 is the atmospheric pressure.

2.2.6 *Prediction of flow characteristics*

Here the first and second laws of thermodynamics might also be described as part of the mathematical description of the thermodynamical changes of the system. However we do not present these principles because very little is known concerning thermal effects in mudflows and this topic is outside the scope of this book.

Setting aside thermal effects, system changes are entirely determined by the set of equations (2.3) and (2.6) (or (2.5) and (2.8)) in conjunction with boundary and initial conditions. The solution is to be found in terms of $\boldsymbol{u}$ and Σ as a function of (x,y,z,t). This involves four scalar equations whereas there are ten unknowns: ρ, u_x, u_y, u_z, and the six components of the symmetrical stress tensor (σ_{xx}, σ_{xy}, etc.). The apparent lack of equations demonstrates that additional relationships between $\boldsymbol{u}$ and Σ are needed for solving the problem. It is here that the constitutive equation of the material appears providing a cure for this problem. The constitutive equation is directly related to material properties. Since it takes the form of equations linking each stress tensor component to velocity components, this will provide us with six additional relations and, theoretically at least, will make it possible to solve the problem. Before giving more information concerning the form and properties of this constitutive equation we wish to discuss in more detail the possibility of solving the complete system of equations.

2.2.7 *Turbulence*

Under some conditions local variables may vary in time. In practice, properties determined by measurements necessarily concern averaged values over a more or less long time. Thus motion and mass equations are valid for describing observed flow properties if these equations are also averaged over a certain duration. However, the averaged flow equations are not directly deduced from local flow equations using the corresponding averaged quantities throughout. Indeed, even if the variations with time are in fact slight fluctuations around mean values, they can significantly affect the solution for the mean variables obtained from the complete system of averaged equations. More precisely, if $\boldsymbol{u}^*$ is the average velocity, the fluctuating velocity, which can be written $\boldsymbol{u}' = \boldsymbol{u} - \boldsymbol{u}^*$, has an average value equal to zero. When averaging over the time, any linear function of $\boldsymbol{u}$ (proportional to $\boldsymbol{u}$ or to one of its derivatives) in (2.4) transforms into a linear function of $\boldsymbol{u}^*$. An analogous result is obtained for linear functions of other variables. However additional terms appear in the averaged momentum equations because of the average of the product $\boldsymbol{u}.\nabla\boldsymbol{u}$ which is equal to $\boldsymbol{u}^*.\nabla\boldsymbol{u}^* + (\boldsymbol{u}'.\nabla\boldsymbol{u}')^*$ in which the second term is *a priori* different from zero. Since this term is also *a priori* unknown, additional equations would be needed to solve the set of equations.

A useful approach is to point out that these terms can be considered as additional stress components that can be included in the averaged stress tensor. The new stress tensor includes viscous components ("laminar") originating in mean motions and additional ("turbulent") stress components originating in fluctuations. However, in spite of considerable research efforts, current knowledge of turbulence is rather poor, especially for quantitative predictions. Thus, practically, turbulent stress components and turbulence occurrence criterion (see below) can only be determined empirically. This holds for any fluid type. Turbulence has recently motivated a large number of works. Further general information on this subject may be found for example in Batchelor (1953), Landau and Lifshitz (1959), or Tritton (1988).

When the stress terms arising from these fluctuations are significant, fluid elements are agitated in various directions and vortices develop. From a physical point of view turbulence occurs when momentum transfers due to local velocity fluctuations become significant in comparison with momentum transfers due to viscous forces corresponding to mean motions. This qualitative criterion has been expressed mathematically with the help of the Reynolds number which is the ratio of a reference kinetic energy amount to a reference viscous energy dissipation. This dimensionless number also appears naturally in dimensionless flow equations (see Chapter 7). Before studying any flow it is necessary to determine whether turbulence is significant or not (see for example 3.1.4 and 8.1).

At this stage we think it may be useful to emphasize the additional complexity introduced by turbulence when predicting flow characteristics. For laminar flows and when the constitutive equation has been determined, flow characteristics may be theoretically determined by solving the complete system of equations. It provides a solution to the problem as long as the particular local flow conditions remain within the range of validity of the constitutive equation. This range of validity is generally wide enough though the constitutive equation may have been determined under completely different macroscopic flow conditions. In contrast, when turbulence is significant, in order to solve the complete system of equations, an exact expression is needed for the turbulent stress components which holds specifically under these particular flow conditions. Thus an empirical approach can be useful both with turbulent and laminar flows (for the determination of the "constitutive equation") but in addition, turbulent flow characteristics are still fundamentally unpredictable unless exactly the same macroscopic flow has been previously experienced.

2.3 BASIC RHEOLOGY

We are looking for the material constitutive equation which reflects its mechanical behaviour in terms of a relationship between the actual stress tensor and motion history. It is an intrisic characteristics of the fluid under consideration within a given range of laminar flow experiments and is intimately linked to fluid microstructure.

2.3.1 *Generalities*

The most general form of a constitutive equation can be expressed as follows (Truesdell 1966): the stress tensor at the instant t at a point M within the fluid can be written as:

$$\Sigma(M,t) = \mathcal{F}_t\,[\chi_{\tau<t}\,(X), M] \tag{2.10}$$

where $\mathcal{F}_t$ is a functional of the previous configurations χ_τ of the fluid elements (whose positions at a reference instant (τ=0) are given by the set of $\boldsymbol{X}$) up to the instant t. This expression must satisfy various principles (determinism, local action, material frame indifference) which makes it possible to simplify the form of the above expression (see Noll (1955)). Moreover, for a so-called "simple material" (see Truesdell (1966)), $\mathcal{F}_t$ is assumed to depend only on $\boldsymbol{F}^t(\boldsymbol{X})$ which is the strain gradient tensor defined as $\nabla\chi_t(\boldsymbol{X})$, so that finally the general formal expression for the stress tensor is:

$$\Sigma = -p\boldsymbol{I} + \boldsymbol{T} \tag{2.11}$$

$$\text{with } \boldsymbol{T}(M,t) = \mathcal{G}[\boldsymbol{F}^t(\boldsymbol{X})] \text{ ; tr}\boldsymbol{T} = 0 \text{ ; } p = -\frac{1}{3}(\text{tr}\Sigma) \tag{2.12}$$

Here, a scalar, p, which is the pressure, and the extra-stress (deviatoric) tensor, $\boldsymbol{T}$, have been used. This tensor is of particular interest since, for liquids, no mean motion results from isotropic pressure, and thus essentially $\boldsymbol{T}$ plays a role in motion. It should be borne in mind that in (2.6) $\nabla.\Sigma$ can now be replaced by $(-\nabla p + \nabla.\boldsymbol{T})$. Noll (1955, 1958) then distinguished "fluids" and "solids" as described in 2.1.1 because this corresponded to materials for which the function $\mathcal{G}$ exhibited particularly simple mathematical properties. Nevertheless this kind of mathematical approach finally appeared limited in practice since it was incapable of providing more information for a classification of the various other possible behaviour types.

In simple practical cases the constitutive equation is more commonly sought in the form of a relationship between the extra-stress tensor and the strain rate tensor, $\boldsymbol{D}$, which is the symmetrical part of $\nabla\boldsymbol{u}$, the velocity gradient, and may be written:

$$\boldsymbol{D} = \frac{1}{2}(\nabla\boldsymbol{u} + \nabla\boldsymbol{u}^T) \tag{2.13}$$

The simplest constitutive equation is then obtained for an isotropic, homogeneous, Newtonian fluid:

$$\boldsymbol{T} = 2\mu\,\boldsymbol{D} \tag{2.14}$$

where μ is the fluid viscosity.

Even when the constitutive equation is given in that way, one last additional equation is required to solve the set of equations (2.4) and (2.7). This is the state equation of the material which relates the isotropic pressure to ρ and θ, the temperature. As a first approximation, for liquids or suspensions of solid particles in a liquid, ρ can be considered as constant. This is a special form of state equation.

A lot of behaviour types have been observed or proposed (see the review of Bird (1976)) and, far from being exhaustive, we will present the main constitutive equation types which appear to be of interest for studying mudflow rheology. Since at this stage these constitutive equations often correspond to phenomenological observations it now appears useful to present some elements of rheometry which is the technical way to determine them.

2.3.2 *Rheometry*

Rheometry comprises the techniques which aim to measure the rheological properties of materials. As has been already pointed out, under the same boundary and initial conditions, flow characteristics only depend on material behaviour. This led people to develop empirical tests such as the penetrometer, the slump test (Abrams' cone), etc., in which the material is subjected to forces or motions and reacts in different ways depending on its behaviour. These reactions in terms of length, force, etc., are recorded and used to compare the behaviour of different materials. These are standards techniques for example in fresh concrete or the food industry. However these methods do not provide any relevant information concerning the constitutive equation of the material because the flow conditions are complex and far from controllable and because only one flow characteristic (velocity, fluid height, etc.) is generally recorded at a particular time after starting the test. For example, in these tests, similar experimental data may be obtained with two fluids exhibiting completely different constitutive equations. These tests only make it possible to compare roughly the viscous properties of different materials. Other techniques involve measuring directly the "viscosity" of the material. This is the case of the cone test or the falling bead viscometer. Since the involved flow is in general complex, these apparati yield some relevant information concerning the constitutive equation when the latter only depends on one parameter, i.e. is linear (Newtonian). For non-Newtonian fluids these techniques may only provide a rough estimate of the apparent viscosity of the fluid under these particular flow conditions.

The aim of rheometry is to determine fluid characteristics from measurements concerning simple (from a mathematical point of view) and controlled flows (the experimenter is able to initiate a well defined flow). Viscometric flows, which are of particular importance, roughly correspond to this definition (see Coleman et al. (1966) or Piau (1979) for more details). Within these flows the fluid is strained in a simple way so that only a few components of its stress tensor are different from zero. Some clear information concerning the overall constitutive equation is thus obtained through the equations relating these stress and strain rate components. In particular, the case of simple shear flow is of great practical interest, since, in a Cartesian coordinate frame, we find:

$$u = \gamma y \; ; \; v = w = 0 \tag{2.15}$$

where γ is a constant called the shear rate. It follows that:

$$\boldsymbol{D} = \frac{\gamma}{2}\begin{pmatrix} 0 & 1 & 0 \\ 1 & 0 & 0 \\ 0 & 0 & 0 \end{pmatrix} \tag{2.16}$$

Then, whatever the constitutive equation in the form (2.11), $\boldsymbol{T}$ is expressed as (Coleman et al. 1966):

$$\boldsymbol{T} = \begin{pmatrix} \tau_{xx} & \tau_{xy} & 0 \\ \tau_{xy} & \tau_{yy} & 0 \\ 0 & 0 & \tau_{zz} \end{pmatrix} \quad (2.17)$$

We define the shear stress as τ_{xy}, referred to as τ in the following, and the first and second normal stress differences as:

$$N_1 = \tau_{xx} - \tau_{yy} \; ; N_2 = \tau_{yy} - \tau_{zz} \quad (2.18)$$

The apparent viscosity in simple shear is defined as:

$$\eta = \frac{\tau}{\gamma} \quad (2.19)$$

The functions $\tau(\gamma)$, $N_1(\gamma)$, $N_2(\gamma)$ represent the behaviour (or constitutive equation) of the material under simple shear conditions. Other simple flow types are presented in rheology textbooks (Tanner 1988). At best, conventional rheometrical measurements provide us with the value of the various material functions. The complete three-dimensional constitutive equation must be inferred from these results in a more or less speculative way. This problem is not encountered with pure, straightforward, theoretical approaches to constitutive equations (for example, kinetic theory).

In practice, in order to impose viscometric flow, the fluid must be confined between some tools which fix specific boundary conditions (through a tube, between parallel disks, between concentric cylinders, etc.). The choice of the appropriate geometry for complex materials like mud suspensions is a crucial point (see Chapter 4). In practice, for any fluid, it is fairly hard to find a geometry which makes it possible to have a perfectly homogeneous shear rate between the tools. In most cases there remains some shear rate heterogeneity while conventional rheometrical data only correspond to mean flow characteristics (torque, flow rate, pressure, rotation velocity, etc.) at boundaries. Even in these cases, some calculations make it possible to deduce the shear rate-shear stress data from as complete as possible a set of torque (or pressure loss or force) vs rotation velocity (or flow rate or velocity) values. The corresponding detailed calculations for main geometries may be found in Appendix 2. It is worth noticing that these calculations rely on the assumption that there exists a unequivocal relationship between τ and γ and they are consequently not valid for time-dependent fluids.

2.3.3 *Main simple shear behaviour types*

2.3.3.1 *Basic constitutive equations*

Here we present some useful mathematical models designed to represent types of rheological behaviour commonly observed in practice. We shall only consider their expressions under simple shear conditions since this corresponds to the situation easily studied with rheometers and also to the predominant flow type in simple free surface flows. In that case only the relationship ($\tau(\gamma)$) between shear stress and shear rate, referred to as flow curve, will be considered. Once again it is worth noting that the phenomenological models presented below only aim to represent some observable trends within a certain range of flow conditions, not universal physical properties. Indeed, at least for extremely low or high shear rates, deviations may appear due for example in Brownian motion, elastic properties or turbulence. More about rheological constitutive equations may be found for example in Bird (1976), Bird et al. (1982), Tanner (1988) or Ferguson and Kemblowski (1991).

The simplest case of the Newtonian fluid has already been presented. In simple shear (2.14) in conjunction with (2.16) gives:

$$\tau = \mu\gamma \tag{2.20}$$

This type of constitutive equation is observed with air, water, sirups, glycerin, oils, alcohols, etc., under usual conditions. A slightly more complex model is the power-law model (Otswald 1925):

$$\tau = a\,\gamma^{n} \Rightarrow \mu = a\,\gamma^{n-1} \tag{2.21}$$

where a and n are fluid parameters, which makes it possible to represent simple changes in viscosity at increasing shear rates. For some fluids, for shear rates approaching zero or infinity the value of μ approaches a constant finite value (respectively μ_0 and μ_∞). Then it is preferable to use a model of the following type:

$$\frac{\mu-\mu_\infty}{\mu_0-\mu_\infty} = f(\gamma) \tag{2.22}$$

where $\mu_0>\mu_\infty$ and the function f is a decreasing function of γ which has to be chosen in such a way that $f(\gamma) \to 1$ when $\gamma \to 0$ and $f(\gamma) \to 0$ when $\gamma \to \infty$. The Powell-Eyring, Carreau or Ellis models (see Tanner (1988)) are examples of this. A more complex situation, of great interest within the context of this book, is encountered with yield stress fluids whose flow curve could take the form:

$$\tau = \tau_c + f(\gamma) \tag{2.23}$$

where f is an increasing function of γ which tends to zero when γ tends to zero. Few such models are listed below but a more complete presentation may be found in Bird et al. (1982).

$$\tau = \tau_c + \mu_B\gamma \qquad \text{(Bingham and Green 1920)} \tag{2.24}$$

$$\tau = \tau_c + K\gamma^{n} \qquad \text{(Herschel and Bulkley 1926)} \tag{2.25}$$

$$\sqrt{\tau} = \sqrt{\tau_c} + \sqrt{K\gamma} \qquad \text{(Casson 1959)} \tag{2.26}$$

In these models parameters τ_c, μ_B or K differ from one model to another and must be seen as empirical model parameters. Roughly speaking this behaviour type is observed for fresh concrete, greases, concentrated suspensions, mud, lavas, foodstuffs, paint, inks, filled polymers, etc.

2.3.3.2 *Main behaviour trends*

Now we wish to present some general behaviour types expressing specific forms of variation of the apparent viscosity with shear rate or time. For shear-thinning materials such as yoghurt, mud, blood, and many other concentrated suspensions, the apparent viscosity decreases with the shear rate:

$$\frac{d\eta}{d\gamma} < 0 \tag{2.27}$$

For shear-thickening fluids, such as some granular suspensions, some water-silicium or clay suspensions, the apparent viscosity increases with shear rate. Equations (2.22) or (2.24)-(2.26) are examples of constitutive equations of possible shear-thinning or shear-thickening materials.

Another class includes viscoelastic fluids which are able to store elastic energy when they are deformed. Then for sufficiently rapid initial flows they can behave as elastic bodies and for sufficiently progressive initial flows they can behave more or less as simple fluids. The simplest representation of this behaviour type in simple shear is based on the Maxwell equation:

$$\frac{1}{G'}\frac{d\tau}{dt} + \frac{1}{\eta}\tau = \gamma \tag{2.28}$$

where G' is the elastic modulus of the fluid. Equation (2.28) shows that elastic effects are not observed for flow durations much larger than a characteristic time of the fluid, η/G'.

The last class includes fluids whose apparent viscosity changes with time under given macroscopic boundary conditions:

$$\frac{d\eta}{dt} \neq 0 \tag{2.29}$$

This property is now commonly referred to as thixotropy though some authors still distinguish anti-thixotropy (or rheopexy) from thixotropy depending on whether the apparent viscosity increases or decreases with time. It is essential to distinguish thixotropy from shear-thinning. In the first case the apparent viscosity of the material may decrease with time whereas in the second case its steady state, apparent viscosity value decreases with imposed shear rate. Thixotropy is for example observed with some clay-water mixtures or paints, which happen to be also shear-thinning in general. More generally all fluids are viscoelastic and thixotropic but the characteristic times associated with these phenomena vary in size so that the corresponding effects may not be observed under usual flow conditions. In practice thixotropy is not observed if a characteristic time of viscosity changes is much less than a characteristic time of flow duration.

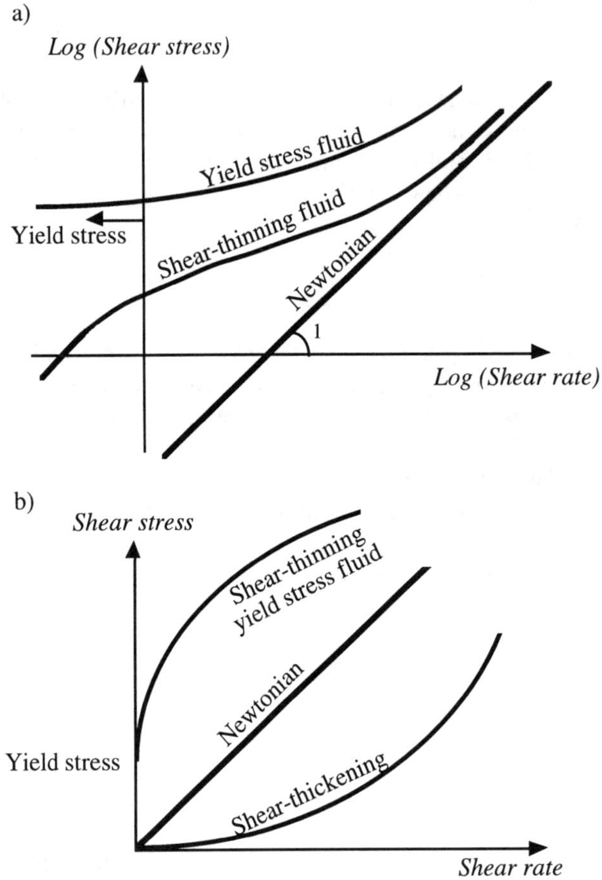

Figure 2.5: Main steady state behaviour types. a) Logarithmic diagram b) Linear diagram.

So far, inspite of many attempts, there are very few complete explanations of the various behaviour types described above on the basis of microstructural considerations concerning changes in particle interaction during flow. Such approaches in the particular case of suspensions and clay-water mixtures are reviewed in Chapters 3 and 5.

Finally, it is worth noticing that the presentation of the steady state data or of a theoretical constitutive equation in linear scale and logarithmic scale shows very different aspects of the flow curve (Fig. 2.5). The choice of a scale type is not neutral. On a logarithmic scale no more importance is attached to any particular shear rate or shear stress range than to another: all the parts of the curve corresponding to ranges whose boundaries have the same ratio occupy the same space in the curve and are directly comparable. Additionally, on a logarithmic scale, one may easily notice the possible tendency of the flow curve to tend to a finite shear stress value when the shear rate tends to zero. For most fluids this corresponds to their yield stress. One cannot proceed in a similar, relevant way with a flow curve plotted on a linear scale. The range of observation of the flow curve corresponds to particular boundaries. This has led many authors to fit a Bingham model to their data, whereas this model is clearly inappropriate when data corresponding to a wide shear rate range are plotted on logarithmic scale (see Chapter 5).

2.4 REFERENCES

Astarita, G., et Marrucci, G. 1974. *Principles of non-Newtonian fluid mechanics*. McGraw Hill, New York.

Barnes, H.A., Hutton, J.F., and Walters, K. 1989. *An Introduction to Rheology*. Elsevier, Amsterdam.

Batchelor, G.K. 1967. *An Introduction to Fluid Dynamics*. Cambridge University Press, Cambridge.

Batchelor, G.K. 1953. *The Theory of Homogeneous Turbulence*. Cambridge University Press, Cambridge.

Batchelor, G.K. 1989. A brief guide to two-phase flow. in *Theoretical Applied Mechanics*. P. Germain, M. Piau and D. Caillerie (Ed.), Elsevier, Amsterdam, 27-41.

Bingham, E.C., and Green, H. 1920. Paint, a plastic material and not a viscous liquid; the measurement of its mobility and yield value. *Proc. Am. Soc. Test. Mat.*, 20, 640-675.

Bird, R.B., Armstrong, R.C., Hassager, O., and Curtiss, C.F. 1977. *Dynamics of Polymeric Liquids*. John Wiley, New York.

Bird, R.B., Dai, G.C., and Yarusso, B.J. 1982. The rheology and flow of viscoplastic materials. *Rev. Chem. Eng.*, 1, 1-70.

Bird, R.B. 1976. Useful non-Newtonian models. *Ann. Rev. Fluid Mech.*, 8, 13-34.

Casson, N. 1959. A flow equation for pigment-oil suspensions of the printing ink type. in *Rheology of disperse systems*. C.C. Mill (Ed.), Pergamon Press, London.

Chapman, S., and Cowling, T.G. 1952. *The Mathematical Theory of Non-uniform Gases*. Cambridge University Press, Cambridge.

Cherimissinoff, N.P. 1988. *Encyclopedia of Fluid Mechanics*. Gulf Publishing Company, Paris.

Coleman, B.D., Markowitz, H., and Noll, W. 1966. *Viscometric Flows of Non-Newtonian Fluids*. Springer Tracts in Natural Philosophy, Volume 5, Springer Verlag, Berlin.

Doi, M., and Edwards, S.F. 1986. *The Theory of Polymer Dynamics*. Oxford University Press, Oxford.

Drew, D.A. 1983. Mathematical modeling of two-phase flow. *Ann. Rev. Fluid Mech.*, 15, 261-291.

Ferguson, J., and Kemblowski, Z. 1991. *Applied Fluid Rheology*. Elsevier, Amsterdam.

Glasstone, S., Laidler, K., and Eyring, H. 1941. *The theory of rate process*. Mc Graw-Hill, New York.

Herschel, W.H., and Bulkley, R. 1926. Über die viskosität und Elastizität von Solen. *Am. Soc. Test. Mat.*, 26, 621-633.

Landau, L.D., and Lifshitz, E.M. 1959. *Fluid Mechanics*. Pergamon, Oxford.

Midoux, N. 1993. *Mécanique et rhéologie des fluides en génie chimique*. Tec&Doc, Lavoisier, Paris.

Munson, B.R., Young, D.F., and Okiishi, T.H. 1994. *Fundamentals of Fluid Mechanics.* John Wiley & Sons, New York.

Noll, W. 1958. A mathematical theory of the mechanical behavior of continuous media. *Arch. Rational Mech. Anal.*, 2, 197-226.

Noll, W. 1955. On the continuity of the solid and fluid states. *J. Rational Mech. Anal.*, 4, 13-81.

Otswald, W. 1925. Uber die Geschwindigkeitsfunktion der viskosität disperser systeme. *Kolloid Zeistricht*, 36, 99-117.

Piau, J.M. 1979. Fluides non-newtoniens, *Techniques de l'Ingénieur.* A710, 1-16, A711, 1-24.

Tanner, R.I. 1988. *Engineering Rheology.* Clarendon Press, Oxford.

Tritton, D.J. 1988. *Physical Fluid Dynamics.* Clarendon Press, Oxford.

Truesdell, C. 1974. *Introduction à la mécanique rationnelle des milieux continus.* Masson, Paris.

Truesdell, C. 1966. *Six Lectures on Modern Natural Philosophy.* Springer-Verlag, Berlin.

CHAPTER 3

Introduction to suspension rheology

Suspensions are heterogeneous two-phase systems made of buoyant solid particles distributed in a continuous liquid. In order to ensure the continuity of the suspension and liquid phase around particles it is necessary (but not sufficient) that the order of magnitude of the suspended particle diameter should be intermediate (mesoscopic) between that of typical interstitial fluid elements (microscopic) and a characteristic length (macroscopic) of the suspension sample as a whole. Here, we intend to review basic elements concerning the rheology of suspensions from a general point of view. It is conventional to start by calculating the viscosity of very dilute suspensions of particles. It is worth noting that historically this is also the first approach to the rheology of suspensions and this remains almost the unique, complete, theoretically sound one, which has in addition been confirmed by experiments. Even if they have provided interesting qualitative concepts and new theoretical tools for a better general approach to suspension rheology, subsequent works have essentially extended this initial viscosity expression to less dilute suspensions or developed partly speculative constitutive equations for various types of suspensions. Finally, with regard to the diversity and complexity of suspension rheology, very little is actually known concerning this field. For example, Einstein's (1956) formula (equation 3.1) or related subsequent works must not hide the widescale, persistent ignorance regarding suspensions for which hydrodynamic effects are not predominant during flow. In particular, in the following it will be shown that the separation between dispersions, which contain colloidal particles, and other suspensions does infer a strict correspondence in fluid behaviour. However this separation is not sufficient and we shall propose a more comprehensive review of suspension rheology.

Rheological (macroscopic) properties result from momentum transfers due to interaction at a mesoscopic scale (the scale at which basic interactions are found). The intensity of these interactions may change with time and flow characteristics, which induce changes in resistance to flow. In the following, we shall try to relate these rheological properties as closely as possible to the basic interactions at an appropriately small scale. Five types of effect leading to momentum transfer will be considered: Brownian, hydrodynamic, packing, colloidal and inertial. The general rheological properties induced in each specific case, assuming that the corresponding effect is larger than the others, will be reviewed. Then we propose some approximate criteria for determining which of these effects prevails for a given suspension under particular flow conditions.

3.1 MAIN MOMENTUM TRANSFERS DURING FLOW

3.1.1 *Brownian motion effects*

Brownian motion was presented in Chapter 1. These random motions are superimposed on the mean motions of particles resulting from the flow of the suspension. As with turbulence, the first effect of Brownian motion is to increase energy dissipation because of additional, local, relative motions between the different elements of the suspension. However, since they do not vary much with flow intensity, these effects only become significant for very slow flows, when energy dissipation due to other types of interaction is small enough. In that case, the suspension viscosity increases when shear rate decreases. As a consequence, Brownian effects may explain the shear-thinning trend of some dilute suspensions (Ferguson and Kemblowski 1991). A second effect of Brownian motions is their tendency to randomize the spatial and orientational distribution of particles. For anisotropic particles, ordering effects due to flow can thus vanish at low shear rates which can also induce a shear-thinning trend.

A final, important effect results from the time-dependent character of the Brownian diffusion process. Diffusion may bring the dispersion back to a possible initial random state corresponding to equilibrium. Induced elastic effects are reported by Mewis and Spaull (1976). Brownian motions might also explain the ability of concentrated suspensions to recover their initial properties after a sufficient period of rest, as is commonly observed with thixotropic clay-water systems (Cf Chapter 5). Due to Brownian diffusion and rotation, small particles are able to slowly migrate towards a position where they are more strongly linked to other particles (via double-layer or van der Waals interactions). The exact role of Brownian motions on other thixotropic properties of these suspensions under rapid flow conditions remains unclear. The reader can find more information about these phenomena in the review by Russel (1980).

3.1.2 *Hydrodynamic effects*

Hydrodynamic effects are the first consequence of the presence of solid particles in the suspending liquid medium. In the absence of solid particles, fluid flow under given macroscopic boundary conditions results in a particular strain field. The flow of a suspension under the same macroscopic boundary conditions obviously gives rise to a different strain field. Particles now take the place of a certain fluid volume. From a general point of view, if the surrounding liquid does not slip at the surface of the particles, the average liquid strain increases with the solid concentration, all other things being equal. As a consequence, the macroscopic viscosity of the suspension increases with the solid fraction. The viscosity even tends towards infinity when the solid fraction (ϕ) tends towards the maximum packing fraction

(ϕ_m). It is worth noting that, when hydrodynamic effects are predominant and when the particle configuration remains statistically isotropic, a suspension of spheres in a Newtonian liquid keeps a Newtonian behaviour (Batchelor 1970). A large number of works have proposed formulae (see below) for the Newtonian viscosity of these suspensions but recent works have also emphasized some non-Newtonian characteristics of sufficiently concentrated suspensions (see 3.1.3).

3.1.2.1 *Dilute suspensions*

Historically, the calculation of the viscosity of very dilute suspensions of non-interacting spheres or ellispoids (Einstein 1956, Jeffery 1922) in a Newtonian fluid was the first serious approach of suspension rheology, later confirmed by experiments. Unfortunately this remains almost the only, exact, theoretical approach. This is because, for a low solid fraction, particles induce only localized perturbations to the fluid strain field which would take place in their absence under the same external conditions (see for example the complete mathematical development in Batchelor (1967) or the review of Schowalter (1978) which also concerns suspensions of deformable particles). The expression for the viscosity (μ) of a dilute suspension of spheres (ϕ<2%) in a Newtonian fluid whose viscosity is μ_0, is therefore:

$$\mu = \mu_0 (1 + 2.5\phi) \quad (3.1)$$

For a larger solid fraction, the strain field of the interstitial fluid becomes complex and hard to predict. The expression for the second order suspension viscosity was also established (Batchelor and Green 1972):

$$\mu = \mu_0 (1 + 2.5\phi + 7.6\phi^2) \quad (3.2)$$

Unfortunately this expression is only valid for elongational flows and for a solid fraction smaller than a few percent (say 10%). For simple shear no expression exactly equivalent to (3.2) has been established.

3.1.2.2 *Moderately concentrated and concentrated suspensions*

Many empirical or semi-empirical formulae giving the viscosity of Newtonian suspensions as a function of solid fraction may be found in the literature [see reviews of Kamal and Mutel (1985), Utracki (1988), Blanc (1983) or Metzner (1985)]. These formulae approximately predict the non-linear viscosity increase with solid concentration within a limited solid fraction range (below 45%). An interesting example is the empirical formula proposed by Chong et al. (1971) fitted to various data corresponding to mono- and poly-disperse sphere suspensions:

$$\mu=\mu_0 \left[1 + 0.75/\left(\frac{\phi_m}{\phi} -1\right)\right]^2 \quad (3.3)$$

It is worth noting that, as is the case with many other semi-empirical formulae, equation (3.3) predicts both Einstein's formula (3.1) for low ϕ (if we take ϕ_m=60.5%, which approximately corresponds to the mean value obtained for a disordered ensemble of spheres (Utracki 1988) and the singularity of viscosity when ϕ tends towards ϕ_m.

For moderately concentrated or concentrated suspensions theoretical approaches have been faced with a major obstacle, namely the prediction of configuration changes during flow. For example, for moderately concentrated suspensions, Simha (1952) presented a "cage" model in which, at each instant, each solid particle is assumed to be surrounded by a spherical liquid enclosure. He then deduced the exact theoretical expression for suspension viscosity from an analysis similar to Einstein's, calculating the theoretical flow field within the liquid enclosure:

$$\mu = \mu_0 \left[1 + 2.5\lambda\phi\right] \tag{3.4}$$

where λ is a function of the ratio (ζ) of the particle radius to the spherical enclosure radius, defined as:

$$\lambda = \frac{4(1-\zeta^7)}{\left[4(1+\zeta^{10})-25\zeta^3(1+\zeta^4)+42\zeta^5\right]} \tag{3.5}$$

The main problem of this approach concerns the size of the liquid enclosure (ζ). Indeed it depends on solid fraction in a quite uncertain manner. Should one consider that the particles are in a crystal configuration (but which one?) and then compute the corresponding uniform spherical liquid enclosure or should one consider that, whatever the configuration, all the liquid more or less participates in the liquid enclosures and then use a mean liquid enclosure whose volume would be computed by dividing the total fluid volume by the particle number?

For concentrated suspensions several other models have been proposed on the assumption that hydrodynamic effects were predominant. When the solid concentration is high enough, say above 40%, the mean gap between two close particles can be small enough for specific hydrodynamic forces within the corresponding liquid layers to be the main source of energy dissipation. For small separating distance to particle diameter ratios, these forces may be estimated from lubrication theory (see Chapter 1). Starting from this idea, Frankel and Acrivos (1967), and later more precisely Goddard (1977), estimated the suspension viscosity by considering that each particle is at the centre of a local cubic network and by taking into account viscous dissipation due to the squeezing of interstitial fluid when particles approach each other. The most simple and popular model is the one due to Frankel and Acrivos (1967). They started from the comparison of the first order values of the various forces acting on close spheres in relative motion in a Newtonian fluid (see 1.2.3.1.1). When ϕ tends to ϕ_m, ε tends towards zero, so that f_2 and f_3 become negligibly small compared to f_1. Frankel and Acrivos (1967) then computed the energy dissipated within a cell surrounding a particle during a pure strain motion and compared it to the energy dissipated within the same volume in the absence of particles. They finally obtained the following equation for the Newtonian (elongational) viscosity of the suspension, which they proved to be in agreement with some avalaible data:

$$\mu = \frac{9}{4} \mu_0 \frac{1}{\varepsilon} \tag{3.6}$$

For reasons that are not explained, Goddard (1977) obtained a significantly different result from a more complete theoretical treatment. Other authors (Tanaka and White 1980, Jarzebski 1981) used a similar approach for power-law interstitial fluids. All these approaches, like the one used by Simha (1952), at least provided theoretical explanations why the viscosity tends towards infinity when the concentration tends towards the maximum packing fraction.

Some assumptions at the basis of these models were contested by Marrucci and Denn (1985) and Adler and al. (1985) who claimed that viscous dissipations cannot be sufficient to explain this singularity. The difference between these two results was later shown (Brady and Bossis 1985, Wang and Cheau 1988) to be due to the fact that, in the latter approach, a periodic configuration composed of particle layers was assumed. Then, during shear, particles only periodically come close to each other. In contrast, in the former approach a constant configuration is assumed. The latter physical scheme is much more realistic especially for high solid concentrations or when the grain size distribution is poorly sorted since in these cases the instantaneous configuration is probably random at any time during flow and any relative motion of particle is balanced by an opposite motion between two other particles, which was not considered in the other works.

A completely different approach prevailed in Mooney's model (1951) later corrected by Krieger and Dougherty (1959) (see details in Krieger (1972)). There it is assumed that the suspension's viscosity exists as a function of the solid fraction and the suspending medium viscosity, $\mu=\mu_0 f(\phi)$ where f is an increasing positive function, and that this function is valid even when one considers the mixture of some particles with the liquid as the suspending medium. Then the viscosity of a suspension may be obtained from μ by either of these procedures:

1) One first adds a solid fraction ϕ_1 of particles to the liquid then a solid fraction ϕ_2 so as to obtain the final suspension with a solid fraction ϕ.

2) One directly adds a solid fraction ϕ to the liquid.

It is worth noting that in the first procedure, ϕ_2 to be taken into account when using f is not simply equal to ϕ-ϕ_1, since the suspending medium is already formed by the mixture of particles with the liquid. In fact we have:

$$\phi_2 = \frac{\phi - \phi_1}{1 - \phi_1} \tag{3.7}$$

Then, equating the final viscosities obtained using the two possible process gives:

$$f(\phi_1) f(\phi_2) = f(\phi) \tag{3.8}$$

Here, in order to take into account the packing effect and the impossibility of having a concentration larger than the maximum packing fraction, it is conventional (see Quemada (1985)) to modify the above approach by using ratios of solid fraction to maximum packing fraction (ϕ/ϕ_m) instead of simple solid fractions (ϕ). Under these conditions, equation (3.8) may be solved taking into account (3.7) giving:

$$f(\phi) = (1-\frac{\phi}{\phi_m})^{-\alpha} \tag{3.9}$$

The coefficient α has been found by many authors (see Quemada (1985)) to be close to 2. However its value depends on that of ϕ_m. Moreover this equation does not make it possible to reproduce experimental results within the whole solid fraction range.

For concentrated suspensions all the above models must be seen as good practical tools for obtaining a rough estimation of the apparent viscosity. However this should not mask the fact that more complex rheological properties may be observed with these suspensions. This mainly originates from the fact that the orientation of anisotropic particles (Jeffery 1922) or the particle distribution may vary with shear and time whereas none of these phenomena are taken into account in the above models. For dilute and moderately concentrated suspensions these effects may induce slight non-Newtonian trends of the steady state properties and thixotropic effects. For highly concentrated suspensions the consequences of these effects may be much stronger and are therefore reviewed in the next section.

3.1.3 *Packing effects*

Here we consider effects due to configuration changes during flow of highly concentrated suspensions of non-colloidal particles. Packing effects may also occur with colloidal suspensions for much smaller solid fractions but the resulting flow phenomena are greatly influenced by the specific nature of colloidal interaction.

3.1.3.1 *Particle configuration*

For a given fraction of granular material, different particle configurations may be found for which the particles are more or less close to each other. For concentrated suspensions the subsequent effects on the nature of the configuration and corresponding main interaction types may be drastic. A critical example of this argument is as follows: consider a square surface (side length: 4 r) on which we put four disks whose diameter is r; they may be placed either in each corner, and then will be separated by a distance equal to $2r$ which may be filled with interstitial fluid so that essentially hydrodynamic effects will take place during flow, or along a cross line (parallel to one side) and then will be in direct contact so that friction may prevail. At a macroscopic level the above remark is also supported by the fact that the maximum random packing fraction depends on the conditions under which it was obtained and depends a great deal on the current particle configuration (distribution) (Onoda and Liniger 1990, Utracki 1988, Barnes et al. 1991). In order to take this into account in the semi-empirical expressions for viscosity (3.3)-(3.9) some authors suggested that the maximum packing concentration to be used for calculations should be a function of shear rate intensity (Quemada 1977, Wildemuth and Williams 1984, Dabak and Yucel 1986). Though these models are based on qualitative realistic assumptions they lead to more speculative expressions for the viscosity because of the difficulty to predict the configuration change as a function of flow and materials characteristics.

The fact that the configuration can change during flow was confirmed by some experimental observations (Nadim 1988, Petit and Noetinger 1988, Petit 1991, Abbott et al. 1991, Graham et al. 1991) which demonstrated that, within monodisperse suspensions, particles can slowly migrate towards specific positions during flow. They can therefore form particular ordered structures which depend on the applied shear rate (Utracki 1988, Petit 1991). For these suspensions the effect of configuration change on behaviour, which obviously depends on the nature of interaction between particles and fluid, may be shear-thickening or time-dependency (Barnes et al. 1991, Gadala-Maria and Acrivos 1980). For example, it has been shown that concentrated suspensions of nearly monodisperse non-aggregating solid particles in a low viscosity Newtonian liquid exhibit a clear (reversible) viscosity increase with increasing shear rate (shear-thickening) (see for example Hoffman 1972, 1974, Barnes 1989, or Boersma et al. 1991). This phenomenon takes the form of a rapid change in viscosity level over a short shear rate range (one decade) but it may be even faster, then appearing for a critical shear rate. This effect has been attributed to the transition from an ordered (possibly two-dimensional layered (Barnes 1989)) structure of particles to a less ordered (possibly random three-dimensional) one. With increasing shear rate the ordered structure becomes unstable and the viscosity goes to the higher level of a less ordered state. Since this effect could have a physical origin similar to those observed by Reynolds (1885) for granular packings, it is sometimes referred to as dilatancy.

3.1.3.2 *Contact network*

When the solid fraction is large direct contacts between particles may occur. This is possible only if sufficient forces act on the particles since normal forces arising from their relative motion (lubrication) will generally tend to push them away (see 1.2.3.1). Direct, frictional contacts between particles are then possible when the solid fraction is so large that particles are more or less packed against each other and forced to remain in contact. Within granular packing, there exists a specific pressure due to particle contacts and it is conceivable that a finite force is required to break the contact network.

Though it is hard to determine its exact value with conventional rheometers, recent experimental results (Kytomaa and Prasad 1993, Coussot and Piau 1994, Barnes et al. 1991, Husband et al. 1993) tend to demonstrate that there exists a concentration threshold above which a concentrated force-free particle suspension exhibits yield stress (Cf Figs 3.1-3.2). There is not yet a general agreement concerning the critical solid fraction value above which this yield stress appears. The value of Coussot and Piau (1994) and Husband et al. (1993), approximately 47%, is different from the one suggested by rough estimations from Barnes & al (1991) results: 56%; and from the one that may be deduced from Kytomaa and Prasad (1993) results: 50%. The difference could be due either to differences in materials or in the way of estimating yield stress (extrapolated from results in a particular shear rate range or obtained from controlled shear stress tests within the accuracy of the rheometer).

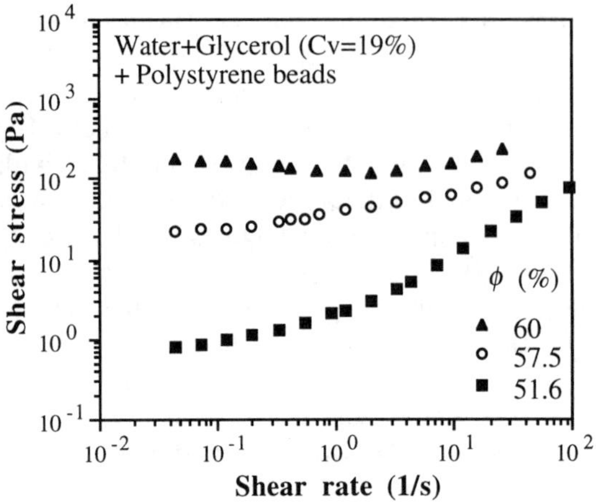

Figure 3.1: Water-glycerol mixture (volume fraction Cv=19%) + different volume fractions of polystyrene beads: Flow curves. Tests were made with a parallel plate rheometer with rough surfaces (disk diameter = 5 cm; gap: 3mm). (from Coussot and Piau (1994))

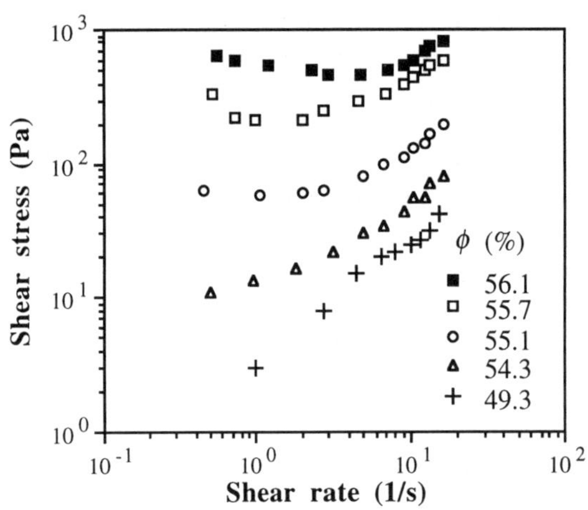

Figure 3.2: Water-glycerine mixture + different volume fractions of glass spheres (Diameter = 2 mm): Flow curves. Tests were made with a cylindrical shear cell (the torque is applied by the lower plate but the sample is confined between two vertical cylinders) whose dimensions were: cylinder height, about 30 cm; inner diameter, 21.9 cm; outer diameter, 43.5 cm. The bottom rotating wall was covered with glass beads at a mean interval of 5 mm. (from Kytomaa and Prasad (1993))

Coussot and Piau's (1994) data (Fig. 3.3) showed that the suspension yield stress seems to tend towards infinity when the solid concentration tends to a finite value.

This is analogous to the phenomenon of viscosity singularity and can be associated with the existence of a concentration threshold, i.e. a "maximum packing concentration", for which no deformation can take place without breakage or volume change (Cf Reynolds (1885)). It is remarkable that the two changes in suspension behaviour, i.e. appearance of yield stress for a moderate volume concentration and large yield stress increase for very high concentrations, is similar to the two successive behaviour changes (Travers et al. 1986) in the compression of a mixture of 'hard' and 'soft' cylinders for the increase of the hard element fraction. This suggests that a similar physical process is at the basis of both phenomena.

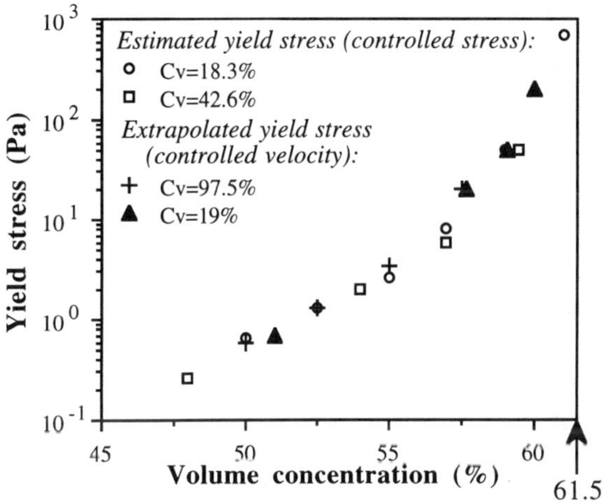

Figure 3.3: Yield stress value estimated from controlled shear rate or controlled shear stress tests for water-glycerol mixtures (volume fraction: Cv) + different concentrations of polystyrene beads. (from Coussot and Piau (1994))

The yield stress appearance above a certain solid fraction represents a sharp change of suspension physical properties. The transition from a fluid to solid state associated to this yield stress was for example compared to glass transition (Shapiro and Probstein 1992). By analogy with percolation theory, it has been suggested that this originates in the formation of a particle contact network (de Gennes 1979). This statement is supported by experimental observations (Bouillot et al. 1982, Blanc et al. 1983) of two-dimensional systems and numerical simulations (Brady and Bossis 1985). This type of network must be broken, at least momentarily, for flow to take place.

For concentrated suspensions the reality might be more complex: depending on the average configuration, the predominant interaction type during flow can either be direct contacts or viscous dissipation within the interstitial fluid. In steady regime, for increasing shear rate, there could be a progressive change from a "close configuration" within which particles may essentially interact via solid frictions to a "loose configuration" within which particles mainly interact hydrodynamically (Cf

Fig. 3.4). Since these changes in configuration are not instantaneous, thixotropic effects may then be observed.

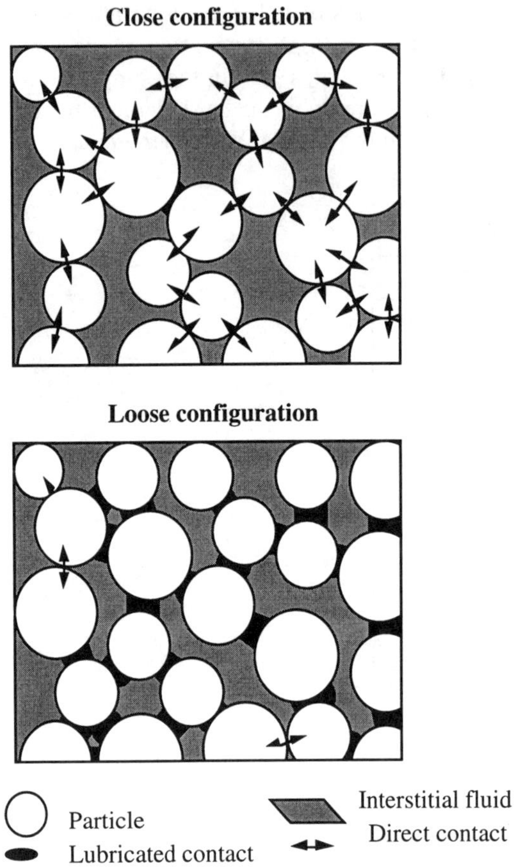

Figure 3.4: Two possible planar configurations of particles for the same solid concentration. In the close configuration, direct contact is mainly observed whereas for the same particle concentration essentially lubricated contacts are observed in the loose configuration.

Finally, it should be noted that a suspension of strictly force-free particles, for which changes in configuration may occur, cannot exhibit a "true yield stress", i.e. recoverable after rest. Indeed, since Brownian motion is in general negligible with grain suspensions, the yield stress will essentially depend on the configuration which was reached just before rest and which is linked to the deformation history, except if some disturbing effects induce slow particle migration after apparent stoppage of the overall flow.

3.1.3.3 *Flow instability*

For very high solid concentrations a shear stress minimum in flow curve was observed (see Figs 3.1-3.2) (Kytomaa and Prasad 1993, Coussot and Piau 1994, 1995): the flow curve decreases for low shear rates and subsequently increases beyond a critical value. This original trend (similar to the one observed for some thixotropic clay-water systems; as in Chapter 5) can easily be explained by a sufficiently rapid change in configuration for increasing shear rates along with a low viscosity of the interstitial fluid. Indeed, at low shear rates, due to "solid friction", normal forces and jamming structures within the close configuration, high stress is necessary to induce a slow strain rate, whereas, in the loose configuration, the shear stress needed to impose relative motion by particles interacting via lubricated contacts can be much smaller (Coussot and Piau 1994).

With this type of very concentrated granular suspension many experimental disturbing effects might foster, increase, or simply explain this effect of shear stress minimum in flow curve: small, unsheared fluid parts near the tools at the periphery of the sample (see Chapter 4) or slight suspended phase dilatancy. These two perturbating effects have similar consequences: they lead to an artificial change in configuration due to a solid concentration decrease of the sheared part of the sample. Considering the large sensitivity of shear stress to configuration changes, very small changes may lead to large experimental errors. Nevertheless, even if the "true" rheological behaviour of this kind of suspension is so far hardly determinable and might be different from what has been observed until now, it remains that their macroscopic flow characteristics can be qualitatively well explained with the help of such a curve type (with a minimum).

3.1.4 *Inertia effects*

3.1.4.1 *Generalities*

As for any fluid type, inertia effects may become significant within a suspension for sufficiently rapid flows. However, for suspensions, two classes of inertia effects must be distinguished. First, one may observe usual (macroscopic) turbulence within the suspension considered as a homogeneous fluid with a given viscosity (Cf Chapter 2). Then the constitutive equation determined in laminar conditions is no longer valid and energy dissipations or flow resistance should be determined from a large number of experiments under different conditions (flow velocity, roughness, etc.). The second class includes three types of inertia effects originating from the granular structure of suspensions (Batchelor 1970) and which are *a priori* independent of the macroscopic turbulence:

1) Turbulence within the interstitial fluid in pores between particles (here the solid fraction is assumed to be such that pore volumes are of the order of particle volume)
2) Grain collisions
3) Grain velocity fluctuations around mean motions

Experimentally, each effect can hardly be distinguished from the other and almost no study has attempted to progress in this field. However it is remarkable that, at least theoretically, the flow of a suspension might remain “macroscopically” laminar, whereas a kind of "microscopic turbulence" develops within the flow due to the particulate structure of the fluid. Within this framework some models (see below) have been developed which make it possible to deduce the (laminar) constitutive equation from a good description of the "microscopic" inertial phenomena. Though these models are theoretically satisfactory, an important question remains open: when local particulate inertia effects (for example collisions) are so large that the macroscopic constitutive equation depend on them, can the flow remain macroscopically laminar? This is what occurs with gas because the mean free path is large compared to particle (molecule) size. Then local collisions do not influence other regions of the flow. This is not the case for liquids: when turbulence occurs, local inertia effects have an influence throughout the material and thus structures (eddies) form at different scales up to that of the sample. Conversely, macroscopic boundary conditions influence the local form of energy dissipation through inertial processes. [This is why, so far, no general, macroscopic, "rheological" behaviour has been found in the case of turbulent flows.] The least that one may conclude from these considerations is that the range of validity of the following models is certainly quite limited.

3.1.4.2 *Bagnold's model*

The pioneering, specific, constitutive equation for a suspension where collisions are predominant was proposed by Bagnold (1954). Indeed, in the so-called "inertia" regime, Bagnold assumed that momentum transfers were mainly produced by elastic collisions between particles of parallel layers in relative motion. The momentum transmitted through each collision is proportional to the relative velocity between the two colliding particles which is proportional to γ because they belong to two adjacent layers in relative motion. For the same reason the collision frequency is also proportional to γ. The shear stress originating in collisions is therefore proportional to γ^2. The momentum transfer process also yields a normal stress proportional to shear stress.

These trends were found to be in agreement with some experimental results (Bagnold 1954, Savage and Sayed 1984) but we still do not have a complete experimental validation of this theory either for suspensions or for dry granular flows (see the review and discussion of Chen and Ling (1996)). In Bagnold's approach the main uncertainty concerns the collision process: particle rotation, velocity angle, fluctuations, etc. These phenomena can be taken into account using the coefficient of proportionality between τ and γ^2. In particular, since it is assumed that each relative motion between two neighbouring particles gives rise to one collision, the coefficient of proportionality increases with the number of particles in each layer and thus with the solid fraction.

3.1.4.3 *Kinetic theory*

Extending Bagnold's model a three-dimensional constitutive equation derived from the gas kinetic theory was proposed for the behaviour of dry granular flows in "inertia" regime. The basic model was first presented by Savage and Jeffrey (1981) then corrected and improved by Jenkins and Savage (1983) and Lun et al. (1984)]. This model involves all sophisticated mathematical tools used for the gas kinetic theory. Its two fundamental assumptions are that momentum transfers via collisions prevail and that only binary collisions exist. This implies that the granular phase must be sufficiently dispersed which, for dry granular flows subject to gravity, seems possible only for very rapid motion. As for Bagnold's model, the exact range of validity of this approach remains unclear. Some simpler models considered other types of energy dissipation (Ogawa et al. 1980, McTigue 1981, Shen and Ackerman 1982, Shahinpoor and Siah 1981) occurring specifically during collisions, but here the latter process (collisions) was still considered to be predominant. Unfortunately most flows of practical interest probably fall into the intermediate regime where both frictional contacts and particle-particle collisions are significant. For example the experimental results of Johnson et al. (1990) and Ancey et al. (1996) clearly demonstrated that dry granular flows exhibit complex properties and that their behaviour cannot in general be described with the kinetic theory alone. With a view to providing a more realistic approach capable of predicting both slow and rapid granular flows, Johnson & Jackson (1987) proposed adding a Coulomb frictional yield stress term to the stress found in the Lun et al (1984) development. Nevertheless a complete, theoretical treatment of slow or moderately rapid, channelled, granular flows taking into account both friction and collisions has yet to be performed.

3.1.5 *Colloidal effects*

3.1.5.1 *Generalities*

For sufficiently small suspended particles and slow flows, various types of electrochemical (colloidal) forces can become larger than common hydrodynamic interactions. So called colloidal effects are mainly van der Waals forces, double-layer interactions, Born repulsion and steric effects (see the review of Russel (1980) and Chapter 1). The particularity of these forces is to give rise to interaction between particles at relatively long distances even at rest. Before studying the rheology of dispersions, a crucial preliminary point is whether the distribution and intensity of forces are such that the suspension is stable, i.e. capable of staying in a homogeneous situation, or whether for example aggregates or other local heterogeneities will tend to form under some flow conditions (see Chapter 1). In the following we shall only consider stable, homogeneous suspensions (dispersions) for which colloidal effects are predominant.

Various force distributions make it possible to obtain a stable suspension: repulsion between close particles, weak flocculated state, etc. Anyway, in equilibrium, this suspension is certainly in a minimum potential energy configuration. When this configuration is broken, a certain amount of energy, corresponding to the height of a local barrier of potential, will be needed to move two close particles definitely far away from each other (Mewis and Spaull 1976). During motion, initially distant particles may move closer so that they will interact significantly within the new instantaneous configuration. It then looks as if one of these particles has fallen into a potential well created by the other particle. The reality is in fact much more complex since the local potential barriers are created by all the surrounding particles. Anyway, it is as if there existed some bonds between particles, which might be reversibly ruptured or restored during flow.

Since under given flow conditions, a certain amount of time is required for equilibrium to be reached, thixotropic effects are obviously induced by particle configuration changes. In addition, for sufficiently small deformations of the suspension, close particles may move relatively to each other without overcoming the potential barrier. This will induce elastic effects.

Therefore these systems have more complicated rheological properties than conventional (moderately concentrated) suspensions of non-colloidal particles [Mewis and Spaull 1976). However the diversity of possible behaviour types for colloidal suspensions is great and it would be hard and outside the scope of this book to exhaustively review them. Here we simply propose a general review of the main physical explanations for the basic rheological trends of dispersions. The following qualitative, physical concepts have often been used in the literature (Moore 1959, Hunter 1982, Firth and Hunter 1976, Patel and Russel 1985, Doi and Chen 1989, Tsenoglou 1990, Coussot et al. 1993) to develop microscopic and macroscopic descriptions of rheological phenomena in dispersed systems. Some models developed for clay-water systems are presented in detail in Chapter 5.

3.1.5.2 *Yield stress*

Dispersions often exhibit yield stress (Navickis and Bagley 1983, Coussot et al. 1993, Mewis and Spaull 1976) as soon as the particle concentration is sufficiently large. The reason is that bonds between particles may form a continuous network which breaks only when a minimum force is applied. The structure of these dispersed systems may be described using percolation theory (Stauffer 1985, de Gennes 1985): at a small concentration ϕ, the particles create a cluster structure and, when $\phi > \phi_c$, where ϕ_c is the percolation threshold, they produce a typical gel or network structure. In this case (Coussot et al. 1993), the concentrated ($\phi > \phi_c$) dispersed systems will behave like elastic solids (see below), when external forces applied to the systems are small enough. When external forces are sufficiently high or when a critical strain rate is applied to the system, it is considered that the network has broken and the suspension flows.

3.1.5.3 *Shear-thinning*

Dispersions are often shear-thinning because the energy dissipated via bond ruptures and restorations does not increase significantly with shear rate. As far as we know there exists no microstructural, theoretical approach concerning slow flows of concentrated suspensions when colloidal effects remain predominant. Thus no relevant explanation of the shear-thinning behaviour of most concentrated dispersions has been given.

Existing theory mainly focuses on the shear-thinning tendency which appears for increasing shear rates when energy dissipated through electrochemical interactions attains the same order as hydrodynamic dissipations (see Section 3.2). In that case it was postulated that locally some bonds are not ruptured so that there remain "flocs" of bonded particles within the suspension and the average (in time and space) size of flocs decreases with flow intensity. A possible explanation for the shear-thinning behaviour is consequently that the volume of jammed, unsheared water (within flocs) decreases with shear rate, which induces a decrease of the solid concentration of the equivalent suspension of non-interacting flocs (Tsenoglou 1990). This explanation is based on the assumption that hydrodynamic interactions within the equivalent suspension of non-interacting flocs can be considered as predominant. This model type could thus be essentially valid for very low solid fractions or high shear rates (see Section 3.2).

In parallel, for dispersions of uniform, colloidal spheres, shear-induced ordering effects may also be observed (Ackerson 1990) which were interpreted as phase transitions. This rather particular effect may also lead to shear-thinning.

3.1.5.4 *Thixotropy*

These suspensions are often thixotropic (Mewis 1979), because the breakage and restoration of the network described above, and the subsequent possible decrease in floc size, are not instantaneous. When for example a steady shear is imposed, as long as the equilibrium structure has not been reached, the apparent viscosity differs from the final one. Attempts to predict changes of floc size as a function of time and shear were proposed for example by Ruckenstein and Mewis (1973) or Mercer and Weymann (1974). Nevertheless, from a general point of view, as for non-colloidal suspensions (see Section 1.3), the problem of the description of configuration changes during flow remains complete.

3.1.5.5 *Elasticity*

The preceding qualitative considerations also result in the appearance of viscoelastic behaviour in the dispersed systems (Onogi et al. 1973). As already mentioned elastic effects result from the ability of bonds to store energy before rupture. In simple shear the simplest approach to taking account of elastic effects is to consider that as long as the yield stress is not overcome, the fluid behaves as a solid and stress is proportional to strain (Doraiswamy et al. 1991):

$$\tau = G'\gamma^* \tag{3.10}$$

where γ^* is the deformation and G' the elastic modulus of the suspension, as in linear elasticity. However for dispersions a constant modulus is only to be expected in the limiting case of infinitely small deformations (Mewis and Spaull 1976) and thus (3.10) should be seen as an approximation. A basic model for estimating G' as a function of particle interactions and the form of the network was proposed by van der Tempel (1961). For a particle with a shape between a sphere and a flat plate, its relation, derived from a potential energy expression, is:

$$G' = \frac{5A\phi R^{1/2}}{12\pi d_0^{7/2}} \tag{3.11}$$

where d_0 is the equilibrium particle distance. However this expression must only be used for rough estimations of G' since for example more complex dependencies on particle size and concentration have been reported or proposed (Mewis and Spaull 1976).

3.2 PREDOMINANT EFFECTS

In the following, we shall compare the effect on the behaviour of the suspension of the various effects reviewed in the previous section. We consider a simply sheared, Newtonian liquid in which solid particles of a given type are suspended. Criteria for the predominance of each effect are established and the resulting (generally approximate) limit curves are reported in Figure 3.5 in a (γ,ϕ) diagram. In this diagram the other material parameters (particle shape and radius, interstitial fluid viscosity and density, temperature, etc.) are fixed.

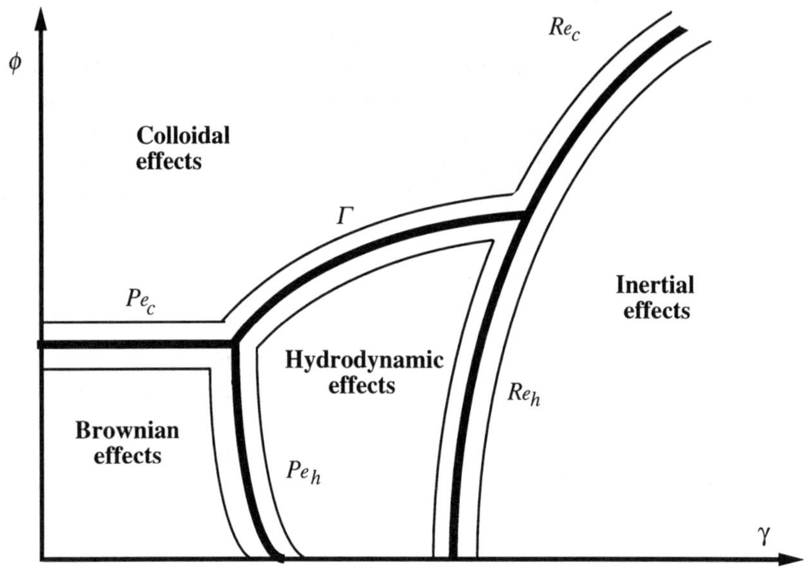

Figure 3.5: Conceptual diagram of predominance of the main momentum transfer effects within a suspension as a function of solid fraction and shear rate. This suspension contains a sufficiently large colloidal fraction and is thus assumed to exhibit a yield stress for solid fractions far below the maximum packing fraction. Packing effects have been neglected. Inertia effects due to the granular structure of the suspension have not been distinguished from usual inertia effects (turbulence). All other parameters (particle size distribution, temperature, interstitial fluid viscosity, fluid density, etc.) are constant. The transition ranges (see text) are often approximately known and may depend on boundary conditions. Here they are represented as two more or less parallel thin lines situated above and below a thick line which should correspond to a constant value of the locally relevant, dimensionless number.

3.2.1 *Colloidal and hydrodynamic effects*

Let us consider a suspension in simple laminar shear. Because of the lack of knowledge concerning the exact microstructural effects occurring during flows of dilute suspensions it is a priori hard to propose a serious, general criterion for the transition from a colloidal regime, for which colloidal effects would dominate, to a hydrodynamic regime, for which hydrodynamic effects would dominate. In the case of suspensions which exhibit a yield stress for moderate solid fractions this appears possible. Indeed in that case, the yield stress provides an estimation of the "amount" of colloidal forces within the suspension. Assuming that the network breakage requires the rupture of almost every bond, the yield stress roughly expresses the number of bonds times the mean bond force. With this physical interpretation, when flow is not too rapid, the order of magnitude of the stress originating in colloidal effects is equal to that of the yield stress value. The amount of hydrodynamic effects may be estimated through the stress of the equivalent suspension of non-interacting particles. Then the ratio of hydrodynamic to colloidal effects is readily expressed as:

$$\Gamma = \frac{\mu\gamma}{\tau_c} \tag{3.12}$$

Here $\mu(\phi)$ is the viscosity of the corresponding force-free particle suspension which may be estimated with one of the formulae proposed in Section 3.1.2 [formally Γ^{-1} is a "Bingham number"]. In general τ_c rapidly increases with ϕ. For concentrated clay-water mixtures τ_c was found to follow an exponential law as a function of ϕ (see Chapter 5). For fine mud suspensions the range of Γ for the transition from a colloidal regime to a hydrodynamic one was found to be [0.3; 50] (Coussot 1995) (see Chapter 5). The transition range for other materials should also be around Γ=1. It should be noted that Γ also corresponds to the ratio of a material characteristic time (μ/τ_c) to a characteristic time of the flow ($1/\gamma$).

3.2.2 *Viscous and Brownian effects*

We must now determine whether Brownian effects play a role in the behaviour of the suspension in a laminar regime. For this purpose, the relative importance of macroscopic viscous effects (either due to hydrodynamic or colloidal effects) and Brownian effects can be estimated by comparing two corresponding characteristic times. Let us consider a dispersion undergoing simple shear (Cf Chapter 1). From (1.15) it results that the mean time taken by the particle to move by R through the suspension because of Brownian motion alone is $R^2/2D$. Consider two neighbouring particles in a line parallel to the direction of shear (perpendicular to the direction of fluid velocity). During flow these particles will move in two parallel planes with a relative velocity equal to $2R\gamma$, where γ is the macroscopic shear rate. The approximate time taken by a suspended particle to move by R apart from the other particle because of externally imposed shear is $1/2\gamma$. In a rheological study for which one intends to relate shear intensity to stresses, translational Brownian motions are negligible if the time they need to affect particle motion is large compared to that needed by flow. As a first approximation this means that the ratio of the two above described characteristic times must be large compared to 1. This ratio, Pe, is the Peclet number:

$$P_e = \frac{\gamma R^2}{D} = \frac{\gamma R^2 f}{k\theta} \tag{3.13}$$

The value of f as a function of the rheological parameters of the surrounding fluid has been discussed in Chapter 1, Section 1.3.2.1. However, in order to determine dimensionless numbers able to characterize the transition between regions of the diagram (γ,ϕ) for which predominant flow effects change, it is sufficient to have orders of magnitude of the functions involved in (3.13) as soon as the expressions we have for these functions include all important parameters.

When the solid fraction is sufficiently small and the shear rate sufficiently large the surrounding fluid may be considered as Newtonian and the Peclet number writes:

$$P_{e_h} = \frac{3\pi R^3 \mu\gamma}{k\theta} \tag{3.14}$$

When the solid fraction is sufficiently large and the shear rate sufficiently small colloidal effects become predominant. Since the value of f is *a priori* unknown for non-Newtonian fluids, it seems difficult to use the expression (3.14) in a straightforward way. However it is noticeable that, for Newtonian suspensions, P_e may be written as the ratio of the viscous stress $(\mu\gamma)$ to $k\theta/3\pi R^3$, a characteristic stress value which only depends on temperature and particle radius. We suggest extending this formal result to non-Newtonian fluids and in particular to yield stress suspensions for which colloidal effects are predominant. This means that Brownian effects are negligible compared to viscous effects (which in the present case are essentially what we called colloidal effects) when P_{e_c} is much larger than 1, where:

$$P_{e_c} = \frac{3\pi R^3 \tau_c}{k\theta} \tag{3.15}$$

3.2.3 *Inertia and viscous effects*

3.2.3.1 *"Macroscopic" inertia effects*

From a scale analysis of motion equations along with stability considerations, a general result is that inertia effects cannot be neglected when the following dimensionless (Reynolds) number is much larger than 1:

$$R_e = \frac{\rho V^2}{\sigma} \tag{3.16}$$

where V is a characteristic fluid velocity and σ a characteristic stress. The Reynolds number can take different, precise forms depending on flow and fluid type. In simple shear we can use γh, where h is the sample thickness in the direction of shear, instead of V. In addition, for Newtonian suspensions, as a first approximation, when hydrodynamic effects dominate at sufficiently low shear rates, σ is equal to $\mu\gamma$ and (3.16) becomes:

$$Re_h = \frac{\rho h^2 \gamma}{\mu} \tag{3.17}$$

For yield stress suspensions, when colloidal effects dominate, it is necessary to compare inertia effects to viscous effects due to colloidal interactions, obtaining immediately:

$$Re_c = \frac{\rho \gamma^2 h^2}{\tau_c} \tag{3.18}$$

The exact transition range for which inertia effects are significant depends on boundary conditions and should be determined experimentally. For example for Newtonian flows through tubes, if we use L instead of h and V/L instead of γ in (3.17) where L is the cylinder diameter and V the mean velocity over a cross-section, the transition range varies slightly around [40; 2500] depending on the exact boundary conditions. For non-Newtonian fluids, and in particular for yield stress fluids, it is only apparent from existing data that the critical range of Re corresponding to the transition between a laminar and a turbulent flow regime in general has much larger boundaries than for Newtonian fluids (see Midoux (1993)). In Chapter 8 a criterion for transition to turbulence in the case of free surface flows of yield stress fluids is presented. Anyway it is likely that, in practice, as a first approximation turbulence is always negligible for Re smaller than 10.

3.2.3.2 *Particle inertia effects*

Following Batchelor's (1970) indications, general, formal, criteria (based on the ratio of the various induced forces) can be easily established (Coussot 1992) for the predominance of the three possible phenomena specifically induced by the presence of particles (Cf 3.1.4.1). Each of them involve a dimensionless number which is formally similar to the Reynolds number (eq. (3.16)), whose corresponding general expression is obtained from the ratio of a stress now related to the particle inertia to a stress related to macroscopic viscous dissipations within the flow:

$$I = \frac{\rho_s (R\gamma)^2}{\sigma} \tag{3.19}$$

where R, the particle radius is now used instead of fluid thickness and σ should either be taken to $\mu\gamma$ or τ_c depending on the type of predominant viscous effects. However, as for the "macroscopic" turbulence criterion for the whole suspension, the critical value of I for which each of the phenomena cited above becomes significant has yet to be determined empirically. One can only be sure that this occurs when I is much larger than 1.

Bagnold's (1954) work aimed to provide a criterion for the transition between a so-called macroviscous and an inertia regime. However, this author only referred to collision effects, did not provide extensive data with quite different viscous fluids and did not consider possible non-Newtonian behaviour of the interstitial fluid. Additionally Bagnold provided data corresponding to inertia (collisional) regimes which were probably affected by turbulence. Indeed, for a typical point on the part of the curve referred to as the collisional regime (due to Bagnold's data and analysis), we have σ=12 Pa and γ=75 s^{-1}. The generalized Reynolds number (eq. (3.16)) is therefore equal to 735 (for fluid elements close to the outer cylinder). For a suspension of this sort, this could be in the transitional regime towards turbulence. In conclusion, Bagnold's results do not provide a general criterion for the transition to a collisional regime.

At least, we may retain from Bagnold's work, as confirmed by the data of Coussot and Piau (1994), that collisions should be negligible for Bagnold numbers under 40. Furthermore, because of our inability to distinguish the different particle inertia effects, we suggest that Bagnold's criterion could in fact predict the absence of any of these three effects. However, in order to take into account possible non-Newtonian effects, we suggest to use, instead of the Bagnold number, a generalized Bagnold number:

$$Ba = \frac{\sqrt{2}\rho_s(2R)^2\gamma^2}{\sigma\sqrt{\varepsilon}} \qquad (3.20)$$

To compute ε we recommend using the formula (1.12) with the real maximum random loose packing fraction, instead of the maximum packing fraction corresponding to a particular crystal configuration.

3.3 REFERENCES

Abott, J.R., Tetlow, N., Graham, A.L., Altobelli, S.A., Fukushima, E., Mondy, L.A., and Stephens, T.S. 1991. Experimental observations of particle migration in concentrated suspensions: Couette flow. *J. Rheol.*, 35, 773-795.

Ackerson, B.J. 1990. Shear induced order and shear processing of model hard sphere suspensions. *J. Rheol.*, 34, 533-590.

Adler, P.M., Zuzovsky, M., and Brenner, H. 1985. Spatially periodic suspensions of convex particles in linear shear flows. *Int. J. Multiphase Flow*, 11, 387-417.

Ancey, C., Coussot, P., and Evesque, P. 1996. Examination of the possibility of a fluid-mechanics treatment of dense granular flows. *Mech. Cohesive-Frictional Mat.*, 1, 385-403.

Bagnold, R.A. 1954. Experiments on a gravity free dispersion of large solid spheres in a Newtonian fluid under shear. *Proc. R. Soc. London*, A 225, 49-63.

Barnes, H.A. 1989. Shear-thickening ("dilatancy") in suspensions of non-aggregating solid particles dispersed in newtonian liquids. *J. Rheol.*, 33, 329-366.

Barnes, H.A., Jomha, A.I., Lips, A., Merrington, A., and Woodcock, L.V. 1991. Recent developments in dense suspension rheology. *Powder Technol.*, 65, 343-370.

Batchelor, G.K. 1967. *An introduction to fluid dynamics*. Cambrige University Press, Cambridge.

Batchelor, G.K., and Green, J.T. 1972. The determination of the bulk stress in a suspension of spherical particles to order c^2. *J. Fluid Mech.*, 56, 375.

Batchelor, G.K. 1970. The stress system in a suspension of force free particles. *J. Fluid Mech.*, 41, 545-570.

Blanc, R., Belzons, M., Camoin, C., and Bouillot, J.L. 1983. Cluster statistics in a bidimensional suspension: comparison with percolation. *Rheol. Acta*, 22, 505-511.

Blanc, R. 1983. Hydrodynamique et rhéologie des colloïdes. in *Colloïdes et Interfaces*, Les Editions de Physique, Les Ulis (France).

Boersma, W.H., Baets, P.J.M., Laven, J., and Stein, H.N. 1991. Time-dependent behavior and wall slip in concentrated shear thickening dispersions. *J. Rheol.*, 35, 1093-1120.

Bouillot, J.L., Camoin, C., Belzons, M., Blanc, R., and Guyon, E. 1982. Experiments on 2D suspensions. *Adv. Colloid Interface Sci.*, 17, 299-305.

Brady, J.F., and Bossis, G. 1985. The rheology of concentrated suspensions of spheres in simple shear flow by numerical simulation. *J. Fluid Mech.*, 155, 105-129.

Chen, C.L., and Ling, C.H. 1996. Granular-flow rheology: role of shear-rate number in transition regime. *J. Eng. Mech.*, 122, 469-480.

Chong, J.S., Christiansen, E.B., and Baer, A.D. 1971. Rheology of concentrated suspensions. *J. Appl. Polymer Sci.*, 15, 2007-2021.

Coussot, P. 1992. *Rheology of debris flows - Study of concentrated dispersions and suspensions*. Ph. D. thesis, INPG, Grenoble. (in French)

Coussot, P., and Piau, J.-M. 1995. A large-scale field coaxial cylinder rheometer to study the rheology of natural coarse suspensions. *J. Rheol.*, 39, 105-124.

Coussot, P., and Piau, J.-M. 1994. Rheology of very concentrated suspensions of force-free particles. *Les Cahiers de Rhéologie* (Groupe Français de Rhéologie), XIII, 266-277. (in French)

Coussot, P., Leonov, A.I., and Piau, J.-M. 1993. Rheology of concentrated dispersed systems in a low molecular weight matrix. *J. Non-Newt. Fluid Mech.*, 46, 179-217.

Coussot, P. 1995. Structural similarity and transition from Netwonian to non-Newtonian behavior for water-clay suspensions. *Phys. Rev. Lett.*, 74, 3971-3974.

Dabak, T., and Yucel, O. 1986. Shear viscosity behavior of highly concentrated suspensions. *Rheol. Acta*, 25, 527-533.

de Gennes, P.-G. 1979. Conjecture on the transition from Poiseuille flow in suspensions. *J. Phys.*, 40, 783-787.

de Gennes, P.G. 1985. Continuum and discontinuum: the percolation. in *Encyclopedia Universalis*, Symposium: Les Enjeux, 313-318. (in French)

Doi, M., and Chen, D., 1989. *J. Chem. Phys.*, 90, 5271.

Doraiswamy, D., Mujumdar, A.N., Tsao, I., Beris, A.N., Danforth, S.C., and Metzner, A.B. 1991. The Cox-Merz rule extended: a rheological model for concentrated suspensions and other materials with a yield stress. *J. Rheol.*, 35, 647-685.

Einstein, A. 1956. in *Investigation of the Brownian movement*. Dover publication, New-York. [English translation of Ann. Physik, 19, p.286 (1906), et 34, p.591, (1911)].

Ferguson, J., and Kemblowski, Z. 1991. *Applied Fluid Rheology*. Elsevier, Amsterdam.

Firth, B.A., and Hunter, R.J. 1976. Flow properties of coagulated suspensions. *J. Colloid Interface Sci.*, 57, 248-275.

Frankel, N.A., and Acrivos, A. 1967. On the viscosity of a concentrated suspension of solid spheres. *Chem. Eng. Sci.*, 22, 847-853.

Gadala-Maria, F., and Acrivos, A. 1980. Shear-induced structure in a concentrated suspensions of solid spheres. *J. Rheol.*, 24, 799-814.

Goddard, J.D. 1977. An elastohydrodynamic theory for the rheology of concentrated suspensions of deformable particles. *J. Non-Newt. Fluid Mech.*, 2, 169-189.

Graham, A.L., Altobelli, S.A., Fukushima, E., Mondy L.A., and Stephens, T.E. 1991. Note: NMR imaging of shear-induced diffusion and structure in concentrated suspensions undergoing Couette flow. *J. Rheol.*, 35, 191-201.

Hoffman, R.L. 1972. Discontinuous and dilatant behavior in concentrated suspensions. I. Observation of a flow instability. *Trans. Soc. Rheol.*, 16, 155-172.

Hoffman, R.L. 1974. Discontinuous and dilatant behavior in concentrated suspensions. II. Theory and experimental data. *J. Colloid Interface Sci.*, 46, 491-506.

Hunter, R.J. 1982. The flow behavior of coagulated colloidal suspensions. *Adv. Colloid Interface Sci.*, 17, 197-211.

Husband, D.M., Aksel, N., and Gleissle, W. 1993. The existence of static yield stresses in suspensions containing noncolloidal particles. *J. Rheol.*, 37, 215-235.

Jarzebski, G.J. 1981. On the effective viscosity of pseudoplastic suspensions. *Rheol. Acta*, 20, 280-287.

Jeffery, G.B. 1922. The motion of ellipsoidal particles immersed in a viscous fluid. *Proc. R. Soc. London*, A102, 161-179.

Jenkins, J.T., and Savage, S.B. 1983. A theory for the rapid flow of identical, smooth, nearly elastic, spherical particles. *J. Fluid Mech.*, 130, 187-202.

Johnson, P.C., and Jackson, R. 1987. Frictional-collisional constitutive relations for granular materials, with application to plane shearing. *J. Fluid Mech.*, 176, 67-93.

Johnson, P.C., Nott, P., and Jackson, R. 1990. Frictional-collisional equations of motion for particulate flows and their application to chutes. *J. Fluid Mech.*, 210, 501-535.

Kamal, M.R., and Mutel, A. 1985. Rheological properties of suspensions in Newtonian and non-Newtonian fluids. *J. Polymer Eng.*, 5, 293-382.

Krieger, I.M., and Dougherty, T.J. 1959. A mechanism for non-Newtonian flow in suspensions of rigid spheres. *Trans. Soc. Rheol.*, III, 137-152.

Krieger, I.M. 1972. Rheology of monodisperse latices. *Adv. Colloid Interface Sci.*, 3, 111-136.

Kytomaa, H.K., and Prasad, D. 1993. Transition from quasi-static to rate dependent shearing of concentrated suspensions. in *Powders & Grains*, C. Thornton (Ed.), Balkema, Rotterdam, 281-287.

Lun, C.K.K., Savage, S.B., Jeffrey, D.J., Chepurniy, N. 1984. Kinetic theories for granular flow: inelastic particles in Couette flow and slightly inelastic particles in a general flowfield. *J. Fluid Mech.*, 140, 223-256.

Marrucci, G., and Denn, M.M. 1985. On the viscosity of a concentrated suspension of solid spheres. *Rheol. Acta*, 24, 317-320.

McTigue, D.F. 1981. A nonlinear constitutive model for granular materials: Application to gravity flow. *J. Appl. Mech.*, 49, 291-296.

Mercer, H.A., and Weymann, H.D. 1974. Structure of thixotropic suspensions in shear flow. III. Time-dependent behavior. *Trans. Soc. Rheol.*, 18, 199-218.

Metzner, A.B. 1985. Rheology of suspensions in polymeric liquid. *J. Rheol.*, 29, 739-775.

Mewis, J., and Spaull, A.J.B. 1976. Rheology of concentrated dispersions. *Adv. Colloid Interface Sci.*, 6, 173-200.

Mewis, J. 1951. Thixotropy - A general Review. *J. Non-Newt. Fluid Mech.*, 6, 1-20.

Midoux, N. 1993. *Mécanique et rhéologie des fluides en génie chimique*. Tec&Doc, Lavoisier, Paris.

Mooney, M., 1951. *J. Colloid. Sci.*, 6, 162-170.

Moore, F. 1959. The rheology of ceramic slips and bodies. *Trans. Brit. Cer. Soc.*, 58, 470-484.

Nadim, A. 1988. The measurement of shear-induced diffusion in concentrated suspensions with a Couette device. *Phys. Fluids*, 31, 2781-2785.

Navickis L.L. and Bagley, E.B. 1983. Yield stresses in concentrated dispersions of closely packed, deformable gel particles. *J. Rheol.*, 27, 519-536.

Ogawa, S., Unemura, A., Oshima, N. 1980. On the equations of fully fluidized granular materials. *J. Appl. Math. Phys.*, 31, 483-493.

Onoda, G.Y., and Liniger, E.G. 1990. Random loose packings of uniform spheres and the dilatancy onset. *Phys. Rev. Lett.*, 64, 2727-2730.

Onogi, S., Matsumoto, T., and Warashina, Y. 1973. Rheological properties of spherical particles in polymer solutions. *Trans. Soc. Rheol.*, 17, 175-190.

Patel, P.D., and Russel, W.B. 1985. *Colloids and Surface*, 31, 355.

Petit, L. 1991. Ecoulement et mise en structure de suspensions macroscopiques. *Annales Phys.*, 16, 155-191.

Petit, L., and Noetinger, B. 1988. Shear-induced structures in macroscopic dispersions. *Rheol. Acta*, 27, 437-441.

Quemada, D. 1977. Rheology of concentrated disperse systems and minimum energy dissipation principle - I Viscosity-concentration relationship. *Rheol. Acta*, 16, 82-94.

Quemada, D. 1985. Phenomenological rheology of concentrated dispersions - I. Clustering effects and the structure-dependent packing fraction. *J. Theor. Appl. Mech.*, Numéro Spécial, 267-288.

Reynolds, O. 1885. On the dilatancy of media composed of rigid particles in contact. With experimental illustrations. *Phil. Mag. J. Sci.*, Series 5, 20, 469-481.

Ruckenstein, E., and Mewis, J. 1973. Kinetics of structural changes in thixotropic fluids. *J. Colloid Interface Sci.*, 44, 532-541.

Russel, W.B. 1980. Review of the role of colloidal forces in the rheology of suspensions. *J. Rheol.*, 24, 287-317.

Savage, S.B., and Jeffrey, D.J. 1981. The stress tensor in a granular flow at high shear rates. *J. Fluid Mech.*, 11, 255-272.

Savage, S.B., and Sayed, M. 1984. Stresses developed by dry cohesionless granular materials sheared in an annular shear cell. *J. Fluid Mech.*, 142, 391-430.

Schowalter, W.R. 1978. *Mechanics of non-Newtonian fluids*. Pergamon press, Oxford.

Shahinpoor, M., and Siah, J.S.S. 1981. New constitutive equations for the rapid flow of granular materials. *J. Non-Newt. Fluid Mech.*, 9, 147-156.

Shapiro, A.P., and Probstein, R.F. 1992. Random Packings of spheres and fluidity limits of monodisperse and bidisperse suspensions. *Phys. Rev. Lett.*, 68, 1422-1425.

Shen, H., and Ackerman, N.L. 1982. Constitutive relationships for fluid-solid mixtures. *J. Eng. Mech. Div.*, 108, 748-763.

Simha, R. 1952. A treatment of the viscosity of concentrated suspensions. *J. Appl. Phys.*, 23, 1020-1024.

Stauffer, D. 1985. *Introduction to Percolation Theory*. Taylor and Francis, Philadelphia.

Tanaka, H., and White, J.L. 1980. A cell model theory of the shear viscosity of a concentrated suspension of interacting spheres in a non-Newtonian fluid. *J. Non-Newt. Fluid Mech.*, 7, 333-343.

Travers, T., Bideau, D., Gervois, A., Troadec, J.P., and Messager, J.C. 1986. Uniaxial compression effects on 2D mixtures of 'hard' and 'soft' cylinders. *J. Phys., A: Math.Gen.*, 19, L1033-L1038.

Tsenoglou, C. 1990. Scaling concepts in suspension rheology, *J. Rheol.*, 34, 15-24.

Utracki, L.A. 1988. The rheology of two-phase flows. In *Rheometrical Measurement*, A.A. Collyer and D.W. Clegg (Ed.), Elsevier Applied Science, Amsterdam, Chapter 15, 479-594.

van der Tempel, M. 1961. Mechanical properties of plastic-disperse systems at very small deformations. *J. Colloid Sci.*, 16, 284-296.

Wang, M.-L., and Cheau, T.-C. 1988. Shear viscosity of slightly-elastic concentrated suspensions at low and high shear rates. *Rheol. Acta*, 27, 596-607.

Wildemuth, C.R., and Williams, M.C. 1984. Viscosity of suspensions modeled with a shear-dependent maximum packing fraction. *Rheol. Acta*, 23, 627-635.

CHAPTER 4

Shear rheometry with concentrated suspensions

The mud suspensions under consideration belong to the category of concentrated suspensions which are systems generally exhibiting most of the more complex rheological properties such as thixotropy and yield stress. Accurately determining their characteristics requires a wide rheometrical investigation and careful analysis of obtained data.

Because concentrated suspensions can contain coarse particles and exhibit particular rheological properties, a number of experimental difficulties may be encountered with those materials during shear rheometry. With this type of material it appears rather dangerous just to make a blind record of some torque (or pressure) and velocity data in order to deduce the constitutive equation of the suspension. When doing so one may easily interpret experimental data erroneously. A typical example is wall slip which occurs with sufficiently smooth tool surfaces. In that case, the viscosity directly inferred from torque-velocity measurements may be significantly lower than the real viscosity because the suspension is much less sheared than expected from data.

Rheometry aims to determine the rheological properties of materials when they are in such conditions that, during flow, they can be considered at each instant as homogeneous systems undergoing continuous deformations at a macroscopic scale. This means that discontinuity in strain field and material heterogeneities should be avoided. One can therefore study the suspension in the field of continuum mechanics. Furthermore, the rheological behaviour of the material is deduced from rheometrical measurements assuming a theoretical (viscometric) flow within the rheometer. The experimenter's task will be to check that these hypotheses are respected or even, ideally, to control them. Under these conditions the stress-strain rate couple deduced from measurements will effectively corresponds to the true behaviour of the homogeneous suspension. In practice shear rheometry with concentrated suspensions requires specific attention to various phenomena such as: observable strain field, changes of free surface form with time, evaporation rate, wall slip, particle migration and any slow or rapid changes of recorded shear stress or shear rate. These precautions are crucial since they will avoid interpreting some disturbing effects as real material properties and should make it possible to obtain relevant measurements of some peculiar properties of these suspensions. However this is not so easy because the different phenomena which may occur are hardly quantifiable and a great number of tests are often necessary before it is possible to distinguish them. The developments and quantifications of disturbing phenomena and their relation with suspension properties remains widely unexplored. Here we shall only review techniques and precautions for shear rheometry of concentrated suspensions to obtain data reflecting intrinsic fluid behaviour.

The first part reviews the advantages and disadvantages of the main rheometer geometries for testing concentrated suspensions. The various possible disturbing effects occurring specifically with these materials are reviewed along with the means to control them. In the second part, we shall examine the specific procedures for determining the main rheological properties of concentrated suspensions.

4.1 WHICH GEOMETRY

In the following we shall only consider the usual, simplest geometries in shear rheometry: capillary rheometer, parallel plates, cone and plate and coaxial cylinders. In addition we shall consider a specific geometry, the inclined plane, which has been rarely used but appears quite useful for mud suspensions. All these geometries along with the corresponding adequate theoretical viscometric calculations to be used are described in more detail in Appendix 2. For a more general description of rheometrical techniques the reader is referred to current textbooks (Coleman et al. 1966, Tanner 1988, Collyer and Clegg 1988).

In Table 4.1 the advantages and disadvantages of these geometries for testing concentrated suspensions are summarised according to the following detailed considerations. It is worth noticing that, although we distinguish different points and discuss them one after another, they are often more or less dependent: shear rate homogeneity depends on tool scale, choice of tool scale depends on the possible disturbing effects, etc.

4.1.1 *Tool scale*

4.1.1.1 *Continuum assumption*

In order to respect the continuum hypothesis (see 2.2.1) the tool scale must be sufficiently large compared to the characteristic particle size. Typically the ratio (r) of the smallest gap between tools (or flow depth for inclined plane tests) to the maximum particle diameter must be large. However, although the theoretical, possible, minimum value of this ratio is 10*, the practical value depends on the macroscopic and microscopic characteristics of the suspension in a currently unknown manner. In any case, the continuum hypothesis must be validated empirically *a posteriori* from direct observations, local measurements or tests with different geometries. Furthermore, when the suspension contains rigid particles, r must be sufficiently large to avoid the formation of a jammed structure of grains

* This is in fact the minimum theoretical value which makes it possible to respect the continuum hypothesis within 10% for a two-phase material. Indeed, for such a material the ratio of the difference between average and local particle velocity to the average velocity is at least of the order of 10%.

between tools. Nevertheless this condition is generally weaker than the continuum condition except when this jamming structure develops within the whole sample. This reflects peculiar material properties (typical of dense granular suspensions) which must be studied specifically. Finally, in Table 4.1, we compute the minimum tool dimensions on the basis of r=30 which seems to be a reasonable value based on our own experience with various suspensions. The systematic computations of Table 4.1 should make it possible to prevent many problems but, with some materials, relevant data might nevertheless not be obtained under these conditions.

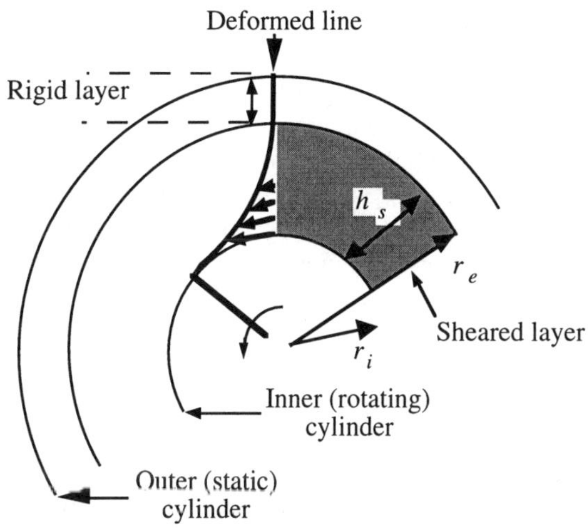

Figure 4.1: View of the free surface of a fluid sheared in a coaxial cylinder rheometer. Because of shear stress heterogeneity a fluid region close to the outer cylinder remains rigid. A line drawn along one radius before starting the test is deformed in the sheared region and remains static elsewhere.

4.1.1.2 *Specific effects for suspensions*

Because of the particular behaviour of concentrated suspensions (see Chapter 3), the above general theoretical arguments, though usual in fluid mechanics, may not work in some cases. When the shear stress is heterogeneous within the geometry, it may be smaller than the yield stress in some areas. These areas therefore remain rigid whereas the rest of the suspension flows. This effect typically occurs with a large gap coaxial cylinder rheometer, a capillary rheometer and an inclined plane (see Chapter 8). For a coaxial cylinder rheometer, the thickness (h_s) of the sheared area may be expressed as:

$$h_s = \sqrt{\frac{C}{2\pi H_0 \tau_c}} - r_i \qquad (4.1)$$

where H_0 is the cylinder length, C the applied torque and r_i the inner cylinder radius. From (4.1) it is clear that the sheared thickness decreases with C and is not necessarily equal to the gap (r_e-r_i) where r_e is the outer cylinder radius (see Fig. 4.1). This is the same for other geometries: the sheared thickness may be smaller than the gap (or flow depth for the inclined plane). As a consequence, during flow, obeying the continuum hypothesis can no longer simply be expressed as a sufficiently large gap to particle diameter ratio. The validity of the continuum assumption would at least require a large ratio of sheared thickness to maximum particle size. Since this ratio depends not only on tool dimensions but also on material yield stress and imposed flow rate or shear stress, no general rule can be proposed. To avoid any problems, the experimenter should either devise a new rheometer for each given material in a given shear rate range or use another geometry for which this effect is negligible. Since this is often not possible in practice, this effect remains a constant source of possible uncertainty.

Tool dimensions must also be such that edge effects (see 4.1.5.1.1) be minimized. For this reason, it is important to have a large ratio of characteristic length of sheared surfaces to characteristic edge length. For parallel plates and the cone-plate this is the ratio of plate diameter to peripherial gap, for a coaxial cylinder the ratio of depth to gap, for a capillary the ratio of tube length to diameter, for an inclined plane the ratio of plane width to flow depth. In Table 4.1 we systematically assume a minimum value of 10 for these ratios.

4.1.1.3 *Tool dimensions*

Making allowance for the above constraints the tool dimensions were calculated as a function of the particle characteristic diameter (φ) (see Table 4.1). Then the required volume of suspensions was deduced in each case. The bulk of the final geometry may be estimated from these values. For clay-water mixtures or silt-and-clay-water mixtures one can use conventional laboratory rheometers since the required volume is in general less than 1 cm^3. For sand-silt-and-clay-water mixtures, a special large device (O'Brien and Julien 1988) may be used. For coarser suspensions special large-scale rheometers must be built (Major and Pierson 1992, Phillips and Davies 1991, Coussot and Piau 1995). In practice, current large-scale rheometers are only capable of testing suspensions containing particles whose maximum diameter is less than few centimeters. For natural coarse debris flow mixtures, which contain particles whose diameter is up to few meters, one would need a rheometer testing a volume of at least one million cubic meters of material.

Finally, it should be remembered that a basic technique for detecting disturbing effects and ensuring the validity of viscometry measurements is to test the same material with different geometries. For two different rheometers, shear conditions (related to volume) change whereas disturbing effects (related to gap, free surface, etc.) change in an *a priori* different way. Thus data obtained by different means should in general be in agreement only if the material has, in each case, been homogeneously sheared as predicted by theory.

4.1.2 *Homogeneity of the theoretical shear rate field*

For suspensions containing coarse particles it is necessary to use rheometers with large gaps (see 4.1.1), which generally increases shear rate heterogeneity. In that case it is also necessary to use the appropriate rheometrical calculations which, even for heterogeneous shear rate, make it possible to derive the shear rate-shear stress curve from a set of torque or pressure difference vs velocity or flow rate data (see Appendix 2). However, since they assume a simple relation between τ and γ, these calculations are not necessarily valid for suspensions exhibiting time-dependent behaviour. Indeed, if τ also depends on the instantaneous internal state of the material, it will depend on the local flow history and thus may be quite different from one point to another. Conventional measurements, for example torque vs velocity data, do not make it possible to deduce the complete function $\tau(\gamma,t)$, since they do not provide any information concerning the structural state at different points of the material. They only give the force and the velocity resulting from a particular, instantaneous, structural state distribution which may change in time. Moreover, two different structural state distributions may yield the same macroscopic data in terms of force and velocity. In fact, thixotropy should be studied in rheometers providing homogeneous shear rate fields (Mewis 1979). Indeed, in that case, γ and τ are geometrically deduced from rotation velocity (or flow rate) and torque (or pressure difference) and thus $\tau(\gamma,t)$ is found directly from instantaneous measurements. In addition, shear rate heterogeneity may enhance the development of disturbing effects (see 4.1.5). In Table 4.1 the homogeneity of shear rate for each geometry is estimated.

4.1.3 *Use*

As rheometrical tests are difficult with concentrated suspensions, it is common to repeat the same tests many times. For this reason the apparatus should be devised with a wide space around the testing geometry giving the experimenter easy access with sample and tools. Additionally after each test it is important to be able to clean the sample easily and rapidly. It must be possible to separate the geometry, with plenty of clearance around tools. Since the above constraints are not directly related to geometry type but depend on the aspect and size of the whole device, they are not addressed in Table 4.1. In addition, sample set up within tools must be as easy as possible. Set up is made difficult in particular by the possibility of yield stress and the tendency of the fluid to develop large normal stresses when squeezed. In Table 4.1 set up difficulty is also estimated for each geometry.

4.1.4 *Observation of the strain field*

When possible it can be enlightening to observe the strain field in order to check shear rate distribution. Some disturbing effects can thus be discovered (see 4.1.5). For that purpose, an appropriate geometry presents large, sheared, free surfaces. For a coaxial cylinder rheometer this is the horizontal free surface, for a cone and plate the peripherial free surface, etc. Along these surfaces the strain field can be directly observed by making a mark on the sample (Magnin and Piau 1987, 1990, Boersma et al. 1991, Coussot et al. 1993, Persello et al. 1994, Aral and Kalyon 1994). However, one cannot be sure that when the shear rate measured from these direct observations corresponds to the theoretical one, it is also the case inside the sample. With a transparent material, one can first disperse or inject a coloured material in the sample and thus directly observe the internal strain field (El Kissi and Piau 1990, Pignon et al. 1994). With an opaque material, in the peculiar case when the geometry dimensions are large enough, it is possible to partly insert a small tool into the sample without significantly affecting flow. The motion of the external part of this tool provides information concerning internal material deformations (Coussot and Piau 1995). Finally, one can place a thin coloured layer of material in the sample before beginning the test. Analysis of the deformation of this layer can give some information concerning the internal strain field (Coussot 1992). However, in practice, this method is rather difficult to use. Indeed it requires to stop the flow after a limited deformation and then to cut the dried or frozen sample in slices upon which the deformations of the coloured material can be observed. Table 4.1 sums up the difficulty of carrying out direct observations with opaque materials.

Recent techniques for observing of suspended particle motions or structures (Nuclear Magnetic Resonance (Altobelli et al. 1991, Chow et al. 1994), X-ray diffusion (Keates et al. 1994), micro-radiography (Akay 1983), etc.) have been efficiently developed to relate microstructure changes to rheological properties (thixotropy, yield stress, etc.). However, if the techniques are to be subsequently related to macroscopic rheological properties, it is preferable that they be used when rheometrical problems have been controlled or understood from previous, repeated, rapid tests combined with simple direct observations.

4.1.5 *Experimental problems*

The experimental problems considered take the form of disturbing phenomena which typically occur with concentrated suspensions and affect the expected strain field or material homogeneity thus leading to erroneous interpretation of data in terms of the constitutive equation. They are generally due to both material characteristics and geometry features. Here, we shall not consider the numerous sources of errors which may occur with all fluid types (inertia effects, secondary flows, Weissenberg effect, flow geometry, viscous heating, equipment imperfections, surface tension, etc.) (see the review of Powell (1988)) except if they have already known, significant, particular consequences with concentrated suspensions. In Table 4.1 all the effects described below are taken into account along with their possible consequences on apparent viscosity measurements. When a viscosity change is mentioned, it refers either to a change compared to the one that should be obtained in steady state conditions or to a temporal change during a particular test under given stress or velocity when the disturbing effect develops during this particular test. As an example, in Figure 4.2, the main potential disturbing effects with a coaxial cylinder rheometer are represented.

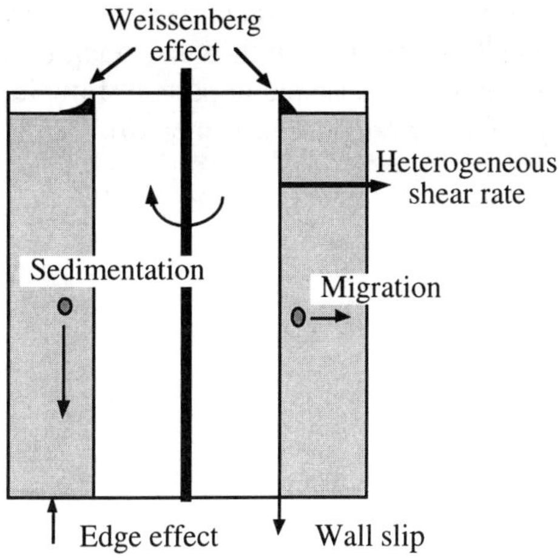

Figure 4.2: Main experimental problems with concentrated suspensions in a coaxial cylinder rheometer. Other problems such as secondary flows, viscous heating, etc., which may occur with other fluids, are not mentioned.

4.1.5.1 *Disturbing effects affecting strain field*

4.1.5.1.1 Edge effects

Material shear is imposed by tools which necessarily have limited lengths. Here we are concerned with alterations induced by shear geometry boundaries or edges to the expected (theoretical) strain field. Edge effects have been estimated theoretically in the case of Newtonian fluids (Oka 1960) or second-order fluids (Griffiths and Walters 1970) for different geometries. For yield stress fluids such as concentrated suspensions it is generally recognized that these effects can be significant (Highgate and Whorlow 1969), but no substantial corresponding theory exists.

Edge effects are highly dependent on the interface boundary conditions. Two main cases can be distinguished: a sample edge in contact either with a fluid or with a solid. In the first case, the main risk is a material free surface change during a test (for example crack (see 4.1.5.1.2)), essentially because of inertia (centrifugal force), normal stresses or/and secondary flows. With rigid boundary interfaces, one may essentially fear disturbance of the expected strain field within the geometry and/or additional stresses which lead to an increase in the apparent viscosity. The magnitude of edge effects also depends on material type and shear rate (Highgate and Whorlow 1969) and thus a general theory can hardly be proposed. Practically, it is essential to minimize their influence by choosing the smallest ratio of boundary surface to shear surface (see 4.1.1.2). Other methods can be proposed for coaxial cylinder rheometer (Coleman et al. 1966, Highgate and Whorlow 1969)) which for example consist in modifying the form of the inner cylinder bottom. In some cases (smooth coaxial cylinder bottom), one may expect concentrated suspensions to tend to slip along these walls (Coussot and Piau 1995), limiting the increase in apparent viscosity. Entrance or end effects in capillary rheometers (Tanner 1988, Senouci and Smith 1988) may be enhanced by particle migration (see paragraph 4.1.5.2.2). In inclined plane tests lateral edges impose additional flow resistance which may become significant when the ratio of flow depth to channel width is too small (see Chapter 8). Moreover the uniform flow depth must be measured in parts of flow far from channel extremities where the flow is gradually varied (see Chapter 10).

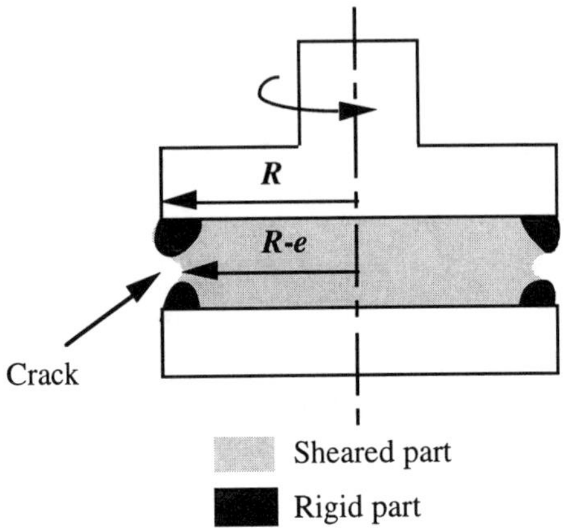

Figure 4.3: Axial cross-section of sample and tools for a parallel plate test: development of crack at the periphery. Cracks during shear take the form of an effectively sheared volume (approximately a cylinder of radius *R-e*) smaller than the theoretical one. The extent of the rigid regions at the periphery depends on material and flow characteristics.

4.1.5.1.2 Cracks

Under some conditions, the peripherial free surface of the sample, originally planar and vertical, bends so as to penetrate the former sample volume (Cf Fig. 4.3). This phenomenon has been essentially observed with torsional geometries. It was originally called fracture (Hutton 1963, 1975, Vinogradov et al. 1975) but we shall keep the latter word for the case when a "crack" really occurs within the sample without significant deformation of the free surface (see 4.1.5.1.4). The crack effect could be due to normal stress difference exerting more influence than surface tension (Hutton 1963, Tanner and Keentok 1983, Lee et al. 1992). In the present case, though the internal strain field may be slightly affected by this crack, deformations remain continuous within the sample. Furthermore, the free surface is separated into a central part, where shear remains visible on the peripherial free surface, and upper and lower parts which slightly jut out of the sample and remain almost rigid during flow (Hutton 1975, Coussot et al. 1993). The extent of the latter parts is rather limited and some authors have considered it possible (Hutton 1975, Coussot et al. 1993, Coussot and Piau 1994) to proceed to relevant rheometrical calculations by simply assuming that a smaller volume than the theoretical one is really sheared in the expected way (without edge effects). For example for a cone and plate rheometer the shear stress may be directly derived from the usual formula (see Appendix 2):

$$\tau(\dot{\gamma}) = \frac{3C}{2\pi((R\text{-}e)^3)} \tag{4.2}$$

where C is the torque, R the cone radius and e the equivalent thickness of the lost volume of material (see Fig. 4.3).

In many cases, this crack develops rapidly at the beginning of the test but remains limited, and then varies slightly during subsequent tests with the same, set up sample. However, since, as can be deduced from equation (4.2), the largest part of the torque is applied at the periphery of the sample, even slight changes in crack extent may seriously affect results. For example this may make the experimenter think that the suspension is thixotropic (Hutton 1963). This must have been the case for some experiments presented in literature, in particular with clay-water mixtures, which consisted in imposing long duration flow without specific control of free surface changes. To minimize this effect, it is possible to impose long, rapid pre-shearing in order, ideally, to stabilize the crack in an aspect which will not change too much during subsequent tests (Coussot and Piau 1994). Nevertheless this method is not always sufficient and the experimenter is advised to remember that this effect is often the subsisting, possible cause of disturbances. Furthermore, when testing materials containing coarse solid particles, the thickness of the central sheared part at the periphery may reach the same order of magnitude of the particle size. Under these conditions, the continuum hypothesis is not respected in this part and particle jammings may occur, which may affect significantly results (Coussot 1992).

With some highly concentrated suspensions exhibiting very large yield stress, this crack may extend rapidly within the sample volume so as to separate the material into two distinct parts. Then, after having separated the tools, one can observe concentric circles of different material depths at the free surface of the two sample parts stick to each tool. These circles originate from the last stage of contact between a few points of the two surfaces which move in relation to each other. This crack effect seems to occur after a critical deformation and thus tests at low shear rates may remain possible. At high shear rates for which the size of this crack increases during a given test, a possible (approximate) solution for deducing the shear rate-shear stress curve is to take into account the changes of the effectively sheared volume as in the above method for less severe cracks. Then alternating short measurements at various shear rates and measurements at a fixed low shear rate level using the appropriate equation (for example Equation 4.2 for a cone and plate) will make it possible to deduce the sheared volume changes between each new shear rate level and thus the flow curve.

4.1.5.1.3 Wall slip

With any viscometric flow, when using smooth tool surfaces, wall slip may occur. It takes the form of a significant velocity difference between a tool surface and suspension elements close to this surface (Cf Fig. 4.4). The importance of this effect varies with material type (particle size and type, interstitial fluid viscosity) and flow velocity (Hutton 1975, Boersma et al. 1991), geometry dimensions (Hutton 1975), wall shear stress (Yilmazer and Kalyon 1989, Kalyon et al. 1993), delay time before test and particle concentration (Buscall et al. 1993), fluid yield stress (Yoshimura et al. 1987) and time (Aral and Kalyon 1994). A general review of wall slip effects (cause, character and cure) was proposed by Barnes (1996). Here we shall focus on its particular features in the case of concentrated suspensions.

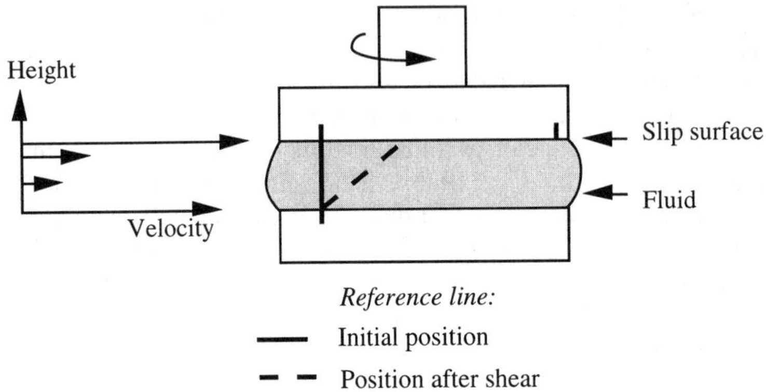

Figure 4.4: Wall slip during parallel plate test. A reference line parallel to the geometry axis has been drawn on sample and tools. One can see its initial position (continuous line) and its position after a finite shear (dashed line). Without slip, the dashed line should not be separated in two independent parts.

Wall slip with concentrated suspensions probably originates in two phenomena in conjunction: the yield stress of the fluid (or its large viscosity) which tends to maintain it as a rigid plug, and the possibility of a thin layer, with a smaller suspended particle concentration (Powell 1988), forming close to the wall. As this layer has a lower viscosity than the rest of the fluid, it is preferentially sheared, which leads to an apparent discontinuity in shear rate. Wall slip is naturally enhanced by the inevitable decrease in particle concentration when approaching a wall. However it is possible to get round this obstacle and even avoid wall slip by using tool surfaces with reasonable roughness, either by grooving (Vinogradov et al. 1975, Magnin and Piau 1990) or blasting (Mas and Magnin 1994) tool surfaces, or gluing particles (Kao et al. 1975, Major and Pierson 1992), sandpaper (Magnin and Piau 1990, Coussot et al. 1993, Collias and Prud'homme 1994) or adhesive material (Navickis and Bagley 1983, Magnin and Piau 1987). The decrease in particle concentration occurs essentially within the roughness whereas the effectively sheared volume is limited by a surface passing directly through the top points of the roughness. This roughness must be large enough compared to particle size for these particles to significantly penetrate the holes but must remain small enough compared to tool size in order to avoid significant disturbance of strain field. For yield stress fluids it has also been proposed to use a conventional test in soil mechanics for measuring the shear strength of cohesive soils, the vane-test (see Appendix 4) (Keentok et al. 1985, Nguyen and Boger 1983, 1985, 1992, Alderman et al. 1991). With this technique the surface of contact between the sheared fluid and the tools is quite limited and thus slip cannot develop. A disadvantage is that the velocity distribution is not quite controlled.

Some authors also proposed to make rheological measurements inspite of wall slip by estimating the intensity of this effect. The original method of Mooney (1931) which required an analysis of data obtained with three different rheometer sizes has recently been improved (Yoshimura and Prud'homme 1988, Kiljanski 1989). A more thorough physical analysis of wall slip has been proposed by Vinogradov et al. (1975) for plastic disperse systems.

In some cases, with extremely concentrated suspensions, even when many precautions are taken, the shear remains restricted to a thin plane near the moving surface (Cheng and Richmond 1978, Kirby 1988). In this case the corresponding measurements cannot be used to deduce the constitutive equation of an equivalent homogeneous material.

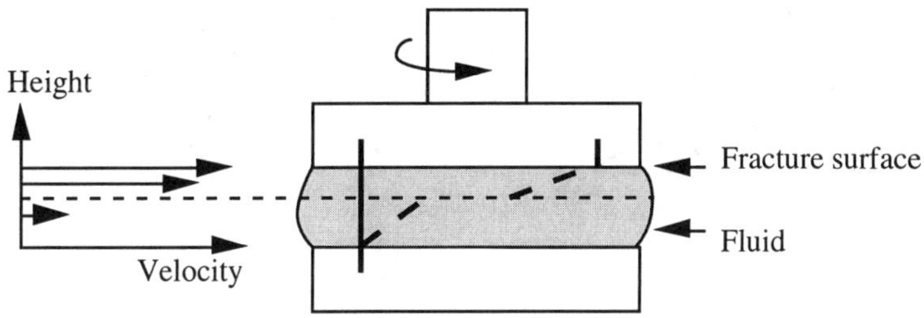

Figure 4.5: Fracture within sample during parallel plate or cone and plate tests. Line drawn at the periphery as in Figure 4.4.

4.1.5.1.4 Fracture

With some materials one may observe a localization of shear within a thin layer of material which extends within the sample (Dealy and Giacomin 1988, Navickis and Bagley 1983, Magnin and Piau 1990, Coussot et al. 1993) (Cf Fig. 4.5). Remarkable observations of this phenomenon in torsional flows have been made by Magnin and Piau (1990) and Persello et al. (1994) along the peripherial free surface and Pignon et al. (1994) within a transparent suspension. This effect should thus be preferentially called fracture (Magnin and Piau 1987) in analogy with rupture in solids or soils. As with wall slip, it is probably enhanced by material yield stress and slight internal heterogeneities. However this phenomenon and its origins remain poorly understood. Persello et al. (1994) called it "internal slip" and suggested that it could originate in a self-lubricating effect: the release of interstitial fluid into the fracture lets solid regions pass over each other. Furthermore, for a given material it seems to occur only beyond or below some shear rate values or solid concentrations. Thus, for the tests, it may constitute a limiting factor in terms of particle concentration and shear rate since no actual means exists to avoid it.

Table 4.1: Advantages and disadvantages of various rheometer geometries and experimental problems

Shear geometry	*Minimum dimensions and volume*	*Ease of use:* *Set up*	*Cleaning*	*Shear rate field*	*Strain field observation (opaque fluid)*	*Experimental problems*	*Effects on measurements*
Coaxial cylinders * Large gap 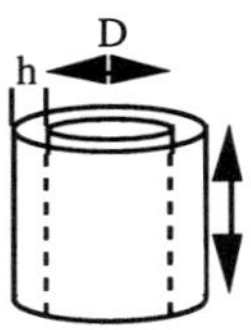	$h = 30\ \varphi$ $H = 10h = 300\ \varphi$ $D = 4h = 120\ \varphi$ $V \approx 4{,}2.10^6\ \varphi^3$	Relatively difficult: - Fluid injection if tools are already installed - Jamming or fluid in excess during insertion of inner cylinder	Difficult if tools cannot be moved apart Relatively easy otherwise	Heterogeneous: varies radially because of shear stress variation from inner to outer cylinder	Possible at the top free surface	Edge effect due to the rigid bottom wall - Wall slip at the inner or outer cylinder - Fracture (generally localized around the inner rough cylinder) Evaporation through the top free surface Shear-induced (outward) radial migration Centrifugation	Apparent viscosity increase (weaker effect if bottom wall slip) Apparent viscosity decrease Apparent viscosity increase Apparent viscosity decrease Apparent viscosity decrease when particles move outwards, viscosity increase otherwise
* Small gap	$D = 30h = 900\ \varphi$ $V \approx 2{,}6.10^7\ \varphi^3$	ditto, but more difficult because of the smaller gap	ditto	Roughly homogeneous	ditto but more difficult because of the smaller gap	Sedimentation	Variable, depends on the new mean viscosity of the heterogeneous suspension, additionally possible jamming of particles in the bottom part
Capillary 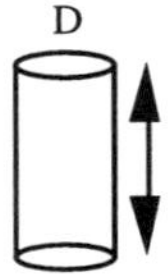	$h = D = 30\ \varphi$ $L = 50h$ V (within the pipe) $\approx 10^6\ \varphi^3$ Vtotal (needed for recirculation) $\approx 10^7\ \varphi^3$	Easy, as part of the flow	Relatively easy	Heterogeneous: varies radially because of shear stress decrease inwards	Impossible no free surface	Wall slip Possible fracture Radial migration ($Re>10^{-3}$), outwards or inwards and possibly "regimentation" Segregation at the entrance or shear-induced axial migration: filtration and formation of particle rich plug Sedimentation	Apparent viscosity decrease Apparent viscosity decrease Apparent viscosity decrease Apparent viscosity increase Variable, depends on the installation

Shear geometry	*Minimum dimensions*	*Ease of use: Set up*	*- Cleaning*	*Shear rate field*	*Strain field observation*	*Experimental problems*	*Effects on measurements*
Parallel plates 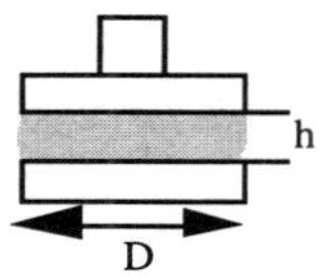 	$h = 30\ \varphi$ $D = 10h = 300\ \varphi$ $V \approx 2.10^6\ \varphi^3$	Generally easy but requires care with highly concentrated materials: large normal stresses or jamming structure in plate axis direction	Easy if plates can be move apart sufficiently	Heterogeneous: varies radially because of shear rate decrease inwards	Possible at the lateral free surface	Edge effects due to fluid in excess at the periphery Crack - Wall slip - Possible shear-induced radial migration - Fracture Evaporation Centrifugation Sedimentation	Apparent viscosity increase, due to additional resistance Variable, depends on relative extent of rigid and sheared parts Apparent viscosity decrease Generally apparent viscosity increase, except when peripheral volume decrease is predominant Apparent viscosity increase when particles move outwards, viscosity decrease otherwise Variable, depends on the new mean viscosity of the heterogeneous suspension
Cone and-plate 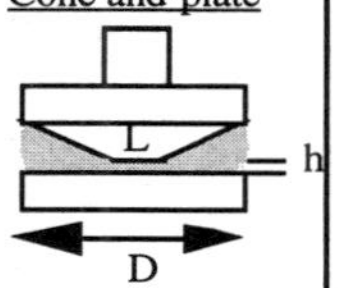 	$\alpha<6°$; $h = 30\ \varphi$ $L = \dfrac{2h}{\text{tg}\alpha} = 575\ \varphi$ $D = 5L = 2875\ \varphi$ $V \approx 6.10^8\ \varphi^3$	ditto, but more difficult because of the central smaller gap	ditto	Homogeneous	ditto	ditto, Negligible radial shear-induced migration because of shear rate homogeneity	ditto
Inclined plane 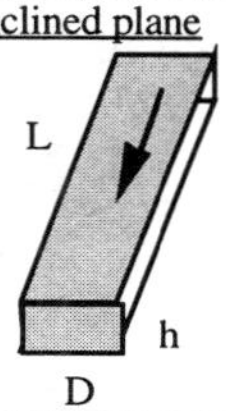 	$h = 30\ \varphi$ $D = 10\ h = 300\ \varphi$ $L = 100\ h = 3.10^3\ \varphi$ V (in channel) $\approx 2.7\ 10^7\ \varphi^3$ Vtotal (needed for recirculation) $\approx 10^8\ \varphi^3$	Easy, as part of the flow	Easy	Heterogeneous: varies radially because of shear stress decrease upwards	Impossible: no free surface in main shearing plane; Edge effect visible at the free surface	- Wall slip - Possible fracture Segregation at the entrance: filtration and formation of particle rich plug Possible migration or sedimentation	Apparent viscosity decrease Apparent viscosity increase Variable, depends on the relative importance of induced heterogeneities

4.1.5.2 *Phenomena affecting the material*

Here we assume that the sample is prepared in a homogeneous and reproducible way before any rheometrical test.

4.1.5.2.1 Evaporation

The consequence of evaporation of the suspending (interstitial) fluid is that the mean solid concentration of the sample increases. This generally gives rise to an apparent viscosity increase except when the corresponding decrease in the volume of the total sample has a more significant effect.

Evaporation is all the more crucial as the ratio of a characteristic surface of the sheared volume to the free surface is smaller. Moreover, with concentrated suspensions, evaporation is enhanced (Powell 1988) because, at the free surface, very thin layers of interstitial fluid surrounding solid particles can evaporate quickly. As a whole, the rate of evaporation increases with solid concentration. Methods to minimize or avoid this effect involve carrying out the test in a saturated atmosphere (Magnin and Piau 1990) or coatting the sample/air interface with a layer of a low volatility, relatively non-viscous oil (Powell 1988) or with the interstitial fluid (Magnin and Piau 1990).

4.1.5.2.2 Particle migration

Suspended particles may migrate through the suspension during flow. This effect generally induces irreversible heterogeneities in particle distribution. Inertia effects or/and shear heterogeneities are the main causes of this phenomenon. In powder flows the observed segregation ("granular sieving") (Savage and Lun 1988, Stephens and Bridgwater 1978) is a conceptually similar effect since it is due to local stress heterogeneities. Particle migration theory is complex in the case of a non-Newtonian interstitial fluid or when colloidal forces between particles are significant. Only a few works have provided answers to this problem in the case of unyielding fluids (Highgate and Whorlow 1969, Caswell 1977, Leal 1979). Particle migration through yield stress fluids in rheometers remains an open problem but it is probable that its qualitative trends are similar to those found with Newtonian fluids, reviewed below.

Particle migration in capillary flows of non-colloidal particle suspensions has been widely studied either experimentally or theoretically (see reviews of Cox and Mason (1971) or Yaras et al. (1994) or the basic works of Segré and Silberberg (1962), Oliver (1962) and Karnis et al. (1963)). In this case migration characteristics depend essentially on particle size (in relation to tube radius), form and concentration, flow direction, Reynolds number and the density difference between fluid and particles. This effect has mainly been considered as an inertia effect which occurs for Reynolds numbers above 10^{-3} but it was also argued that particles could diffuse in the absence of inertia effects (Koch 1989). Radial particle migration in coaxial cylinder geometries with neutrally buoyant, non-colloidal particle suspensions have also been observed (Halow and Wills 1970, Gadala-Maria and Acrivos 1980, Leighton and Acrivos 1987, Nadim 1988, Graham et al. 1991, Abbott et al. 1991, Altobelli et al. 1991, Sinton and Chow 1991). The intensity of this effect, which has mainly been considered as a shear-induced phenomenon, essentially depends on shear rate, particle size and concentration but does not seem to be significantly influenced by particle form (Mondy et al. 1994). However, with non-neutrally buoyant particles one can also expect radial migration (centrifugation) due to inertia effects. Shear-induced migration in torsional flows (cone and plate, parallel plates) has scarcely been studied so far, but it appears to be less significant than in flows between coaxial cylinders (Chow et al. 1994). Since in general migration depends on particle size, a mixture of particles of varying sizes can also be separated (Metzner 1985).

Sedimentation (normal or opposite) occurs when particle density is different from the surrounding fluid density. With concentrated suspensions this may not occur at rest because of the surrounding fluid yield stress. However, as soon as shear is induced, no sufficient normal stress balances the specific gravity applied to the particle which starts to fall (see Chapter 11). From a practical point of view, this effect must therefore be checked. This can be achieved either by simply comparing the characteristic sedimentation time (computed using an approximate particle velocity through the fluid at rest (see Chapter 11)) to the test duration, or by directly measuring the local density variations within the suspension. Special device have also been proposed which make it possible to carry out (more or less approximate) rheometrical tests despite significant sedimentation. Most of these apparatuses consist in imposing a secondary flow between the tools, which tends to maintain the homogeneity of the suspension (Clarke 1967, Valentyik 1971, Overend et al. 1984, Kemblowski 1988, Reeves 1990). Recently another technique was proposed which consists in imposing and thus measuring the shear within a restricted, central zone which is not significantly affected by sedimentation, that is density remains constant, during test duration (Klein et al. 1995). The position of the bob, which imposed the height of sheared fluid, must be chosen and set before each test, depending on settling properties of the suspension and the desired measurement period.

For capillary or inclined plane flows, the necessary rapid reduction of flow section from the upstream reservoir or chamber to the tube or to the open-channel may lead to a particular type of segregation. Solid particles may be compressed towards each other which leads to a greater particle concentration in the chamber than in the tube or in the channel (Metzner 1985, Yaras et al. 1994, Coussot 1994). This filtration effect must be surveyed since, with concentrated suspensions, a slight change in particle fraction may lead to a large behaviour change. Additionally this effect, which is crucial for mud suspensions, may give rise to the formation in the chamber of a particle rich plug which remains around the entrance and increases flow resistance.

4.1.5.2.3 Shear-thickening

Shear-thickening is not properly speaking an experimental problem but, to obtain relevant data, one must keep in mind and possibly take into account the particular effects resulting from this phenomenon. It has been shown that concentrated suspensions of nearly monodisperse non-aggregating solid particles in a low viscous Newtonian liquid exhibit a clear (reversible) viscosity increase with increasing shear rate (shear-thickening) (see 3.3.1) (Barnes 1989). This phenomenon takes the form of a rapid change in the viscosity level over a short shear rate range (one decade) but it may be even faster, then appearing for a critical shear rate. The nature of the shear-thickening effect not only depends on the parameters of suspended phase (concentration, particle distribution, particle shape, mutual interactions, etc.) but also on flow history. First, the jump in viscosity shows typical features of a phase transition: hysteresis and a metastable low viscosity state above the critical shear rate (Laun et al. 1991). In addition, when a constant stress corresponding to the transition range is applied, the recorded shear rate fluctuates (Laun et al. 1991). Finally, the apparent critical shear rate may depend on rheometer geometry (Laun et al. 1991) or more precisely on gap size (Chow and Zukoski 1995).

4.2 SPECIFIC EXPERIMENTAL PROCEDURES FOR SHEAR RHEOMETRY WITH CONCENTRATED SUSPENSIONS

In order to distinguish the consequences of disturbing effects and true rheological properties, it is necessary to control disturbing effects as much as possible (see above) and to analyze results obtained with similar edge effects and initial internal structures. In this way, the appropriate method consists in preparing the fluid and setting it in the rheometer with a fixed technique, cleaning carefully the fluid in excess, then pre-shearing it during a fixed time at a given velocity and finally leaving it at rest for another given time (Coussot et al. 1993, Mas and Magnin 1994). Then steady state or transient tests can be carried out. In order to avoid some disturbing effects it may be necessary to shorten some tests. This is for example the case when, as for mud suspensions, a crack slowly develops during shear (see Coussot (1994)). Two rheological properties, i.e. yield stress and thixotropy, are often typical characteristics of concentrated suspensions, even if time-dependent effects are generally negligible under common flow conditions for natural mud suspensions. In the following we propose a critical review of the experimental procedures presented in literature to quantify these properties. We shall not focus on other properties such as normal stress differences because they do not seem to be typical of concentrated suspensions.

4.2.1 *Yield stress*

The yield stress of a fluid is a minimum stress to overcome for flow to occur (see Chapters 2 and 3). For concentrated suspensions there often exists a shear stress threshold which limits a sharp transition in behaviour during creep tests. Indeed strain-time curves apparently have two different forms: with saturation or with monotonic increase in strain (flow) depending on the relative position of imposed shear stress and yield stress (Nguyen and Boger 1983, Coussot et al. 1993, Mas and Magnin 1994) (Cf Fig. 4.6). This definition is consistent with physical arguments for the existence of yield stress: the yield stress is related to a continuous network of interactions which behaves more or less elastically below a critical stress value and breaks for larger shear stresses. According to this yield stress definition, our considerations below demonstrate that accurate and complete rheological measurements should not, as considered by some authors (Husband et al. 1993, Bonnecaze and Brady 1992), give different yield stress values depending on the method used to determine it. A review of existing techniques for yield stress determination has been proposed by Nguyen and Boger (1992). Here we review the procedures suitable for using basic rheometrical geometries and specific problems encountered with concentrated suspensions.

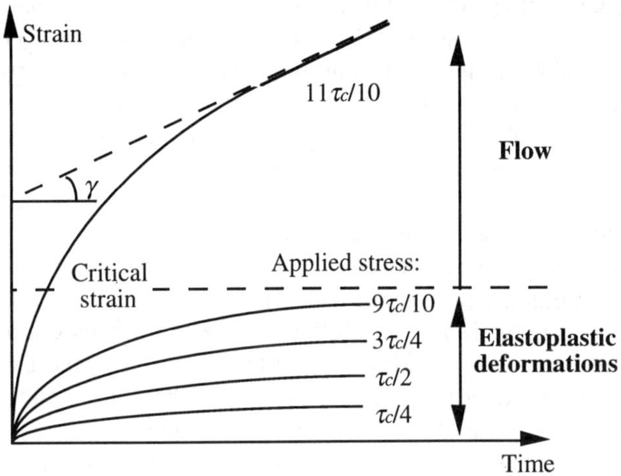

Figure 4.6: Typical aspect of creep test curves: constant levels of shear stress are applied to the material. Results are presented in terms of strain as a function of time. For shear stress smaller than yield stress each curve tends to an asymptotic strain value which depends on stress level. In that case, most of the deformation is recoverable after stress release which demonstrates the elastic behaviour of the material. For shear stress slightly exceeding yield stress there is an irreversible strain and the curve slope tends to an asymptotic value which is the shear rate of the steady flow.

A first method consists in measuring the residual shear stress after relaxation* (Hutton 1975, Keentok 1982, Nguyen and Boger 1983, Magnin and Piau 1990). It is thought that since it reflects the strength of the structure at rest, the yield stress will be recovered after flow stoppage and subsequently recorded by the apparatus. However though initially promising and providing satisfactory results in some cases (Nguyen and Boger 1983, Magnin and Piau 1990), non-reproducible data (Coussot et al. 1993) or yield stress underestimates (Keentok 1982, Mas and Magnin 1994) are often recorded. Possible explanations are the thixotropic properties of the material (Mas and Magnin 1994) or a residual (possibly reversal) shear during stoppage due to imperfections in the apparatus (Coussot et al. 1993). Nevertheless, the particular relaxation test which consists in suddenly stopping the upstream fluid supply during inclined plane flow and measuring the residual fluid depth after a long time of rest can give relevant results (Coussot and Boyer 1995). With this geometry, it is also the best way to determine yield stress (Cf Chapter 12).

* Sudden flow stoppage under controlled shear rate.

The most common method for determining yield stress consists in extrapolating the experimental stress-rate (flow) curve towards low shear rates. The obtained asymptotic shear stress is thought to correspond to the yield stress. First of all, it should be stressed that the accuracy of this extrapolation depends a great deal on the lowest shear rates for which one has obtained data, the extrapolation method and the fluid behaviour. In this way a frequent error (see 2.3.3.2) for yield stress determination is committed when fitting a (linear) Bingham model to flow curve data in a rather small shear rate range whereas suspensions are often shear-thinning (see 5.1.1.2). It is worth noticing that with many concentrated suspensions disturbing effects are crucial at low shear rates which, sometimes, makes it difficult to interpret data. Additionally this method provides a "model" yield stress which is *a priori* not necessarily related to a specific physical property (Nguyen and Boger 1983, Keentok 1982, Coussot and Piau 1994). However, in general, when it has been carefully determined in a wide shear rate range, the flow curve should provide a rather good estimate of the real yield stress (Coussot and Piau 1994). In the particular case of a material exhibiting a flow curve with a minimum (see 5.1.1.2.4) (Coussot et al. 1993, Mas and Magnin 1994), the flow curve never flattens out at low shear rates. Shear stress extrapolation is consequently tricky and can give bad results, especially when flow curve data have only been obtained for high shear rates, in a region where the decreasing part of the curve is not apparent.

Lastly, a method which avoids the problems described above is directly related to the definition of the true yield stress. It consists in applying successive, increasing, shear stress values until a clear change in the corresponding creep curve is observed (Nguyen and Boger 1983, Coussot et al. 1993, Mas and Magnin 1994). Below this yield stress most of the strain is recovered when the stress is released which reflects the elastic behaviour of the fluid. For sufficiently small deformations the ratio of the shear stress to elastic deformation is constant and corresponds to the elastic modulus of the suspension below yield stress. In the case of a fluid with a minimum (see above), this method is the only one that enables us to determine the true yield stress: below this critical stress value there is no flow, just above it the flow very rapidly reaches relatively high shear rate (Coussot et al. 1993). We pointed out before that, from a more general point of view, this method requires waiting a sufficient length of time before each new creep test in order to let the material recover its initial properties. Indeed, even if the material does not really flow before reaching the yield stress, the repeated slight deformations induced by creep can progressively affect the internal structure of the fluid. As an extreme example the technique of the rapid shear stress increase (ramp), which was formerly rather common, cannot in general provide adequate data for yield stress determination. In the case of inclined plane flows, a test of this type involves progressively increasing the channel slope (see Chapter 12), which increases the volume force and consequently the wall shear stress. However, in this case it is rather hard to carry out this test properly. At each new slope one may wait for a sufficiently long time to ascertain that flow does start or not. Since beyond the critical slope the flow velocity is extremely small, perturbing effects (fracture, settling, etc) have then time to develop (see Chapter 8).

4.2.2 *Thixotropy*

The reviews of Mewis (1979, 1994) provided physical explanations for thixotropic phenomena along with an exhaustive, critical review of corresponding measurement methods. Here, we shall only summarize the main results of these works and give some additional comments.

Basically, determining thixotropic properties requires a flow geometry with homogeneous shear stress. Otherwise exact data analysis is almost impossible (see 4.1.2 and Appendix 2). The conventional test which estimates the extent of the hysteresis loop obtained from a shear rate or shear stress increase-decrease ramp hardly makes it possible subsequent relevant comparisons of the behaviour of different materials (see Mewis (1994)). Indeed from a theoretical point of view it involves complex uncontrolled flows whose data can hardly be used for a quantitative estimation of thixotropic characteristics. Even for a simple, relevant comparison of thixotropic properties of various materials, these hysteresis can hardly be used when they are drawn in linear scale as usual. At least a consistent comparison of these hysteresis would require to draw them in logarithmic scale or to compare relative hysteresis surfaces (dividing their surface by a reference shear stress level for each material). In addition, with the ramp test it is impossible to control experimental perturbating effects. In particular the extent of these effects with materials of this type generally vary with flow intensity and duration.

An improved method is now commonly used to characterize thixotropic properties of fluids. It consists in suddenly applying a given shear rate or shear stress to the material originally in a given known internal state (Coussot et al. 1993, Mas and Magnin 1994). The corresponding shear stress- or shear rate-time curve thus obtained generally exhibits a peak (negative or positive depending on the initial conditions) followed by a slow decrease or increase towards steady state. Though this kind of curve reflects a true material property (Lockyer and Walters 1976)) it is probably affected by disturbing effects or equipment defects. It is more reliable to repeat and compare such tests under different conditions (see Fig. 4.7). Indeed this makes it possible to determine some physical characteristics of the suspension: the characteristic (restructuration) time for complete bond restoring at rest and the characteristic (destructuration) time of bond rupture during flow.

The above method is valid only if the material is not impaired by the flow. Moreover, many disturbing effects (see 4.1.5), typical of rheometry with concentrated suspensions, change with time and thus may yield data trends similar to those obtained with thixotropic materials. Other difficulties may originate in the necessary imperfection of the apparatus: the time delay to impose a given shear rate or shear stress and record the corresponding data, the time delay for relaxation, inertia effects (see Krieger (1990)), and so on. A review of works related to these problems has been proposed by Powell (1988). In practice it is often assumed that at least the data obtained with a single apparatus are reasonably self consistent.

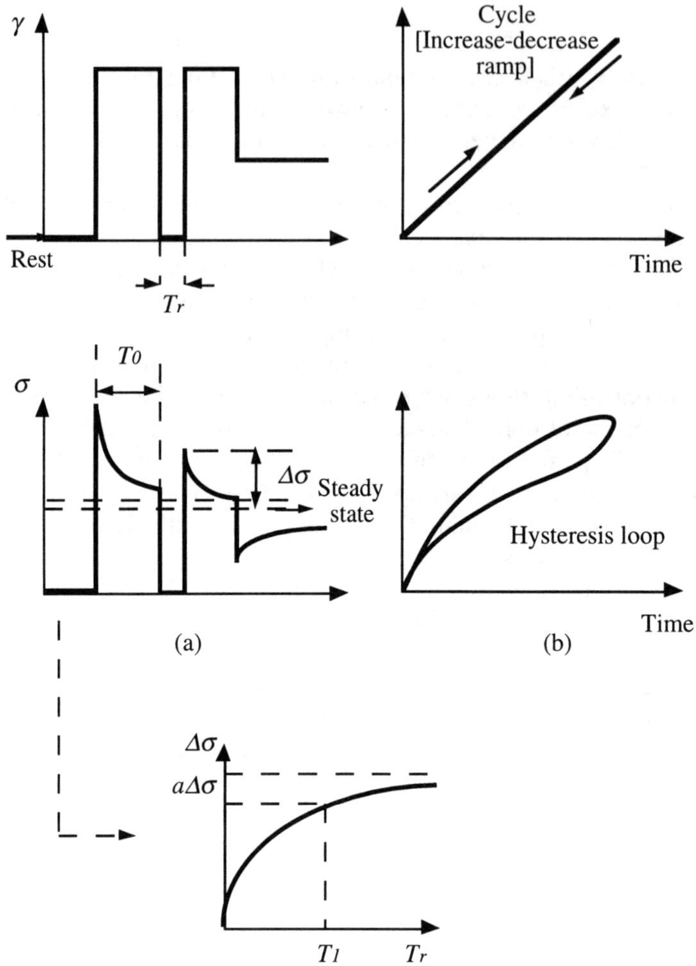

Figure 4.7: Different tests for estimating time properties of fluids. The ramp test is presented in b). In that case the shear stress (or the shear rate) is progressively increased then decreased. The corresponding shear rate (or shear stress) is recorded and yields a hysteresis loop. More valuable tests are presented in a). They consist in imposing various shear stress levels (this may also be done with shear rate levels) after different times of rest (T_r). T_0 is obtained by measuring the time required to reach a given percentage of the steady state value (T_r=∞). T_1 is obtained by measuring the rest time necessary for the overshoot ($\Delta\sigma$) to reach a given percentage (1-a) of its maximum value (after an infinite time of rest). For a suspension, T_0 is a characteristic destructuration time, that is the time necessary for most bonds to be broken during flow. T_1 is a characteristic restructuration time, that is the time necessary for most bonds to be restored at rest.

4.3 REFERENCES

Abott, J.R., Tetlow, N., Graham, A.L., Altobelli, S.A., Fukushima, E., Mondy, L.A., and Stephens, T.S. 1991. Experimental observations of particle migration in concentrated suspensions: Couette flow. *J. Rheol.*, 35, 773-795.

Akay, G. 1983. Unstable capillary flow of reinforced polymer melts. *J. Non-Newt. Fluid Mech.*, 13, 309-323.

Alderman, N.J., Meeten, G.H., and Sherwood, J.D. 1991. Vane rheometry of bentonite gels. *J. Non-Newt. Fluid Mech.*, 39, 291-310.

Altobelli, S.A., Givler, R.C., and Fukushima, E. 1991. Velocity and concentration measurements of suspensions by nuclear magnetic resonance imaging. *J. Rheol.*, 35, 721-734.

Aral, B.K. and Kalyon, D.M. 1994. Effects of temperature and surface roughness on time-dependent development of wall slip in steady torsional flow of concentrated suspensions. *J. Rheol.*, 38, 957-972.

Barnes, H.A. 1989. Shear-thickening ("Dilatancy") in suspensions of nonaggregating solid particles dispersed in Newtonian liquids. *J. Rheol.*, 33, 329-366.

Barnes, H.A. 1995. A review of the slip (wall depletion) of polymer solutions, emulsions and particle suspensions in viscometers: its cause, character and cure. *J. Non-Newt. Fluid Mech.*, 56, 221-251.

Boersma, W.H., Baets, P.J.M., Laven, J., and Stein, H.N. 1991. Time-dependent behavior and wall slip in concentrated shear thickening dispersions. *J. Rheol.*, 35, 1093-1120.

Bonnecaze, R.T. and Brady, J.F. 1992. Yield stresses in electrorheological fluids. *J. Rheol.*, 36, 73-115.

Buscall, R., McGowan, J.I., and Morton-Jones, A.J. 1993. The rheology of concentrated dispersions of weakly attracting colloidal particles with and without wall slip. *J. Rheol.*, 37, 621-641.

Caswell, B. 1977. Sedimentation of particles in non-Newtonian fluids. in *The Mechanics of viscoelastic Fluids*, ASME, AMD, 22, 19-46.

Cheng, D.C.-H., and Richmond, R.A. 1978. Some observations on the rheological behaviour of dense suspensions. *Rheol. Acta*, 17, 446-453.

Chow, A.W., Sinton, S.W., Iwamiya, J.H., and Stephens, T.S. 1994. Shear-induced particle migration in Couette and parallel-plate viscometers: NMR imaging and stress measurements. *Phys. Fluids*, 6, 2561-2576.

Chow, M.K., and Zukoski, C.F. 1995. Gap size and shear history dependencies in shear thickening of a suspension ordered at rest. *J. Rheol.*, 39, 15-32.

Clarke, B. 1967. Rheology of coarse settling suspensions. *Trans. Inst. Chem. Eng.*, 45, T251-T256.

Coleman, B.D., Markowitz H., and Noll, W. 1966.*Viscometric Flows of Non-Newtonian Fluids*, Springer Verlag, New-York.

Collias, D.I. and Prud'homme, R.K. 1994. Inhomogeneous flows of guar/metal ion gels observed by laser Doppler anemometry and rheological measurements. *J. Rheol.*, 38, 217-230.

Collyer, A.A., and Clegg, D.W. 1988. *Rheological measurement*, Elsevier Applied Science, Amsterdam.

Coussot, P. and Piau, J.M. 1995. A large-scale field coaxial cylinders rheometer to study the rheology of natural coarse suspensions. *J. Rheol.*, 39, 105-124.

Coussot, P. and Piau, J.M. 1994. On the behavior of fine mud suspensions. *Rheol. Acta*, 33, 175-184.

Coussot, P., Leonov, A.I. and Piau, J.M. 1993. Rheology of concentrated dispersed systems in a low molecular weight matrix. *J. Non-Newt. Fluid Mech.*, 46, 179-217.

Coussot, P. 1992. *Rheology of debris flows - Study of concentrated suspensions.* Ph.D. thesis, INPG, Grenoble, France. (in French)

Coussot, P. 1994. Steady, laminar, flow of concentrated mud suspensions in open channel. *J. Hydraul. Res.*, 32, 535-559.

Coussot, P., and Boyer, S. 1995. Determination of yield stress fluid behaviour from inclined plane test. *Rheol. Acta*, 34, 534-543.

Cox, R.G. and Mason, S.G. 1971. Suspended particles in fluid flow through tubes. *Ann. Rev. Fluid Mech.*, 3, 291-316.

Dealy, J.M. and Giacomin, A.J. 1988. Sliding plate and sliding cylinder rheometers. in *Rheological measurement*, A.A. Collyer and D.W. Clegg (Ed.), Elsevier Applied Science, Amsterdam, Chapter 12, 383-404.

El Kissi, N., and Piau, J.M. 1990. The different capillary flow regimes of entangled Polydimethylsiloxane polymers: Macroscopic slip at the wall, hysteresis and cork flow. *J. Non-Newt. Fluid Mech.*, 37, 55-94.

Gadala-Maria, F., and Acrivos, A. 1980. Shear-induced structure in a concentrated suspension of solid spheres. *J. Rheol.*, 24, 799-814.

Graham, A.L., Altobelli, S.A., Fukushima, E., Mondy L.A., and Stephens, T.E. 1991. Note: NMR imaging of shear-induced diffusion and structure in concentrated suspensions undergoing Couette flow. *J. Rheol.*, 35, 191-201.

Griffiths, D.F. and Walters, K. 1970. On edge effects in rheometry. *J. Fluid Mech.*, 42, 379-399.

Halow, J.S., and Wills, G.B. 1970. Radial migration of spherical particles in Couette systems. *A.I.Ch.E.*, 16, 281- 286.

Highgate, D.J., and Whorlow, R.W. 1969. End effects and particle migration effects in concentric cylinder rheometry. *Rheol. Acta*, 8, 142-151.

Husband, D.M., Aksel, N., and Gleissle, W. 1993. The existence of static yield stresses in suspensions containing noncolloidal particles. *J. Rheol.*, 37, 215-235.

Hutton, J.F. 1963. Fracture of liquids in shear. *Nature*, 200, 646-648.

Hutton, J.F. 1975. On using the Weissenberg Rheogoniometer to measure normal stresses in lubricating greases as examples of materials which have a yield stress. *Rheol. Acta*, 14, 979-992.

Kalyon, D.M., Yaras, P., Aral, B., and Yilmazer, U. 1993. Rheological behavior of a concentrated suspension: A solid fuel simulant. *J. Rheol.*, 37, 35-52.

Kao, S.V., Nielsen, L.E., and Hill, C.T. 1975. Rheology of concentrated suspensions of spheres. I. effect of the liquid-solid interface. *J. Colloid Interface Sci.*, 53, 358-366.

Karnis, A., Goldsmith, H.L., and Mason, S.G. 1963. Axial migration of Poiseuille flow. *Nature*, 200, 159-160.

Keates, P., Mitchell, G.R., Riti, J.B., Peuvrel-Disdier, E., and Navard, P. 1994. A novel X-ray rheometer for in-situ studies of polymer melts and solutions during shear flow. *J. Non-Newt. Fluid Mech.*, 52, 197-215.

Keentok, M. 1982. The measurement of the yield stress of liquids. *Rheol. Acta*, 21, 325-332.

Keentok, M., Milthorpe, J.F., and O'Donovan, E. 1985. On the shearing zone around rotating vanes in plastic liquids: theory and experiment. *J. Non-Newt. Fluid Mech.*, 17, 23-35.

Kemblowski, V. 1988. Continuous measurements of viscous properties of suspensions with settling particles. in *Progress and Trends in Rheology II*, H. Giesekus et al. (Ed.), Springer, New York, 177-179.

Kiljanski, T. 1989. A method for correction of the wall-slip effect in a Couette rheometer. *Rheol. Acta*, 28, 61-64.

Kirby, J.M. 1988. Rheological characteristics of sewage sludge: a granuloviscous material. *Rheol. Acta*, 27, 326-334.

Klein, B., Laskowski, J.S., and Partridge, S.J. 1995. A new viscometer for rheological measurements on settling suspensions. *J. Rheol.*, 39, 827-840.

Koch, D.L. 1989. On hydrodynamic diffusion and drift in sheared suspensions. *Phys. Fluids*, A1, 1742-1745.

Krieger, I.M. 1990. Bingham Award Lecture-1989. The role of instrument inertia in controlled-stress rheometers. *J. Rheol.*, 34, 471-483.

Laun, H.M., Bung, R., and Schmidt, F. 1991. Rheology of extremely shear thickening polymer dispersions (passively viscosity switching fluids). *J. Rheol.*, 35, 999-1034.

Leal, L.G. 1979. The motion of small particles in non-Newtonian fluids. *J. Non-Newt. Fluid Mech.*, 5, 33.

Lee, C.S., Tripp, B.C., and Magda, J.J. 1992. Does N_1 or N_2 control the onset of edge fracture?. *Rheol. Acta*, 31, 306-308.

Leighton, D., and Acrivos, A. 1987. The shear-induced migration of particles in concentrated suspensions. *J. Fluid Mech.*, 181, 415-439.

Lockyer, M.A. and Walters, K. 1976. Stress overshoot: real and apparent. *Rheol. Acta*, 15, 179-188.

Magnin, A., and Piau, J.M. 1990. Cone-and-plate rheometry of yield stress fluids. Study of an aqueous gel. *J. Non-Newt. Fluid Mech.*, 36, 85-108.

Magnin, A., and Piau, J.M. 1987. Shear rheometry of fluids with a yield stress. *J. Non-Newt. Fluid Mech.*, 23, 91-106.

Major, J.J., and Pierson, T.C. 1992. Debris flow rheology: Experimental analysis of fine-grained slurries. *Water Resour. Res.*, 28, 841-857.

Mas, R., and Magnin, A. 1994. Rheology of colloidal suspensions: Case of lubricating greases. *J. Rheol.*, 38, 889-908.

Metzner, A.B.1985. Rheology of suspensions in polymeric liquids. *J. Rheol.*, 29, 739-775.

Mewis, J. 1994. The thixotropy approach to time dependency in rheology. *Les Cahiers de Rhéologie* (French Rheology Group) XIII, 2-9. (in French)

Mewis, J. 1979. Thixotropy - A general review. *J. Non-Newt. Fluid Mech.*, 6, 1-20.

Mondy, L.A., Brenner, H., Altobelli, S.A., Abott, J.R., and Graham, A.L. 1994. Shear-induced migration in suspensions of rods. *J. Rheol.*, 38, 444-452.

Mooney, M. 1931. Explicit formulas for slip and fluidity. *J. Rheol.*, 2, 210-222.

Nadim, A. 1988. The measurement of shear-induced diffusion in concentrated suspensions with a Couette device. *Phys. Fluids*, 31, 2781-2785.

Navickis L.L. and Bagley, E.B. 1983. Yield stresses in concentrated dispersions of closely packed, deformable gel particles. *J. Rheol.*, 27, 519-536.

Nguyen, Q.D., and Boger, D.V. 1983. Yield stress measurement for concentrated suspensions. *J. Rheol.*, 27, 321-349.

Nguyen, Q.D., and Boger, D.V. 1985. Direct yield stress measurement with the vane method. *J. Rheol.*, 29, 335-347.

Nguyen, Q.D., and Boger, D.V. 1992. Measuring the flow properties of yield stress fluids. *Ann. Rev. Fluid Mech.*, 24, 47-88.

O'Brien J.S. and Julien, P.Y. 1988. Laboratory analysis of mudflows properties. *J. Hydraul. Eng.*, 114, 877-887.

Oka, S., 1960. in *Rheology: Theory and Applications*, F.R. Eirich (Ed.), New York, 3.

Oliver, D.R. 1962. Influence of particle rotation on radial migration in the Poiseuille flow of suspensions. *Nature*, 194, 1269-1271.

Overend, I.J., Horsley, R.R., Jones, R.L., and Vinycomb, R.K. 1984. A new method for measurement of rheological properties of settling suspensions, in *Proc. 9th Int. Congr. Rheology*, Mexico, B. Mena et al. (Ed.), 2, 583-590.

Persello, J., Magnin, A., Chang, J., Piau, J.M., and Cabane, B. 1994. Flow of colloidal aqueous silica dispersions. *J. Rheol.*, 38, 1845-1870.

Phillips, C.J., and Davies, T.R.H. 1991. Determining rheological parameters of debris flow material. *Geomorphology*, 4, 101-110.

Pignon, F., Magnin, A., and Piau, J.M. 1994. Flows of thixotropic clay suspensions. *Les Cahiers de Rhéologie* (French Rheology Group) XIII, 160-169. (in French)

Powell, R.L. 1988. Rotational viscometry. in *Rheometrical measurement*, A.A. Collyer and D.W. Clegg (Ed.), Elsevier Applied Science, Amsterdam, Chapter 9, 247-296.

Reeves, T.J. 1990. On-line viscosity measurement under industrial conditions. *Coal Prep.*, 8, 1-9.

Savage, S.B. and Lun, C.K.K. 1988. Particle segregation in inclined chute flow of dry cohesionless granular solids. *J. Fluid Mech.*, 189, 311-335.

Segré, G., and Silberberg, A. 1962. Behaviour of macroscopic rigid spheres in Poiseuille flow. *J. Fluid Mech.*, 14, 115-157.

Senouci, A. and Smith, A.C. 1988. An experimental study of food melt rheology. II. End pressure effects. *Rheol. Acta*, 27, 649-655.

Sinton, S.W. and Chow, A.W. 1991. NMR flow imaging of fluids and solid suspensions in Poiseuille flow. *J. Rheol.*, 35, 735-772.

Stephens, D.J. and Bridgwater, J. 1978. The mixing and segregation of cohesionless particulate materials. *Powder Technol.*, 21, 17-44.

Tanner, R.I. and Keentok, M. 1983. Shear-fracture in cone-plate rheometry. *J. Rheol.*, 27, 47-57.

Tanner, R.I., *Engineering Rheology*, Clarendon Press, Oxford.

Valentyik, L. 1971. Instrumentation and control of the specific gravity and the rheology of heavy-media suspensions. *Min. Sci. Eng.*, 38-44.

Vinogradov, G.V., Froishteter, G.B., Trilisky, K.K., and Smorodinsky, E.L. 1975. The flow of plastic disperse systems in the presence of the wall effect. *Rheol. Acta*, 14, 765-775.

Yaras, P., Kalyon, D.M., and Yilmazer, U. 1994. Flow instabilities in capillary flow of concentrated suspensions. *Rheol. Acta*, 33, 48-59.

Yilmazer, U. and Kalyon, D.M. 1989. Slip effects in capillary and parallel disk torsional flows of highly filled suspensions. *J. Rheol.*, 33, 1197-1212.

Yoshimura, A., and Prud'homme, R.K. 1988. Wall slip corrections for Couette and parallel disk viscometers. *J. Rheol.*, 32, 53-67.

Yoshimura, A.S., Prud'homme, R.K., Princen, H.M., and Kiss, A.D. 1987. A comparison of techniques for measuring yield stress. *J. Rheol.*, 31, 699-710.

CHAPTER 5

Rheology of clay-water mixtures

Since clay-water suspensions contain small particles they can easily be tested with conventional laboratory rheometers. Moreover these clay-water systems exhibit various behaviour types and are thus often used as model fluids. Clay-water systems have been extensively studied in literature but our knowledge of their behaviour remains partial. For sufficiently low shear rates, the mechanical behaviour of clay-water mixtures or fine mud suspensions containing a large clay fraction may be strongly influenced by the colloidal character of suspended particles (Cf Chapter 1). In the following the suspensions are always considered as stable and homogeneous.

Dilute clay-water suspensions for which hydrodynamic or Brownian effects may play a significant role have rarely been studied. Concentrated suspensions, for which colloidal effects become predominant under usual flow conditions, can exhibit almost all the complex behaviour types generally encountered in the field of rheology: yield stress, shear-thinning, thixotropy, viscoelasticy, etc. We shall start by reviewing these characteristics from a general point of view and examining the influence of some parameters (clay type, pH, electrolyte, concentration, temperature) on the behaviour of the suspensions. In the second part we shall review the theoretical models attempting to explain or predict the rheological behaviour from internal processes.

The following synthesis will certainly appear critical of many previous works. Indeed, it is likely that, out of the large amount of work carried out in this field over the last 50 years, only a relatively small fraction has constituted important or relevant rheological information. We believe that this is due to the difficulty of obtaining reliable and interesting rheometrical data with concentrated clay-water systems. Indeed we must again emphasize that concentrated suspension rheometry requires many precautions (see Chapter 4): various disturbing effects may occur (fracture, slip, crack evolution, sedimentation, etc.) which must be controlled. In addition, considering the acutely non-Newtonian character of most clay-water suspensions and the inherent uncertainty of rheometrical measurements, behaviour transition or evolution with material characteristics (pH, concentration, temperature, etc.) can only be seriously appreciated when, for a given material, a whole range of tests under different conditions has been carefully carried out. For example, in order to have an idea of simple shear behaviour in steady conditions, a simple apparent viscosity measurement at a given shear rate is insufficient. One needs measurements within a wide shear rate range, since apparent viscosity depends on shear rate.

5.1 RHEOLOGICAL PROPERTIES

5.1.1 *Generalities*

Stable homogeneous clay-water suspensions contain colloidal particles and are thus expected to exhibit yield stress when the solid fraction (ϕ) is larger than a critical value (ϕ_c) (see 3.1.5). In the following, clay-water mixtures exhibiting yield stress will be referred to as concentrated suspensions while other mixtures will be referred to as dilute suspensions. The exact limit between these two classes remains hard to determine in practice. Indeed conventional apparati are incapable of detecting yield stress smaller than about 0.1 Pa.

5.1.1.1 *Dilute suspensions*

When immersed in water, clay particles are surrounded by diffuse double-layers and the corresponding suspensions cannot *a priori* be considered as suspensions of rigid non-colloidal particles even if these double-layers do not overlap.

Consider sufficiently dilute suspensions in which clay particles are so far from each other that double-layer overlapping is negligible. This is achieved when ϕ is much smaller than δ^{-3}, where δ is the ratio of the double-layer thickness (see 1.2.2.2) to particle radius. Even if they do not overlap they can deform more or less easily. Then the influence of electrical double-layers on suspension behaviour is negligible when N_e, the ratio of electrical to viscous forces, which may be written:

$$N_e = \frac{\varepsilon_0 \psi_0^2}{6\pi\mu\gamma R^2} \tag{5.1}$$

is much smaller than 1. N_e is in fact the ratio of N_r to N_f (see 1.2.2.3). In that particular case the viscosity of the clay-water system may be computed using the formulae for suspension viscosity (Cf Section 3.1.2) as a function of ϕ. When the order of magnitude of N_e is equal to or larger than 1, electrical effects arise from deformation of the double-layer by the flow (see Saville (1977)). Thus the viscosity of the suspension (μ) may differ from that of the equivalent suspension of rigid bodies. Booth (1950) has proposed a theoretical approach of this problem in which, for low values of P_e, μ is expressed as:

$$\mu = \mu_0 \left[1 + 2.5(1 + q_0^2 P_e Z(\delta))\phi\right] \tag{5.2}$$

with $q_0 = eQ_0 / 4\pi\varepsilon_0 k\theta$ where Q_0 is the total charge at the particle surface. P_e is the Peclet number given by (3.13) where f is replaced by the inverse of the ion mobility through the liquid. Z(δ) is a function which is proportional to δ^4 when δ tends to zero and proportional to δ when δ tends towards infinity. Appropriate expressions taking into account the specific characteristics of the clay material (platelet particles, exchangeable cations in double-layers) have not yet been proposed.

Very little is known concerning the behaviour of dilute suspensions for which double-layer overlapping is not negligible. In general, as yet insoluble experimental problems are encountered with conventional laboratory rheometers when studying dilute suspensions: for large duration tests settling may occur (see 1.3.2); the apparatus should record very small shear stress levels. Thus the rheological transition from dilute to concentrated clay-water suspensions remains an open problem. For example the data of Vali and Bachmann (1988) cannot be relevantly interpreted as demonstrating the rheological transition from Newtonian to viscoplastic behaviour since their experimental shear rates are too large and data are plotted in a linear diagram. We may nevertheless mention the following suggestions of Güven (1992) which result from his experimental observations with dilute clay-water suspensions. At low shear rates, the suspension viscosity level is increased by an isotropic arrangment of particles favored by Brownian motions. This gives rise to an initial Newtonian plateau. For increasing shear rates, the shear forces start breaking flocs and orientating clay particles along lines. Then the viscosity of the suspension decreases and reaches a second Newtonian plateau. For larger shear rates, shear-thickening might result from the randomization of the suspension structure, as for some other suspensions (see 3.1.3.1).

5.1.1.2 *Concentrated suspensions*

We shall now deal with concentrated suspensions exhibiting yield stress, which, according to the above statements, is due to the formation of a continuous network of double-layer overlapping. In the following we shall consider not only simple clay-water mixtures but fine mud suspensions (clay-water mixtures with some fine particles) with a sufficiently large clay fraction that their behaviour can reasonably be considered as being dictated by the clay fraction. For both material types, assuming a typical particle radius of 1 μm and a yield stress of 0.1 Pa, the Peclet number, P_{e_c}, as given by (3.15), is larger than 200 at ordinary temperatures. Thus Brownian effects are negligible compared to colloidal effects. At large shear rates hydrodynamic effects may nevertheless become predominant. Coussot (1995) has shown that for various fine mud suspensions this occurs for Γ larger than 50.

5.1.1.2.1 Yield stress and shear-thinning

In steady-state conditions clay-water mixtures are generally shear-thinning yield stress fluids (see Chapter 2). It is worth noticing that shear-thickening has been reported for some clay-water mixtures (Beazley 1965) but, because of the high concentration of his suspensions and the long duration of the tests, we fear that these data were affected by various perturbating effects (slip, fracture, chemical reactions, etc.). It should also be noted that Rand et al. (1980) observed shear-thickening for Na-montmorillonite-water suspensions.

The viscoplastic character of concentrated natural or pure clay-water mixtures has often been reported in the literature (Qian and Wan 1986, O'Brien and Julien 1988, O'Brien 1986, Phillips 1989, Major and Pierson 1992, Locat and Demers 1988, Migniot 1989, Naik 1983, Johnson 1970, Michaels and Bolger 1962). Some examples of flow curves obtained for various fine clay-water mixtures under controlled shear rates and controlling the various possible perturbating effects (Cf Chapter 4) are presented in Figures 5.1-5.3. In all these diagrams the shear stress clearly tends to a finite, critical value for low shear rates which reflects the yielding behaviour of the suspension. Indeed it is not possible to have flow without imposing a shear stress larger than this critical value. Controlled shear stress experiments (see Section 5.1.2.2)) have been carried out which confirm the existence of this yield stress (Coussot and Piau 1994, Coussot et al. 1993).

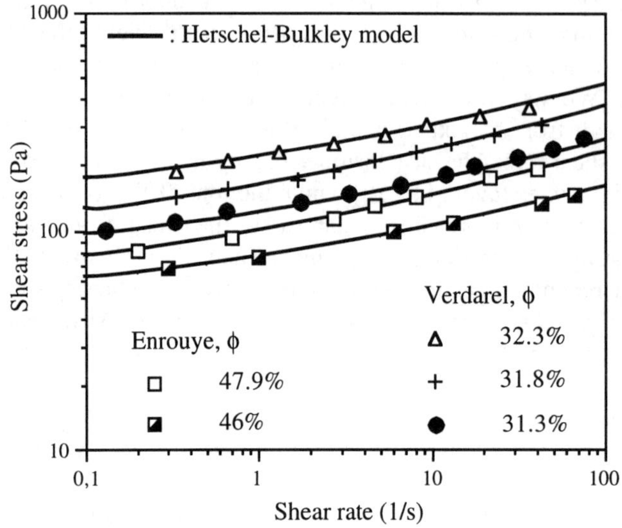

Figure 5.1: Flow curves of various natural fine mud suspensions (fine fraction of natural debris flows). Herschel-Bulkley model fitted to data. [from Coussot and Piau 1994]

The shear-thinning behaviour of these suspensions has also often been noticed in the literature (Street 1956, Weymann et al. 1973, Locat and Demers 1988, Akae and Low 1988, Huifang et al. 1991, Coussot and Piau 1994) and appears more directly visible in a linear scale diagram (Fig. 2.5). Though shear-thinning is often associated with floc breakage there is actually no microstructural theory capable of even predicting these curves qualitatively (Cf Section 5.2). To compare suspension properties or predict flow characteristics, the best that one can do is to fit an empirical model to experimental data and compare the values of the corresponding parameters.

The simplest theoretical model for yield stress fluid, i.e. the Bingham model, cannot be fitted to flow curves in the whole range of tested shear rates. Obviously it is possible to fit this model to experimental data in a narrow range of large shear rates as has been done by most authors. However, very important material properties are overlooked using this method. For example the yield stress deduced from fitting this model to some data (plotted in a linear scale diagram for this fitting) in the range [20; 200 s^{-1}] (Cf Fig. 5.3) is at least twice as high as the real yield stress (for which an estimation is found by extrapolating the shear stress value at very low shear rates in a logarithmic scale). As soon as the Bingham model is used to predict any flow for which another range of shear rates (for example flow initiation) plays a major role, significant errors will be made. In addition the yield stress value obtained by fitting a Bingham model does not specifically represent the strength of the suspension but only accounts for an arbitrary amount of viscous dissipations which depends on the particular shear rate range used.

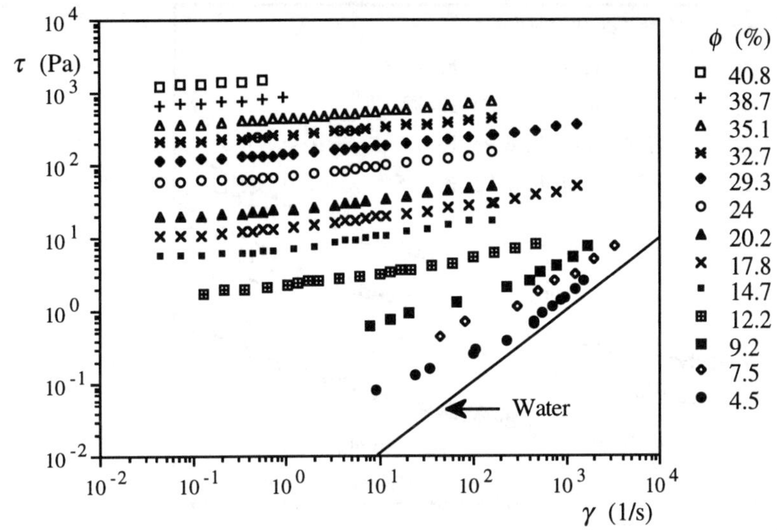

Figure 5.2: Flow curves of natural kaolin-water suspensions for solid fractions within a wide range. [from Coussot, 1995]

As a consequence it was proposed (Nguyen and Boger 1983, Chhabra and Uhlherr 1988, Doraiswamy et al. 1991, Coussot and Piau 1994, Sherwood 1994) to fit a Herschel and Bulkley (1926) model to experimental data for the whole experimental shear rate range (Cf Figs 5.1-5.3):

$$\tau = \tau_c + K\,\gamma^n \qquad (5.3)$$

where τ is the shear stress magnitude, γ the shear rate magnitude, and τ_c, K, and n are positive parameters. Clearly here τ_c is an empirical yield stress. Other models can be used (see the review of Lapasin (1985)) but they contain more parameters or do not represent experimental flow curves in a wide shear rate range. Apart from the usual numerical fitting procedures, the following practical fitting method may be used. One plots $(\tau-\tau_c)$ against γ on a logarithmic scale diagram with different trials for the values of τ_c until it is possible to draw a satisfactory straight line through these points. The slope and position of this straight line give n and K. It is worth noting that the shear stress level corresponding to a shear rate of about 0.01 s^{-1} is in general very close and just above the real yield stress (Coussot et al. 1993, Coussot and Piau 1994, Mas and Magnin 1994, Coussot and Boyer 1995). Finally, using a Herschel-Bulkley model fitted to values in a wide shear rate range including values as small as 0.01 s^{-1} provides a theoretical yield stress which is very close to the real yield stress, with an uncertainty which is sufficiently small for conventional practical applications. However for some suspensions, displaying peculiar behaviour this method can be quite inappropriate (bentonite-water mixture exhibiting a minimum in flow curve (Coussot et al 1993) (Cf Section 5.1.2.4)).

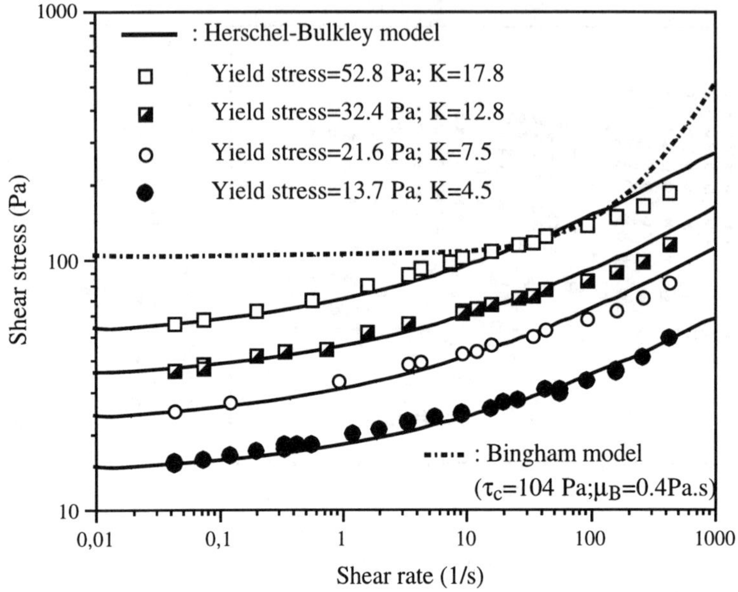

Figure 5.3: Flow curves of Sinard clay-water suspensions at different solid fractions. Herschel-Bulkley model fitted to data. A Bingham model has also been fitted to data corresponding to the highest solid fraction in the range [20-200 s^{-1}]. [from Coussot 1994]

An important question is whether the Herschel-Bulkley model (or the corresponding parameters) does or does not express any physical mechanism occurring within the material during flow. For yield stress the answer was given above: it really reflects the strength of a continuous network of interaction. Concerning the part of the shear stress expression which is shear rate dependent in equation (5.3) the question remains open. It has often been erroneously considered that the Bingham model found for clay-water mixtures reflects the sum of the energy needed to break the particle network, by means of yield stress, and viscous dissipations, using the second term identical to the stress for a Newtonian fluid. Let us try to clarify this problem on the basis of the physical scheme proposed for dispersion in 3.1.5.1. To simplify the description of internal phenomena we can consider that, during motion, there are essentially two sources of energy dissipation: (a) ruptures and restorations of links between clay particles within the material and (b) usual hydrodynamic dissipations due to the shear of interstitial free water. As soon as there is a flow the structure is continuously broken and restored in part, even though each particle obviously occupies a new position. The complete network (with all the links) is only obtained after a sufficiently long (in theory infinite) time of rest. Thus, at each instant, a transient network or a set of clusters exists and the intensity of shear stress to be applied to maintain shear may consequently be not that far from the yield stress value. More probably, since the complete network no longer exists, the stress needed to break and restore a certain number of links is different from yield stress. Thus, physically, during flow, the shear stress is not the sum of a constant force needed to break the static network and an additional force needed to maintain the shear of the suspension of broken flocs, even if a constant fluid "yield stress" appears in the flow curve model (5.3). In fact the total shear stress, either for the Bingham model or the Herschel-Bulkley model, expresses a complex sum of viscous dissipations related to the deformations of the transient network or to the overlapping or friction of flocs. The following critical example clearly demonstrates the above statements. In the case of concentrated dispersed systems in a low molecular weight matrix, it was shown that a minimum in the flow curve can exist (Cf Section 5.1.2.4). This means that, beyond yielding, the shear stress which is necessary to maintain flow can either be higher or lower than the yield stress, depending on the characteristics of the material. Finally, the two terms of equation (5.3) are not independent at all as long as there really is a flow and they both express the two types of phenomena (a and b) (Coussot et al 1993, Coussot and Piau 1994). Consequently, and this is even more noticeable for a Herschel-Bulkley fluid, none of the parameters K and n *a priori* represent a simple, specific property of the fluid, unlike viscosity in Newtonian cases or yield stress.

5.1.1.2.2 Viscoelasticity

Only a few studies have focused on this subject but there is some evidence that concentrated clay-water suspensions exhibit viscoelastic properties. This first appears clearly during creep tests (see 4.2.1). The creep curves (deformation versus time) have two different forms below and above yield stress: with saturation or with monotonic increase in strain (flow) (Coussot et al. 1993). Below this yield value, for small deformations, the suspension has an elastic behaviour.

Other evidence of the viscoelastic behaviour of clay-water suspensions is provided by dynamic tests. Here the sample is submitted to an oscillating shear:

$$\gamma = \gamma^*_0 (\sin \omega t) \tag{5.4}$$

and the recorded stress may be written as:

$$\tau = \tau_0 (\sin(\omega t + \psi)) \tag{5.5}$$

where γ^*_0 and ω are the imposed strain amplitude and frequency, τ_0 is the resulting stress amplitude and ψ is the phase shift between stress and strain. When a stationary oscillating flow has been reached the latter parameters do not change in time and one can compute the respective storage and loss moduli defined as:

$$G' = \frac{\tau_0}{\gamma^*_0} \cos \omega \quad ; \quad G'' = \frac{\tau_0}{\gamma^*_0} \sin \omega \tag{5.6}$$

For a simple elastic material G' is constant and corresponds to the elastic modulus while G'' is equal to zero. For a Newtonian fluid G' is equal to zero while G''/ω is the fluid viscosity. For clay-water suspensions G' is almost constant for small strain amplitudes and significantly decreases for larger values (Cf Fig. 5.4), which is in agreement with the above scheme (elastic response up to a critical strain then flow). Some authors (Sohm and Tadros 1989) considered that the yield stress is reached and then overcome when G' starts decreasing, i.e. beyond a critical strain amplitude (γ_c). However Pignon et al. (1994) demonstrated that for laponite-water suspensions this may result from strain localisation (see Fig. 5.4). Changes in G' and G'' can also be used to estimate the critical concentration for which the suspension starts exhibiting yield stress (Mourchid et al. 1995). For sufficiently small solid fractions G' and G'' remain small reflecting slightly viscous behaviour whereas beyond a critical concentration G' increases markedly which corresponds to network formation.

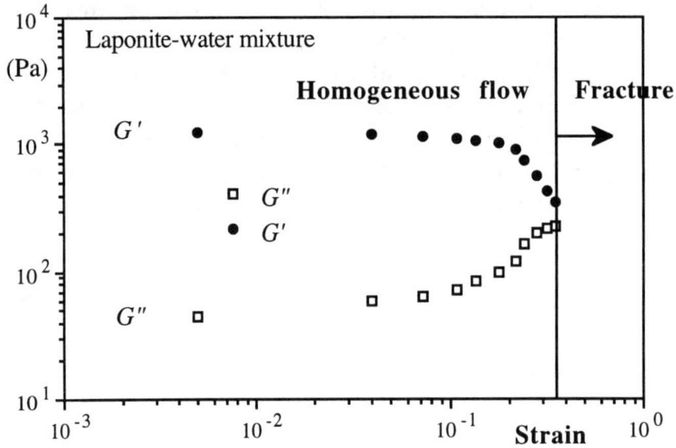

Figure 5.4: Elastic and viscous modulii of Laponite-water mixtures (ϕ=1.1%; [NaCl]=5 10^{-3} mol/l] as a function of strain amplitude (γ^*_0). For γ^*_0 larger than 0.35 the sample fractures and G' and G'' do not correspond to true properties of the homogeneous material. [from Pignon et al., 1994]

5.1.1.2.3 Thixotropy

Some clay-water systems are well-known as typical thixotropic materials (van Damme et al. 1991). Thixotropy (in the general sense of viscosity changes with time) was first noticed in (ramp) tests which involve applying a progressively increasing shear stress up to a certain value then decreasing it to zero (Hu and Fang 1985, Moore 1959, Neumann and Sansom 1971, Wan 1982). In a shear rate-shear stress diagram hysteresis loops are then observed which reflect the difference in material response with flow history (see 4.2.2). Above all thixotropy, or a decrease in viscosity with flow duration, has been reported for clay-water mixtures. Some authors also reported anti-thixotropy (or rheopexy) (viscosity increase with flow duration) for natural clay-water mixtures (Wang et al. 1994). However this ramp (or cycle) test is not quite appropriate for studying fluid thixotropy (Mewis 1979) (see 4.2.2). With appropriate tests (as described in 4.2.2) only a small amount of experimental data is available for clay-water mixtures (Coussot et al. 1993). In addition it is now clear that, with clay-water mixtures, many disturbing effects can develop during a test and lead to stress or rate changes which may be erroneously interpreted as thixotropic effects. Since many works reporting thixotropy did not mention these effects, it is likely that many thixotropic properties reported in the literature on clay-water mixtures did not correspond to the true characteristics of the corresponding homogeneous material.

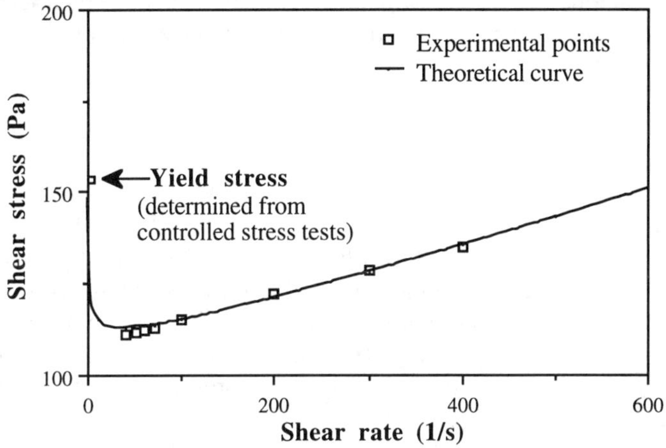

Figure 5.5: Bentonite water mixture (ϕ=4%): flow curve with a minimum. Data were obtained from controlled rate tests except for the yield stress. The theory corresponds to the model presented in 5.2.2.2. [from Coussot et al. 1993]

5.1.1.2.4 Minimum in flow curve

Bentonite-water mixtures and laponite-water mixtures were shown to exhibit a minimum in flow curves (Engelund and Wan 1984, Coussot et al. 1993, Baylocq et al. 1994, Pignon et al. 1996a) (see Fig. 5.5). It takes the form of a shear stress decrease with shear rate for low shear rate values in the flow curve, followed by a subsequent increase at larger shear rates. So far this has not been reported in the literature for other clay-water mixture types but it is probable that the ones that take the form of gels exhibit this property. Experiments carried out at large shear rates or in the form of ramp tests precluded the observation of this minimum. Indeed this requires tests under controlled (progressively increasing) shear stress or under controlled, low shear rates, and this means that perturbing effects must be controlled as effectively as possible (Cf Coussot et al. 1993, Mas and Magnin 1994). A minimum of this sort in the flow curve has been reported for other concentrated suspensions (Coussot (1992) and Coussot and Piau (1994) for granular suspensions, Ducerf and Piau (1994) for alumina suspensions, and Mas and Magnin (1994) for soap fiber suspensions). For materials of this sort, flows are theoretically unstable for the shear rate in the decreasing part of the flow curve (Cf Tanner (1988) and Chapter 9). For this reason, the decreasing part cannot be obtained from rheometrical tests in general but for some materials this was possible for unexplained reasons (Coussot 1992, Mas and Magnin 1994). Pignon et al. (1996a) carried out a careful, systematic, straightforward study of flow regimes by combining rheometry and observation of the strain field for transparent laponite-water suspensions. They demonstrated that the sample is homogeneously sheared only in the increasing part of the flow curve and for very low shear rates up to a critical strain. This means that these suspensions effectively exhibit a minimum in the flow curve. In the decreasing part, shear is localized in a more or less thin layer which demonstrates that rheological properties cannot be determined with conventional methods, which is in agreement with theoretical predictions of instability for such tests (see above).

5.1.2 *Effect of some parameters*

5.1.2.1 *Concentration*

5.1.2.1.1 Flow curve similarity

It has always been reported that, all other things being equal, increasing the solid fraction of a clay-water mixture induces an increase of the overall shear stress level of corresponding flow curves. For clay-water suspensions, in the literature, a complete set of tests in wide shear rate and solid fraction ranges keeping the pH and electrolyte concentration almost constant has only been presented by Coussot (1995). A typical set of results for a given material is presented in Figure 5.2 in a logarithmic diagram. At sufficiently low γ and high ϕ values the flow curves of clay-water suspensions are essentially of the shear-thinning yield stress type which must be attributed to the colloidal interactions. At sufficiently high γ or low ϕ values the Newtonian form of flow curves along with the low viscosity obtained suggests that particles behave as independent force-free particles suspended in water (Cf 3.1.2). It should be noted that for various materials the transition between these two regimes occurs in a common range of Γ, i.e. [0.3; 50]. This suggests the following dimensionless representation of data, using $T=\tau/\tau_c$ as dimensionless shear stress and $\Gamma=\mu\gamma/\tau_c$, where μ is the viscosity of the corresponding suspensions of force-free spheres (see various possible formulae in 3.1.2), as the dimensionless shear rate. The experimental results of Coussot show that, within experimental uncertainty, flow curve data plotted in a diagram (Γ, T) fall on a master curve (see typical example in Figure 5.6) which comprises two main parts: for low values of Γ (<0.3), the flow curve corresponds to shear-thinning yield stress behaviour which could be well represented by equation (5.3) which here takes the following general dimensionless form:

$$T = 1 + K' \Gamma^{n'} \qquad (5.7)$$

where K' and n' are material parameters which depend on characteristic times of rupture and restoration processes but do not depend on ϕ; for high values of Γ (>50), the flow curve corresponds to Newtonian behaviour with viscosity approximately equal to 1.5μ, thus we have:

$$T = 1.5\,\Gamma \tag{5.8}$$

For spherical solid particles one should have $T/\Gamma=1$ instead of (5.8). Since clay particles take the form of thin plates, the present result is not surprising and is consistent with those obtained for suspensions of non-spherical non-colloidal particles (Kamal and Mutel 1985).

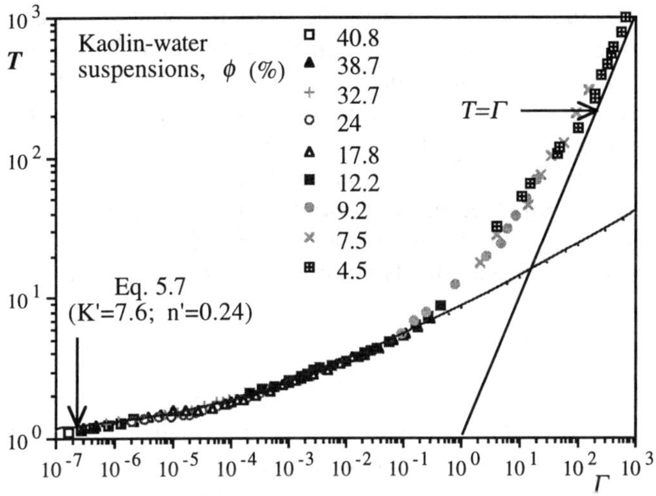

Figure 5.6: Dimensionless flow curve for natural kaolin-water suspensions at various solid fractions. See text for definitions of the dimensionless shear rate, Γ, and shear stress, T. [from Coussot 1995]

It is worth noting that, since rheometrical data can hardly be obtained for very low or large solid fractions (because of settling or fracture) at low shear rates, the empirical master curve (5.7) does not pretend to represent the behaviour of all mixtures of water with a given clay, but, at the very least, a large number of such suspensions within a fairly wide solid fraction range. In particular, the master curve is obviously not valid for suspensions at solid fractions smaller than ϕ_c which do not exhibit yield stress.

5.1.2.1.2 Yield stress

Not far beyond ϕ_c, in the case of a synthetic hectorite-type clay (Laponite), Pignon et al. (1996b) demonstrated by static light scattering that the fractal dimension of micron-sized particle aggregates forming the percolating network increases from 1 to 1.8 around a critical solid fraction (0.55%), which means that the arrangement of aggregates becomes more compact. This structure change also induces a change in yield stress increase as a function of solid fraction: τ_c is proportional to ϕ^2 from 0.35 to 0.48% and is proportional to ϕ^3 from 0.6 to 2%.

For larger solid fractions all authors have noticed the extremely rapid increase in yield stress of different mixtures with solid concentration (Cf Fig. 5.7). From data generally obtained within relatively small solid fraction ranges many authors assumed a power-law yield stress increase with solid concentration (Weymann et al. 1973, Thomas 1963, Michaels and Bolger 1962, Wang et al. 1994). From vane rheometry (see Appendix IV) Alderman et al. (1991) showed that the yield stress of bentonite-water suspensions increases as the third power of the solid fraction for solid fractions in the range 1-10%. From squeeze-film rheometry (the yield stress corresponds to a critical resistance when squeezing fluid between two parallel disks), Sherwood et al. (1991) found that the yield stress of non-uniform mudcakes increases as the power 1.9 of solid fraction in the range 9-60%. Because this experimental procedure does not correspond to viscometric flow with homogeneous materials it is hard to determine whether these results effectively provide a realistic yield stress increase of the mean material.

When results were obtained within sufficiently wide solid fraction ranges within conventional rheometers it appeared more appropriate to fit a straight line to yield stress data plotted in a semi-logarithmic diagram (Cf Fig. 5.7) (Coussot and Piau 1994, Coussot 1995), so that one can finally write:

$$\tau_c = c \exp(s\phi) \qquad (5.9)$$

where c and s are two material parameters. It is noticeable that other authors (Neumann and Sansom (1971) for laponite, Rand et al. (1980) for Na-montmorillonite, Speers et al. (1988) for bentonite, O'Brien and Julien (1988) and Major and Pierson (1992) for natural clays) have noticed the same trend after having fitted a Bingham model to their flow curve data. Rand et al. (1980) also found an exponential increase of the plastic viscosity and the shear modulus (deduced from wave propagation through the material). All the different values of s obtained by the authors (O'Brien and Julien 1988, Major and Pierson 1992, Coussot and Piau 1994) fall between 10 and 60. However it must be recognized that a power-law may also be approximately fitted to data (for example those of Figure 5.7) in a slightly different solid fraction range (10-40%). Then the power varies from 5.5 to 7 depending on the material.

Within the range of solid fraction for which flow curves are similar (see 5.1.2.1.1), it is natural to assume that the networks formed by links between particles which cause yield stresses have some similarities, in particular geometrical. Within this framework the yield stress increase with concentration can be explained by considering two types of phenomena: the increase of the number of interactions (in a fixed volume) and the increase of their intensity. A power-law increase (with a power smaller than 3) of the yield stress could have been explained by an increase of the number of links (see above), but an exponential increase (or a power-law increase with a power larger than 3) is much too rapid for it to be explained by this phenomenon alone. Thus it is likely that the increase in the intensity of interactions is the essential phenomenon. However attempts to compute τ_c from estimations of force intensity have led (Michaels and Bolger 1962, Firth and Hunter 1976) to formulae which do not predict the present exponential increase in a very wide solid fraction range. Quantitative knowledge of interactions in concentrated clay-water suspensions during flow therefore remains insufficient.

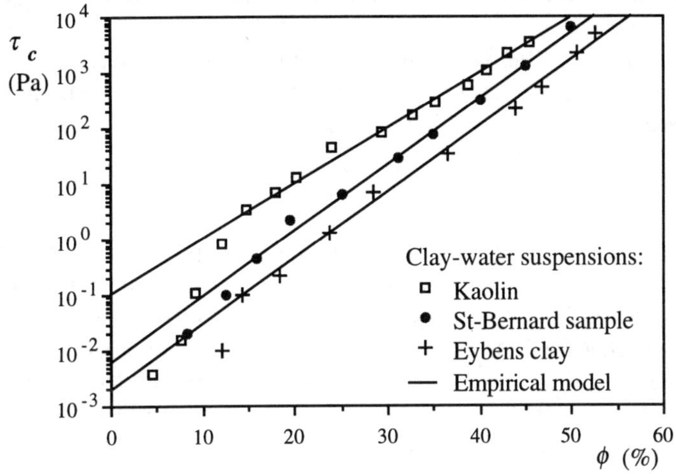

Figure 5.7: Yield stress of fine mud suspensions as a function of solid fraction. [from Coussot 1995]

5.1.2.1.3 Elastic properties

For low solid fractions it has been reported that the elastic modulus significantly increases with solid fraction above a critical value. This should correspond to a percolation process associated to the yield stress appearance (see Section 5.1.5). Above the critical percolating concentration a continuous network of linked particles forms. The number of particles involved in the infinite aggregate should increase as:

$$(\phi-\phi_c)^\alpha \tag{5.10}$$

where α is an exponent related to the fractal dimension of the aggregates. For $\phi>\phi_c$ the elastic behaviour of the suspension is intimately linked to the structure of the network and thus the elastic modulus (G') should increase in a manner similar to (5.10). These considerations have been confirmed by experimental work but since the corresponding rheometrical studies were not always carried out with the precautions required for these suspension types (see Chapter 4), some of them are probably only rough approximations of reality.

Laponite-water suspensions were shown to exhibit a reversible thermodynamical sol-gel transition: above ϕ_c, both the yield stress and G' start growing rapidly as (5.10) with α=2.35 (Mourchid et al. 1995). ϕ_c varies from 0.2 to 2% depending on the ionic strength of the suspension. This transition was confirmed by osmometric measurements. For sepiolite suspensions, Gouami et al. (1994) observed a similar trend for the zero frequency modulus (ω=0) and for the yield stress, and they obtained α=2.7 and ϕ_c around 0.33%. Khandal and Tadros (1988) and Sohm and Tadros (1989) observed the same power-law trend but they did not consider the sol-gel transition and thus took ϕ_c to be equal to zero. For commercial bentonite and Na-montmorillonite-water mixtures these authors found respectively α around 4.5 and 3.4.

Golden et al. (1982) measured the wave rigidity modulus, G^*, for various clay suspensions, by recording the velocity (u) of an electrical pulse propagating through the material. We have:

$$G^* = \rho u^2 \tag{5.11}$$

When viscous dissipations are sufficiently low, G^* is a good approximation of the high frequency limit of the elastic (storage) modulus G'. For sodium kaolinite suspensions they found that G^* varies as predicted by (5.10) with α=2, ϕ_c (=3.8%) here being the lowest volume fraction at which a wave rigidity modulus could be measured. For sodium attapulgite suspensions the same authors found that G^*=9.8 exp (110 ϕ) but, on the basis their data presented in linear diagrams, it is hard to figure out whether this interpretation is relevant or not.

5.1.2.2 *Temperature*

Since water viscosity changes significantly with temperature (see Table 2.1), we should expect a significant change of the apparent viscosity of clay-water suspensions with temperature change. In fact apparently different results have been obtained in the literature for clay-water mixtures within the range [0; 100 °C]: Coussot and Piau (1994) and Lapasin (1985), unlike Ohgaki and Matsuo (1973) (cited in Tsutsumi and Yoshida (1987)), observed a negligible influence of temperature on flow curves. This can be explained in the light of considerations concerning predominant interactions during flow (see 3.2). Indeed, in the first case, the dimensionless number Γ of corresponding flows was smaller than 1 whereas in the second case, Γ, estimated using the actual stress level instead of τ_c (which underestimates its real value), was larger than 100. Thus it appears that when hydrodynamic dissipation mainly originating from shearing of water layers between particles is negligible, that is when colloidal interactions prevail, temperature effects are negligible. Temperature effects become significant only when colloidal effects become negligible. This means that, even during flow, colloidal effects hardly depend on the viscosity of water within double-layers but depend essentially on the electrochemical forces resulting from double-layer interactions. During flow, for sufficiently low shear rates, flow curve values are essentially determined by processes of link restoration and rupture.

A slight viscosity change (in the form of a decrease followed by an increase) with temperature was reported by Chang et al. (1993) for attapulgite-water suspensions. However it is likely that their data were affected by wall slip since they did not mention using rough tool surfaces (Cf 4.1.5.1.3) and their flow curves tend towards a Newtonian plateau at low shear rates. For higher temperatures the viscosity changes could be more complex. Güven (1992) observed viscosity variations which markedly differ when the suspension is heated previously to various temperatures.

In particular for an initial mixture at ambient temperature, with temperature increase, a viscosity increase (four times its initial value) followed by a subsequent decrease (after 280°C) is observed. However, the question remains open whether this trend is due to a behaviour similar to phase transition in liquid crystal as suggested by Güven (1992), irreversible mineralogy changes due to chemical reactions or simply perturbed experimental data (the bentonite-water mixtures used in this study were shown to exhibit rather complex rheological properties (Coussot et al. 1993)).

5.1.2.3 *pH and electrolyte*

The rheological characteristics of clay-water systems significantly change with pH and electrolyte (Pazwash and Robertson 1971; Coussot and Piau 1994) on account of changes in the concentration and type of exchangeable cations in double-layers. Thus a great diversity of changes may be observed depending on clay type and cation types. It is probably for this reason that there is not a general scheme of possible effects of electrolyte and pH changes. Moreover the effect of pH or electrolyte has mainly been studied by observing the resulting change in the (extrapolated) Bingham yield stress of the suspension. It results that the corresponding results and interpretations must be seen as trends which roughly represent physical properties.

For a kaolinitic clay (Huifang et al. 1991, Lagaly 1989) and Na-Kaolinite-water suspensions (Rand and Melton 1977) it was shown that the yield stress essentially decreases with pH at different NaCl concentrations but all curves nearly intersect in a common point (around pH=8). In addition the yield stress decreases with the NaCl fraction at pH values greater than the isoelectric point. Otherwise, it increases.

For illitic and montmorillonitic clay-water mixtures it was shown that the yield stress increases with increasing NaCl fractions (for a fixed pH) and decreases with pH (for a fixed NaCl content) (Huifang et al. 1991). For Na-montmorillonite-water suspensions the yield stress was shown to reach a minimum for a given NaCl content (around 6 mmol/l) and, in parallel, for a given pH (around 8), the viscosity of Ca-montmorillonite-water mixtures was shown to have a minimum for a pH close to 5.5 (van Olphen 1956, Rand et al. 1980, Granquist 1959, M'Ewen and Pratt 1957, Brandenburg and Lagaly 1988, Chen et al. 1990). In addition the storage modulus (G') was shown to increase to a maximum for a pH around 5 for Na-montmorillonite-water suspensions at various solid fractions (Khandal and Tadros 1988). For laponite-water suspensions it has been found that the yield stress increases with electrolyte concentration (Neumann and Sansom 1971).

Wang et al. (1994) observed that, for natural clay-water mixtures, at low solid fractions, clay suspensions with Na+ atoms as exchangeable cations exhibit higher yield stress than those with Ca^{++} and Mg^{++}. At higher clay fractions these suspensions have larger yield stress than those with monovalent cations. Street (1956) also noticed that the yield value of kaolinite-water mixtures increases when the exchangeable cations change from Potassium to Sodium then Lithium. Lagaly (1989) and Permien and Lagaly (1995) also proposed an extensive study of rheological changes along with possible physical interpretations for bentonite and kaolinite suspensions in which dispersing agents, salts, organic compounds, or surfactants are added.

For Sodium montmorillonite dispersions, Brandenburg and Lagaly (1988) proposed a study of the changes in thixotropic behaviour depending on pH, electrolyte and temperature. However the data cannot be consistently interpreted because they involved ramp tests compared on a linear scale (see 4.2.2).

All these trends have been discussed in terms of interaction types (edge-to-edge, edge-to-face, face-to face) (Cf van Olphen (1977)) between clay particles. Since this scheme has not yet been proved to be realistic (see 1.2.2.4), there remains a great uncertainty on the conclusions. These interpretations might nevertheless remain valuable at least concerning the changes in double-layer intensity.

5.1.2.4 *Particle shape and size*

The effect of particle size has been studied by Brandenburg and Lagaly (1988) for sodium montmorillonite dispersions. They observed that the viscosity does not change significantly for a particle size larger than a critical value (around 0.6 μm). Below this value the viscosity increases rapidly. Nitakawi et al. (1981) and Egashira (1977) also suggested that the order of magnitude of some viscosity changes from one clay type to another could be explained by shape changes (see 5.1.2.5). However we must emphasize that the particle size can hardly be considered as a relevant parameter since size changes are often accompanied by changes in shape, specific surface and electrical charge. These changes may induce viscosity changes much larger than those specifically induced by size changes.

5.1.2.5 *Clay type*

The compilation of all the data obtained in the literature could provide qualitative trends for behaviour dependence on clay type. However it would be difficult to give more precise interpretations since the experiments were made with clay suspensions at different pH, electrolyte concentrations, solid fractions and temperatures. As far as we know only one study (Nitakawi et al. 1981) tried to provide a systematic approach to viscosity changes with clay type, all other things being equal (pH, electrolyte, solid concentration). The results are as follows: the apparent viscosity measured for a few shear rate values decreases from imogolite, to montmorillonite, then kaolinite, allophane, halloysite. These trends were interpreted on the basis of particle form changes (Cf 5.1.2.4) considering that, in agreement with some observations, for the above range of clay types, particles change from small fibers to less elongated ellipsoids or thicker plates.

5.2 BEHAVIOUR MODELLING

Here, we intend to provide to a critical review of existing theoretical models designed to represent the rheological behaviour of clay-water mixtures as a function of internal structure changes during flow. Considering the variety of clay types a wide range of models could have been developed but this is not the case. This may be explained by two facts: first the internal structure and its changes during flow have overall features similar from one material to another, which also correspond to those of many dispersions in a low molecular weight liquid (see 3.1.5); secondly, considering present understanding of behaviour as a function of changes in internal structure, it is hardly conceivable also to take into account the particularities of the structure of each clay-water suspension. Since most of the following models have been proved to represent only some qualitative properties of clay-water suspensions and moreover involve many unknown parameters, we do not think it valuable to present them in detail. We shall simply outline these works in order to review possible approaches to modelling and the corresponding difficulties.

From a general point of view the basic framework of all the proposed models is the same as the one assumed for dispersions (see 3.1.5). More precisely it is considered that, in water, clay particles aggregate in flocs. When the solid fraction is large enough, a continuous aggregate (or floc) extends continuously within the sample so that it is necessary to apply a minimum force (the yield stress) to break this network and thus induce flow. During flow there is a continuous process of link ruptures and restorations almost everywhere within the fluid, and at any moment, at least, one might distinguish more or less separate flocs. Many models even tend to consider that flocs remain relatively well identified and separated during flow. These models are more probably valid for low concentrated clay-water mixtures. However the above concept of instantaneous separation in flocs makes it possible to gather in a single physical scheme almost all the proposed models.

Under steady flow conditions the average (in time and space) size of flocs establishes itself at a steady value. A certain time is necessary to reach the steady state and the associated transient responses explain suspension thixotropy. For increasing shear rate, floc size decreases until it reaches the size of a single solid particle. In some cases it is considered that this limiting floc contains many particles (Michaels and Bolger 1962). Finally all models rely on two main physical phenomena: the (instantaneous) behaviour of the suspension of flocs and the law of floc size change as a function of flow history. We shall first consider models that focus on steady states. Then we shall review the models that also seek to take into account time-dependent properties.

5.2.1 *Models for steady flows*

In the following models only the law of floc size as a function of shear rate in steady state is taken into account. In the model of Weymann et al. (1973), it is considered that, at a given shear rate, equilibrium aggregates or flocs of a finite size form, while the rest of the suspension is a suspension of individual particles in water. Within each of the aggregates a fraction of interstitial fluid is trapped and, at least temporarily, particles and fluid embedded in an aggregate may be considered as a rigid coarse particle. The viscosity of the suspension is then given as:

$$\mu_s = \mu_\infty F(\Phi) \tag{5.12}$$

where Φ designates the volume fraction of flocs and μ_∞ the viscosity of the interstitial suspension of water and individual particles given as:

$$\mu_\infty = \mu_0 f(\phi) \tag{5.13}$$

F and f are two functions which may be found from semi-empirical formulae listed in 3.1.2 taking into account as far as possible the mean shape of flocs (spheroidal) and particles (platelets). Knowing these functions the problem of finding the viscosity is reduced to one of finding the volume fraction Φ which is a function of shear rate. To this end Weymann et al. (1973) wrote that the viscous force distribution acting on each spheroidal aggregate balances the cohesive force (F_c) on a particle in the surface of an aggregate. For high shear rates they deduced:

$$\tau = \mu_\infty \gamma + \frac{F_c N^{2/3}}{2.88} \tag{5.14}$$

where N is the number of particles by unit volume.

Tsenoglou (1989) proposed a slightly more sophisticated approach in which he directly considered the suspension as a Bingham fluid. Then each floc embedded in the rest of the medium is subjected to a drag force which must be computed taking into account the behaviour of the whole viscoplastic medium. The drag force was roughly estimated from the work of Ansley and Smith (1967) (Cf Chapter 12) and, balancing this force with interparticle cohesive force, Tsenoglou deduced:

$$\tau = \mu_0 \gamma \left[\frac{1}{1-\phi}\right]^{2+N/2} + \tau_0 \left[\frac{1-\phi_c}{1-\phi}\right]^{1+(7/8)p\sqrt{N}} \tag{5.15}$$

where, according to Tsenoglou (1989), τ_0 is the yield stress value at the onset of its appearance ($\phi=\phi_c$).

It is unlikely that the flocs can be individualized during flow. More probably they are inexorably broken and restored because they come in contact with one another. Thus one should not only take into account the viscous dissipation due to the instantaneous flow of the equivalent suspension of rigid flocs but also the energy dissipated during floc rupture and restoring from one instant to another. In the above models these terms are in fact more or less taken into account by calculating a mean equilibrium size. In other models these two terms of energy dissipation have been clearly identified and separated. For example Michaels and Bolger (1962) considered that, in simple shear, the shear stress is the sum of three terms:

$$\tau = \tau_n + \tau_{cr} + \tau_v \tag{5.16}$$

In this model τ_n corresponds to the energy to be stored in the form of elastic energy within stretched aggregates and subsequently dissipated in viscous form after breakage. It should be noted that for these authors particles form essentially indestructible particle clusters or flocs which may associate to form weakly bonded aggregates, whose size and configuration are governed by the intensity of the local shear field. Thus, here, flocs and aggregates describe different structure units. Estimating τ_n remains problematic. According to Michaels and Bolger (1962) this term should correspond to the suspension yield stress at low shear rates and should tend to zero for high shear rates. In addition it might vary as $(\phi-\phi_c)^3$.

The second term, τ_{cr}, corresponds to the energy dissipated during floc breakage. Indeed it takes energy to pull flocs apart but flocs may be recombined by collision and subsequently broken so that it is necessary to provide a certain amount of energy to ensure constant mean floc size. According to Michaels and Bolger (1962), in steady state, the product $\gamma\tau_{cr}$ is proportional to the number of possible collisions by unit time, to the mean work required to completely separate two close particles and to the number of possible bond formations due to each collision. One difficulty in developing this model further comes from the fact that τ_{cr} can hardly be estimated since it depends on the structure and size of the flocs. Nevertheless, for high and low shear rates, Michaels and Bolger (1962) found:

$$\tau_{cr} = A_0\phi^2 \tag{5.17}$$

where A_0 depends on floc concentration in aggregates, floc to aggregate diameter ratio, particle concentration in flocs and attractive forces between flocs.

Finally τ_v is the shear stress associated to the shear of the suspension of aggregates assumed to be rigid (at least instantaneously). Thus, for sufficently low solid fractions, τ_v is given by Einstein's formula (equation (3.1)) in which ϕ is replaced by the concentration of aggregates.

In the original developments by Michaels and Bolger, ϕ_F, the concentration of flocs in the suspension appeared instead of ϕ in the above reasonings, but since the floc size (and not the aggregate size) is assumed to be constant within the considered shear rate range ϕ_F is proportional to ϕ and thus we have directly used ϕ in the formulae.

Although their demonstration is more detailed the final result of the model of Firth and Hunter (1976), more generally developed for coagulated colloidal dispersions, is similar in some respects to that of Michaels and Bolger (1962). These authors considered different energy terms related to floc or aggregate breakage and coalescence due to collisions but finally only the term related to elastic energy stored in intrafloc bond stretching is significant. In this model, this last term is proportional to the number of collisions by unit time, and thus to the square of the floc volume fraction as in Equation (5.17). Finally the yield stress also increases as ϕ^2.

Though these approaches provide interesting direct computations of the rheological parameters of the suspension, their underlying assumptions appear rather arbitrary and especially devoted to predicting the simple shear behaviour of clay-water suspensions within specific, limited shear rate and solid fraction ranges. In particular, considering their underlying assumptions (separation of well-identified flocs within the medium), these models might be especially appropriate for describing the behaviour of clay-water mixtures at solid fractions just above ϕ_c, since loose aggregation networks form in that case. This is confirmed by the relatively good agreement of (5.17) with experimental results concerning fluid yield stress increase with solid fraction just above yield stress appearance (see 5.1.2). For concentrated clay-water suspensions (5.17) is in complete disagreement with experimental results.

5.2.2 *Models taking into account thixotropic effects*

5.2.2.1 *Generalities*

As has been already pointed out, floc rupture processes are not instantaneous and thus thixotropic effects may occur as a result. In this field the proposed models are even more speculative than models for steady flow only. This is due to our poor knowledge of internal structure changes with flow history and more generally from the absence of reliable general "thixotropy" models based on internal structure changes for fluids (Mewis 1979). There have been some attempts to predict thixotropy on the basis of a continuum treatment (Harris 1967), kinetic arguments (Ruckenstein and Mewis 1973) or thermodynamic arguments (Huang 1972, Leonov 1987) but, compared with reality, none of them has proved to be of practical value so far. Finally, the most common and simple approach to thixotropy is phenomenological, as described in a general form by Cheng and Evans (1965). Within this framework, in simple shear, we have:

$$\tau = h(\lambda;\gamma)\gamma \tag{5.18}$$

Here the viscosity h is not only a function of shear rate, as is usual for non-Newtonian fluids, but also a function of the instantaneous internal structure of the fluid as expressed by the parameter λ. λ is then usually referred to as a structural parameter. The instantaneous value of λ is a function of flow history. A simple expression of this is as follows:

$$\frac{d\lambda}{dt} = f(\lambda;\gamma) \tag{5.19}$$

assuming that λ is only a function of time and instantaneous shear rate. Equation (5.19) is often called the kinetic equation. Under steady state conditions λ is found as a function of γ by equating the second member of (5.19) to zero. If the resulting expression $\lambda(\gamma)$ is introduced in (5.18) one finds the steady state shear stress value as a function of γ only. f and h are two functions which should depend on the material. With the help of equations (5.18) and (5.19) it appears possible to represent the main qualitative aspects observed in experiments. However, in most cases, the determination of h and f is not sustained by direct analysis of the structure but fitted to data.

5.2.2.2 *Basic approach to clay-water systems*

Moore (1959) first proposed a model for clay-water mixtures which was both rather simple and apparently realistic. It is a particular case of the above general model. The complete set of equations is given as:

$$\tau = \eta_0\gamma + c\lambda \tag{5.20}$$

$$a\frac{d\lambda}{dt} = (1-\lambda) - b\lambda\gamma \tag{5.21}$$

where a, b and c are material parameters. In these equations the structural parameter λ is interpreted as the state of build-up, defined as the ratio of the instantaneous number of links to the maximum number of links (obtained after an infinite time of rest). η_0 is consequently the viscosity of the suspension when all bonds are broken. Equation (5.21) in fact states that the rate of change of λ is the difference between the rate of increase in λ, proportional to the number of links remaining to be formed, and the rate of decrease, proportional to the instantaneous number of links and to shear rate. This model was proved to express the basic time-dependent characteristics of some clay-water suspensions, in particular the observed hysteresis (Moore 1959).

A more sophisticated model, which attempts to represent both the viscoelastic and thixotropic properties of concentrated dispersions in a low molecular weight matrix such as clay-water mixtures, was proposed by Coussot et al. (1993) on the basis of previous work concerning filled polymers (Leonov 1990). Considering that a yield stress clay-water suspension at rest has elastic properties, the model of Coussot et al. (1993) essentially assumes that, during flow, a part of elastic energy can be stored within flocs so that finally, the particle matrix can be considered as a simple Maxwell fluid and its shear stress (τ_e) verifies Equation (2.28) in which the elastic modulus G' is assumed to remain constant whereas the instantaneous viscosity η depends on flow history. As for the above models, a stress term due to the viscous dissipation of the equivalent suspension of rigid flocs must be taken into account and the suspension shear stress is finally τ:

$$\tau = \tau_e + \mu_I \gamma \tag{5.22}$$

where μ_I is the viscosity of the equivalent suspension of rigid flocs (here assumed to be constant). In this model η is a key parameter which depends on the concentration of broken bonds (1-λ). When λ tends towards 0, that is when all bonds are broken in steady state, all the particles move relatively to each other and a certain force is needed which depends on the mutual colloidal interaction forces. Then η tends to a limit value referred to as η_0. When, on average, some flocs exist, the energy dissipation arising from these relative motions is larger, so that η increases. Finally it was suggested by Coussot et al. (1993) to write:

$$\eta = \frac{\eta_0}{f_0(\lambda)} \tag{5.23}$$

with $f_0(\lambda) = (1\text{-}\lambda) + 0((1\text{-}\lambda)^2)$ when λ tends towards 1 and $\frac{df_0}{d\lambda} < 0$.

Once again one needs to know the state of the internal structure at any time. For that purpose the variations of λ with t and γ are expressed through a kinetic equation similar to the one due to Moore (1959) (equation (5.21)) but where a is now a function of the instantaneous shear rate, the aim of which is to express the change in material characteristic time with flow intensity. Finally arbitrary, reasonable forms for a and f_0 were proposed by Coussot et al. (1993). A complete three-dimensional expression of the model may also be found in Coussot et al. (1993).

The most remarkable feature of this theoretical model is that it contains no mathematical criterion but predicts the existence of a yield value with a continuous transition from solid state to flow as a bifurcation. The suspension behaviour progressively turns from purely elastic for small deformations to purely viscous for large shear rates. The instantaneous state of the material in terms of the number of links and the shear rate can be positioned in a phase diagram. The state evolution was then drawn by Coussot et al. (1993) within such, quite instructive, phase diagrams, for various initial and boundary conditions.

Although even in simple shear this model is rather complex, its merit is that it appears to be capable of predicting all the main rheological properties of clay-water suspensions: yielding, viscoelasticity, thixotropy. In particular, for a bentonite-water mixture, the model parameters were experimentally determined by some specific tests and then the theory was in reasonable agreement with experimental results obtained under various flow conditions: steady state, elastic response below yield stress, transient tests, dynamic tests. In particular, from (2.28), (5.22-5.23) it results that the suspension yield stress is :

$$\tau_c = \frac{\eta_0}{b} (=G'Z_c) \qquad (5.24)$$

in which the critical strain (Z_c) before yielding (for purely linear elastic motions) was introduced. The above expression makes it possible to find a relation between η_0 and b, the characteristic time of destructuration (appearing in (5.21)), from creep tests before yielding.

This model also predicts the possibility that there exists a minimum in flow curve (Cf 5.1.1.2.4) when the characteristic time of destructuration is sufficiently large and the viscosity of the suspending medium sufficiently low. More precisely this condition is expressed as:

$$\frac{\mu_I}{\tau_c} < b\left(\frac{1}{2}\left(\frac{d^2 f_0}{d\lambda^2}\right)_0 - 1\right) \qquad (5.25)$$

It is worth noting that this condition also involves a characteristic time of the material, μ_I/τ_c, which already appears when determining whether colloidal effects are predominant or not (see Chapter 3). Here this time must be smaller than a constant times the characteristic time b. When (5.25) is true, the resulting flow curve is rather peculiar and theory predicts that flows corresponding to the decreasing part of the curve (at low shear rates) should be unstable (Tanner 1988, and Chapter 9). In fact, in that case, the complete model of Coussot et al. (1993) predicts the formation of sandwich-like structures (alternately sheared and unsheared layers), the stability of which remains to be studied with further thermodynamic arguments. Instability processes are detailed for two similar, theoretical, rheological models proposed by Pearson (1994) and Schmitt et al. (1995).

A rather similar model was proposed by Billingham and Ferguson (1993) with less physical arguments. The resulting set of equations differs from the one due to

Moore (1959) in two ways: it explicitly takes into account the possible existence of a yield stress and it includes a diffusion term in the kinetic equation (5.21). The model is also generalized to three-dimensional flows by these authors. For a pipe flow it is written as:

$$\tau = \lambda\tau + (c+d\lambda)\gamma^n \quad \text{when } \tau>\tau_c\lambda \tag{5.26a}$$

$$\gamma=0 \quad \text{when } \tau<\tau_c\lambda \tag{5.26b}$$

$$a\frac{\partial\lambda}{\partial t} = (1-\lambda) - b\lambda\gamma + \frac{1}{r}\frac{\partial}{\partial r}[rD\frac{\partial\lambda}{\partial r}] \tag{5.27}$$

where τ and γ are the shear stress and shear rate magnitudes, r the distance from the pipe axis, D a coefficient of diffusion which may depend on γ and λ and c and d two parameters. These authors then studied the various possible flow characteristics of a a material of this type under different conditions. Even for simple flows the results are complex. In particular they found that, without the diffusion term, the flow curve which should be obtained from rheometrical data is not monotonous so that flow cannot be stable. Flow characteristics similar to the ones described above would therefore be expected again (sandwich-like structures). They demonstrated that the arbitrary diffusion term introduced within the kinetic equation can eliminate the instability. Another advantage of the constitutive equation (5.26-5.27) is that it makes it possible to describe with a single model both the behaviour of simple non-thixotropic clay-water systems following a Herschel-Bulkley model and the behaviour of complex thixotropic clay-water systems such as bentonite-water mixtures. The disadvantage is obviously that it involves a large number of unknown parameters which, additionally, do not have clear physical meanings yet.

5.3 REFERENCES

Akae, T., and Low, P.F. 1988. Interparticle bond energy and rheological properties of clay suspensions. *J. Colloid Interface Sci.*, 124, 624-631.

Alderman, N.J., Meeten, G.H., and Sherwood, J.D. 1991. Vane rheometry of bentonite gels. *J. Non-Newt. Fluid Mech.*, 39, 291-310.

Ansley, R.W., and Smith, T.N. 1967. Motion of spherical particles in a Bingham plastic. *AIChE J.*, 13, 1193-1196.

Baylocq, P., Cartalos, U., and Piau, J.M. 1994. Rheological characterisation and structural modelling of clay polymer systems. in *Progress and Trends in Rheology IV*, C. Gallegos (Ed.), Steinkopf, Darmstadt, 618-620.

Beazley, K.M. 1965. Factors influencing dilatant behaviour in China clay suspensions. *Trans. Brit. Ceram. Soc.*, 64, 531-546.

Billingham, J., and Ferguson, J.W.J. 1993. Laminar, unidirectional flow of a thixotropic fluid in a circular pipe. *J. Non-Newt. Fluid Mech.*, 47, 21-55.

Booth, F. 1950. The electroviscous effect for suspensions of solid spherical particles. *Proc. R. Soc. London*, A203, 533-551.

Brandenburg, U., and Lagaly, G. 1988. Rheological properties of sodium montmorillonite dispersions. *Appl. Clay Sci.*, 3, 263-279.

Chang, S.H., Ryan, M.E., and Gupta, R.K. 1993. The effect of pH, ionic strength, and temperature on the rheology and stability of aqueous clay suspensions. *Rheol. Acta*, 32, 263-269.

Chen, J.S., Cushman, J.H., and Low, P.F. 1990. Rheological behavior of Na-montmorillonite suspensions at low electrolyte concentration. *Clays and Clay Minerals*, 38, 57-62.

Cheng, D.C.H., and Evans, F. 1965. Phenomenological characterization of the rheological behaviour of inelastic reversible thixotropic and antithixotropic fluids. *Brit. J. Appl. Phys.*, 16, 1599-1617.

Chhabra, R.P., and Uhlherr, P.H.T. 1988. Static equilibrium and motion of spheres in viscoplastic liquids. in *Encyclopedia of Fluid Mechanics*, N.P. Cheremisinoff (Ed.), Gulf Publishing Co., Houston, TX, 1988, Chapter 21, Vol. 7.

Coussot, P. 1992. *Rheology of debris flows - Study of concentrated suspensions*. Ph.D. thesis, INPG, Grenoble, France. (in French)

Coussot, P., and Boyer, S. 1995. Determination of yield stress fluid behaviour from inclined plane test. *Rheol. Acta*, 34, 534-543.

Coussot, P. and Piau, J.M. 1994. On the behavior of fine mud suspensions. *Rheol. Acta*, 33, 175-184.

Coussot, P., Leonov, A.I. and Piau, J.M. 1993. Rheology of concentrated dispersed systems in a low molecular weight matrix. *J. Non-Newt. Fluid Mech.*, 46, 179-217.

Coussot, P., and Piau, J.-M. 1994. Rheology of very concentrated suspensions of force-free particles. *Les Cahiers de Rhéologie* (Groupe Français de Rhéologie), XIII, 266-277. (in French)

Coussot, P. 1995. Structural similarity and transition from Netwonian to non-Newtonian behavior for water-clay suspensions. *Phys. Rev. Lett.*, 74, 3971-3974.

Doraiswamy, D., Mujumdar, A.N., Tsao, I., Beris, A.N., Danforth, S.C., and Metzner, A.B. 1991. The Cox-Merz rule extended: a rheological model for concentrated suspensions and other materials with a yield stress. *J. Rheol.*, 35, 647-685.

Ducerf, S., and Piau, J.M. 1994. Influence des forces inter-particulaires sur la structure et les propriétés rhéologiques de suspensions denses d'alumines. *Les Cahiers de Rhéologie* (Groupe Français de Rhéologie), XIII, 120-129.

Egashira, K. 1977. Viscosities of allophane and imogolite clay suspensions. *Clay Sci.*, 5, 87-95.

Engelund, R. and Wan, Z. 1984. Instability of Hyperconcentrated flow. *J. Hydr. Eng.*, 110, 219-233.

Firth, B.A., and Hunter, R.J. 1976. Flow of coagulated colloidal suspensions. *J. Colloid Interface Sci.*, 57, 249-275.

Golden, S.P., Goodwin, J.W., and Olal, A.D. 1982. The dependence of the elastic properties of clay dispersions on the mode of interaction between particles. *Trans. J. Br. Ceram. Soc.*, 81, 84-87.

Gouami, A., Mourchid, A., Levitz, P., and Bergaya, F. 1994. Rhéologie des suspensions d'argiles fibreuses. *Les Cahiers de Rhéologie* (Groupe Français de Rhéologie), XIII, 170-179.

Granquist, W.T., 1959. Flow properties of dilute montmorillonite dispersions, *Clays and Clay Minerals*, Proc. Nat. Conf. Clays and Clay Minerals, Sixth Conference, NRC publication, 207-219.

Güven, N. 1992. Rheological aspects of aqueous smectite suspensions. *C.M.S. Workshop Lectures*, Volume 4, Clay-water interface and its rheological application, 81-125.

Harris, J. 1967. A continuum theory of time-dependent inelastic flow. *Rheol. Acta*, 6, 6-12.

Herschel, W.H., and Bulkley, R. 1926. Measurement of consistency as applied to rubber-benzene solutions. *Am. Soc. Testing Mater.*, 26, 621-633.

Hu, G., and Fang, Z. 1985. Antithixotropic model for fluid with hyperconcentration of sediment. *Proc. Int. Workshop on Flow of Hyperconcentrations and Sedimentation*, Beijing, China, I.3, IRTCES Publication, 1-9.

Huang, C.R. 1972. A thermodynamic approach to generalised rheological equations of state for time-dependent and time-independent non-Newtonian fluids. *Chem. Eng. J.*, 3, 100-104.

Huifang, Z., Low, P.F., and Bradford, J.M. 1991. Effects of pH and electrolyte concentration on particle interaction in three homoionic sodium soil clay suspensions. *Soil Sci.*, 151 196-207.

Johnson, A.M. 1970. *Physical processes in Geology*, Freeman Cooper and Co, San Francisco.

Kamal, M.R., and Mutel, A. 1985. Rheological properties of suspensions in Newtonian and non-Newtonian fluids. *J. Polymer Eng.*, 5, 293-382.

Khandal, R.K., and Tadros, Th.F. 1988. Application of viscoelastic measurements to the investigation of the swelling of sodium montmorillonite suspensions. *J. Colloid Interface Sci.*, 125, 122-128 .

Kytomaa, H.K., and Prasad, D. 1993. Transition from quasi-static to rate dependent shearing of concentrated suspensions. in *Powders & Grains*, C. Thornton (Ed.), Balkema, Rotterdam, 281-287.

Lagaly, G. 1989. Principles of flow of kaolin and bentonite dispersions. *Appl. Clay Sci.*, 4, 105-123.

Lapasin, R. 1985. Clay/kaolin aqueous suspensions. *J. Theor. Appl. Mech.*, Numero spécial, 239-251.

Leonov, A.I. 1987. On a class of constitutive equations for viscoelastic liquids. *J. Non-Newt. Fluid Mech.*, 25, 1 59.

Leonov, A.I. 1990. On the rheology of filled polymers. *J. Rheol.*, 34, 1039-1068.

Locat, J., and Demers, D. 1988. Viscosity, yield stress, remolded strength, and liquidity index relationships for sensitive clays. *Can. Geotechn. J.*, 25, 799-806.

M'Ewen, M.B., and Pratt, M.I. 1957. The gelation of montmorillonite. Part I : The formation of a structural framework in sols of Wyoming bentonite. *Trans. Faraday Soc.*, 53, 535-547.

Major, J.J., and Pierson, T.C. 1992. Debris flow rheology : Experimental analysis of fine-grained slurries. *Water Resour. Res.*, 28, 841-857.

Mas, R., and Magnin, A. 1994. Rheology of colloidal suspensions: Case of lubricating greases. *J. Rheol.*, 38, 889-908.

Mewis, J. 1979. Thixotropy - A general Review. *J. Non-Newt. Fluid Mech.*, 6, 1-20.

Michaels, A.S., and Bolger, J.C. 1962. The plastic flow behavior of flocculated kaolin suspensions. *Ind. Eng. Chem. Fundamentals*, 1, 153-162.

Migniot, C. 1989. Bedding-down and rheology of muds. *La Houille Blanche*, 1, 11-29, and 2, 95-111, 1989. (in French)

Moore, F. 1959. The rheology of ceramic slips and bodies. *Trans. Brit. Ceram. Soc.*, 58, 470-492.

Mourchid, A., Delville, A., Lambard, J., LeColier, E., and Levitz, P. 1995. Phase diagram of colloidal dispersions of anisotropic charged particles: equilibrium properties, structure, and rheology of laponite suspensions. *Langmuir*, 11, 1942-1950.

Naik, B. 1983. *Mechanics of mudflow treated as the flow of a Bingham fluid*. Ph.D. thesis, Washington State University, Washington.

Neumann, B.S., and Sansom, K.G. 1971. The rheological properties of dispersions of laponite, a synthetic hectorite-like clay, in electrolyte solutions. *Clay Minerals*, 9, 231-243.

Nguyen, Q.D., and Boger, D.V. 1983. Yield stress measurement for concentrated suspensions. *J. Rheol.*, 27, 321-349.

Nitakawi, Y., Wada, K., and Egashira, K. 1981. Particle-particle and particle-water interactions in aqueous clay suspensions. Part II. Viscosity data and interpretation. *Clay Sci.*, 5, 319-331.

O'Brien J.S. and Julien, P.Y. 1988. Laboratory analysis of mudflows properties. *J. Hydraul. Eng.*, 114, 877-887.

O'Brien, J.S. 1986. *Physical processes, rheology and modeling of mud flows*. Ph.D. thesis, Colorado State University, Fort Collins.

Pazwash, H., and Robertson, J.M. 1971. Fluid dynamic consideration of bottom materials. *J. Hydraul. Div.*, ASCE, 97, 1317-1329.

Pearson, J.R.A. 1994. Flow curves with a maximum. *J. Rheol.*, 38, 309-331.

Permien, T., and Lagaly, G. 1995. The rheological and colloidal properties of bentonite dispersions in the presence of organic compounds. V. Bentonite and sodium montmorillonite and surfactants. *Clays and Clay Minerals*, 43, 229-236.

Phillips, C.J., 1989. *Rheology of debris flow material*. Ph. D. thesis, Lincoln College, Univ. Canterbury.

Pignon, F., Magnin, A., and Piau, J.M. 1994. Ecoulements de suspensions thixotropes d'argile. *Les Cahiers de Rhéologie* (Groupe Français de Rhéologie), XIII, 160-169.

Pignon, F., Magnin, A., and Piau, J.M. 1996a. Thixotropic colloidal suspensions and flow curves with minimum: identification of flow regimes and rheometric consequences. *J. Rheol.*, 40, 573-587.

Pignon, F., Piau, J.M., and Magnin, A., 1996b. Structure and pertinent length scale of a discotic clay gel. *Phys. Rev. Lett.*, 76, 4857-4860.

Qian, N., and Wan, Z. 1986. *A critical review of the research on the hyperconcentrated flow in China*. IRTCES publication, China.

Rand, B., and Melton, I.E. 1977. Particle interactions in aqueous kaolinite suspensions, Part I - Effect of pH and electrolyte upon the mode of particle interaction in homoionic sodium kaolinite suspensions. *J. Colloid Interface Sci.*, 60, 308-336.

Rand, B., Pekenc, E., Goodwin, J.W, and Smith, R.W. 1980. Investigation into the existence of edge-face coagulated structures in Na-montmorillonite suspensions. *J. Chem. Soc. Faraday Trans.*, 76, 225-235.

Ruckenstein, E., and Mewis, J. 1973. Kinetics of structural changes in thixotropic fluids. *J. Colloid Interface Sci.*, 44, 532-541.

Saville, D.A. 1977. Electrokinetic effects with small particles. *Ann. Rev. Fluid Mech.*, 9, 321-337.

Schmitt, V., Marques, C.M., and Lequeux, F. 1995. Shear-induced phase separation of complex fluids: The role of flow-concentration coupling. *Phys. Rev. E*, 52, 4009-4015.

Sherwood, J.D. 1994. Thixotropic mud, compacted cake and swelling shale. *Les Cahiers de Rhéologie* (Groupe Français de Rhéologie), XIII, 111-119.

Sherwood, J.D., Meeten, G.H., Farrow, C.A., and Alderman, N.J. 1991. Squeeze-film rheometry of non-uniform mudcakes. *J. Non-Newt. Fluid Mech.*, 39, 311-334.

Sohm, R., and Tadros, Th.F. 1989. Viscoelastic properties of sodium montmorillonite (gelwhite H) suspensions. *J. Colloid Interface Sci.*, 132, 62-71.

Speers, R.A., Tung, M.A., and Williamson, D.T. 1988. Rheological determination of peptizing agents in bentonite clays. *Rheol. Acta*, 27, 561-564.

Street, N. 1956. The rheology of kaolinite suspensions. *Austral. J. Chem.*, 9, 467-479.

Tanner, R.I. 1988. *Engineering Rheology*, Clarendon Press, Oxford.

Thomas, D.G. 1963. Non-Newtonian suspensions - Part I: Physical properties and laminar transport characteristics. *Ind. Eng. Chem.*, 11, 18-29.

Tsenoglou, C. 1989. Viscoplasticity of agglomerated suspensions. *Rheol. Acta*, 28, 311-314.

Tsutsumi, A., and Yoshida, K. 1987. Effect of temperature on rheological properties of suspensions. *J. Non-Newtonian Fluid Mech.*, 26, 175-183.

Vali, H., and Bachmann, L. 1988. Ultrastructure and flow behavior of colloidal smectite dispersions. *J. Colloid Interface Sci.*, 126, 278-291.

Van Damme, H., Lemaire, E., and Levitz, P. 1991. La physique de la boue. *Pour la Science*, 167, 82-90.

Van Olphen, H. 1956. Forces between suspended bentonite particles. *Clays and Clay Minerals*, Proc. Nat. Conf. Clays Clay Minerals, Fourth Conference, NRC publication, 204-224.

Van Olphen, H. 1997. *Introduction to clay colloid chemistry*, Wiley and Sons, New-York,.

Wan, Z. 1982. *Bed material movement in hyperconcentrated flow*. Series Paper 31, Inst. of Hydrodynamics and Hydr. Eng., Tech. Univ. of Denmark.

Wang, Z., Larsen, P., and Xiang, W., 1994. Rheological properties of sediment suspensions and their implications. *J. Hydr. Res.*, 32, 560-580.

Weymann, H.D., Chuang, M.C., and Ross, R.A., 1973. Structure of thixotropic suspensions in shear flow: I. Mechanical properties. *Phys. Fluids*, 16, 775-783.

CHAPTER 6

Rheology of mud suspensions

Here we shall consider mud suspensions in the general sense of mixtures of water, clay, silt, sand, gravel, pebbles and boulders. At first sight this kind of material is very complex since it includes elements of very different types, sizes and shapes giving rise to very different, possible interactions (Cf Chapter 1). As a first approximation (see Chapter 1), it has been possible to roughly separate solid particles into two main types depending on the main type of interactions to which they give rise. In water, fine particles such as clay particles give rise to colloidal interactions and may be subjected to significant Brownian motions. Grains whose diameter is larger than a critical value can be considered as non-colloidal particles which interact hydrodynamically or via direct contacts giving rise to friction or collisions. Once again it is worth noticing that the predominant type of interaction between particles depends on flow conditions (see 3.2). Thus the above particle distinction must be seen as true only under usual flow conditions. Additionally there are particles whose diameter is in a transition range, for which these two types of interaction play a significant role. Nevertheless, for the sake of simplicity, we shall make a rough separation of all solid particles into two types, colloidal and non-colloidal, depending on their diameter compared to 40 μm. This limit is both reasonable from a theoretical point of view since it should be approximately situated at the centre of the transition range for most practical situations and practical from the point of view of usual laboratory grain size analyses. Finally, we shall assume that the clay fraction is significant within any fine solid fraction (< 40 μm) considered below, so that the properties of the water-fine particles mixture is greatly influenced by the clay fraction.

Given these assumptions, the particles of the first type, those with a diameter smaller than 40 μm, referred to as the fine particles, form, when mixed with water, a fine particle-water mixture whose behaviour is similar to a clay-water mixture (see Chapter 5). Moreover this mixture can be considered as the interstitial fluid of a mud suspension. The particles of the second type, those with a diameter larger than 40 μm, referred to as grains, are added to the fine particle-water mixture to finally obtain the whole mud suspension. Thus, from a rheological point of view, the final material is a suspension of non-colloidal particles in a dispersion.

Let us consider a mixture of given amounts of water, fine particles (mainly clay) and grains. We wish to obtain a homogeneous material whose dynamical properties can be described within the framework of fluid mechanics (see Chapter 2). For this purpose the size of the coarsest grains must be small compared to a characteristic macroscopic length of the sample and the initial distribution of grains must be isotropic. The initial sample may easily be prepared so as to respect the above two conditions. Two other important conditions are required: the mixture must not fracture under usual flow conditions, and sedimentation must be negligible within the duration of the flow. These two conditions should make it possible to determine concentration and component limits beyond which the mixture cannot be considered as a fluid under particular flow conditions. In the first part, on the basis of experimental results, we shall propose a classification of mud suspensions according to their overall aspect (fracturing, settling, homogenously deformed, etc) as a function of their components. In the second part we shall review the rheological behaviour of mud suspensions which belong to regions corresponding to homogeneous mixtures in the above classification. This leads to a rough classification of mud suspension behaviour as a function of their components, explained by internal structure. Mud suspension rheology might take advantage of knowledge of the rheology of analogous industrial or natural suspensions. Conversely the present methodology might be used for other coarse suspensions. In the third part we shall review rheological knowledge of more or less natural, coarse suspensions.

6.1 CLASSIFICATION OF NATURAL MUD SUSPENSIONS AS A FUNCTION OF THEIR ASPECT

6.1.1 *Generalities*

If the viscosity of the mixture is not sufficiently large, at least the coarsest particles will settle rapidly. In any flow whose duration is longer than the settling time, sedimentation will break the initial homogeneity of the fluid and we may be confronted with two-phase flows. Obviously the components for which sedimentation occurs also depend on flow conditions (shear intensity, normal stress difference, flow duration, etc.).

If the solid fraction is too large, fracture may occur during flow. In other words there are some specific surfaces along which shear is localized, and the strain field can no longer be considered as continuous. This effect is observed for example during rheometrical tests (see 4.1.5.1.4) and then considered as a disturbing effect. In soil or solid mechanics the rupture (or fracture) occurs so systematically that the value of the critical strength for which it occurs is commonly considered as a basic characteristic of the material. This effect is associated with apparent density changes which depend on initial, apparent, material density. More generally fracture also depends on boundary conditions: confined or free surface flow, external pressure, imposed shear rate, total deformation, etc. A simple, general relation between fracture and material characteristics and boundary conditions cannot yet be proposed. At least, for a given solid material, other things being equal, fracture occurs beyond a critical solid concentration.

6.1.2 *Experimental observations*

Finally mud suspensions for which neither fracture nor sedimentation occur cannot be determined in a systematic way independently of flow conditions. Nevertheless, in order to have an approximate view of mud suspensions that can be considered as homogeneous fluids under usual flow conditions, we shall review the procedure of Coussot (1992), who obtained a classification of mud suspensions according to their aspect on the basis of a simple experimental protocol.

For a given solid particle distribution (fine particles and grains) all mud suspensions will be obtained by mixing different amounts of water with solid particles. Since in that case the solid particle distribution is fixed the ratio of fine particles to (all) solid particles (fine particles + grains) remains constant. Two empirical limit solid volume fractions can be defined (Coussot 1992):

1- ϕ_i, the lowest solid volume fraction without almost immediate (a few seconds after the end of mixing) settling of the coarsest particles. In practice if, after mixing, no grains are visible at the free surface one can presume that ϕ, the actual solid concentration, is lower than ϕ_i. This can be checked by plunging a hand into the mixture.

2- ϕ_s, the largest solid volume fraction for which an open fracture of 5 cm in depth (directly visible) occurs after one revolution (at a speed of 6 rev/mn) during a vane-test in a small cylindrical bob (diameter: 8 cm; depth: 10 cm) with a vane which is 6 cm long and 4 cm high.

We emphasize that this protocol has been chosen arbitrarily for practical reasons and because it is thought to reflect the main changes in material characteristics. It is not intended to be generalized to other materials or used systematically with natural mud suspensions as a standard technique. Roughly speaking, between the two empirical limit concentrations a mud suspension is in general homogeneous during most natural or laboratory flows. Beyond these two limits, settling or fracture generally occurs.

The two limit concentration curves for a solid material were determined and explained physically from a qualitative point of view (see 6.1.3) so that the qualitative results were extrapolated to other materials (Coussot 1992). The procedure was as follows. Given volumes of fine and coarse particles were mixed altogether. Then water was progressively added and mixed by hand: the fine to solid particle ratio thus remained constant and the two corresponding limit concentrations were successively determined by means of the above described tests. The experimental results are presented in Figure 6.1 as a function of the total solid volume concentration and the fine to solid particle ratio. Two zones separated by an "intermediate zone" (narrow but experimentally clear and located between the two vertical lines in Figure 6.1) appear clearly. During these tests, mixtures corresponding to points situated in the "intermediate zone" fractured as soon as the settling limit was overcome. Thus, it is not possible to obtain homogeneous fluids corresponding to points situated in this zone. On the contrary, in the left-hand zone, ϕ_i and ϕ_s are also equal but the corresponding mixtures could flow (at least when sheared by the vane within the bob) without fracturing. Nevertheless, in that case, the shear was localized in a zone 3 cm thick, around the vane, and seemed almost constant in this zone.

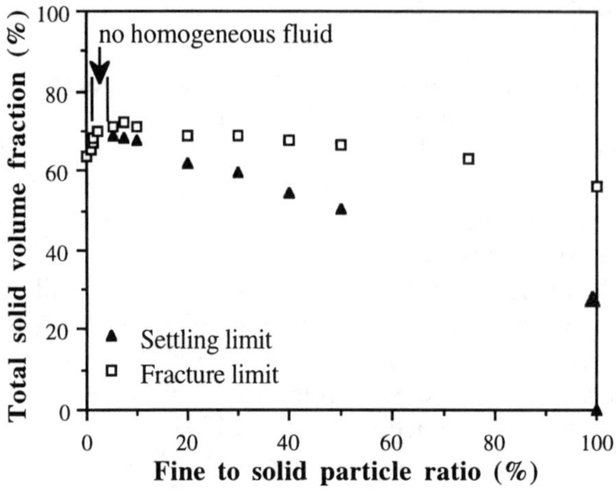

Figure 6.1: Curves defining composition of water-sediment mixtures used in experiments. Limit settling and fracture concentrations as a function of fine (Eybens clay < 40 μm) to solid fraction (Bourgeat sand and Eybens clay) and total solid concentration. [from Coussot 1992].

6.1.3 *Physical explanations*

The form and slope of these curves can be explained from a qualitative point of view:

1- When the concentration of fine particles is sufficiently large, the interstitial fine particle-water mixture exhibits yield stress (see Chapter 5). The range of total solid concentrations for which one obtains a homogeneous fluid becomes wider when the fine to solid particle ratio increases. The reason is as follows: as long as the grain fraction is not too high (say less than 30%), the interstitial clay-water mixture certainly has a significant effect on the yield stress of the whole suspension (see below); a given total solid concentration change (keeping the fine to solid particle ratio constant) has a smaller effect on the solid concentration of the interstitial fluid (fine particle-water mixture) and therefore on its yield stress when the fine to solid particle ratio is larger; the range of solid concentration for which neither settling nor fracture, which are probably related to suspension yield stress, occurs, is therefore wider when the fine to solid particle ratio is larger. The slightly decreasing curves for ϕ_i and ϕ_s from their maximum value (dotted zone) to the right might be explained by the fact that fracture and immediate settling can be associated with critical yield stress values. The corresponding equations for these curves could therefore be deduced from equation (6.3) in which ϕ_a should be expressed as a function of ϕ and the fine to solid particle ratio.

The limit settling curve (ϕ_i) approaches a non-zero value when the fine to solid particle ratio approaches 100% because a certain yield stress level is necessary to maintain coarse particles in suspension. However the ϕ_i value for a pure fine particle-water mixture can be considered to be zero because fine particles do not settle immediately in water at rest. Thus, there is discontinuity at this point.

2- For very low fine to solid particle ratios (close to the ordinate) the interstitial fluid is essentially composed of water and, if it exists, its yield stress is too small to support grains at rest. Thus the only way for these grains to be supported at rest is by a network of grain-to-grain contacts affecting the whole material. Consequently the value of ϕ_i is high because it corresponds to a high grain concentration. Additionally ϕ_s is almost equal to ϕ_i because the corresponding packing network which must be reached at ϕ_i is very close to the maximum packing concentration obtained after mixing. The slope of the common curve (ϕ_i-ϕ_s) starting from the left vertical axis is positive. This is due to the fact that the packing network corresponding to ϕ_i or ϕ_s contains water pores which may easily be replaced by a fine particle-water mixture without any change in grain configuration. This allows an increase in the fine to solid particle ratio in parallel with an increase in the solid fraction.

3- Within the "intermediate zone", coarse grains cannot be supported by the yield stress of the interstitial fluid but must be borne by a grain-to-grain packing network. Nevertheless, when this network exists the total solid fraction (here there is also a significant fine fraction) is too high and fracture occurs as for fine mud suspensions with a large sand concentration.

Assuming again that it is possible to explain the form of these curves from a physical point of view, we can conclude that these curves give an idea of the general properties of mud suspensions: analogous curves should exist for all grain-size distributions and clay types and more generally for suspensions of grains in (colloidal) dispersions.

6.2 RHEOLOGY OF MUD SUSPENSIONS

In the following we shall focus on the behaviour of mud suspensions under steady simple shear (flow curve). Time-dependent properties of these materials have not yet been studied. Nevertheless, when they appear negligible under usual flow conditions for the interstitial clay-water suspension, they should in general be negligible for the corresponding, complete mud suspension.*

6.2.1 *Rheology of fine particle (clay)-water mixture + grains*

When grains are added to a dilute clay-water system they rapidly settle. Consequently we shall focus on the rheology of sufficiently concentrated clay-water suspensions in which grains, when added, do not settle under usual flow

* Except in the case of a highly concentrated granular suspension for which packing effects may be significant (see 3.1.3).

conditions. The behaviour of such mixtures has rarely been studied. Moreover no research work so far was focused on the behaviour of grain suspensions in yield stress fluids. Thus the largely empirical results presented below are only intended to give some basic ideas on this topic.

6.2.1.1 *Sand in a clay-water suspension*

Let us consider a given clay-water (or fine particle-water) suspension. For the sake of simplicity, this mixture will be referred to as the "initial" or "interstitial" fluid. The clay fraction within this fluid is denoted ϕ_a (clay to [water+clay] volume ratio). Let us now consider the suspension obtained by adding grains to this initial mixture. The volume concentration of added sand will be denoted ϕ_S (sand to total mixture volume ratio). The total solid concentration, ϕ, the ratio of the volume of clay+sand to the total mixture volume, may be written:

$$\phi = \phi_S + (1-\phi_S)\,\phi_a \tag{6.1}$$

Observing the change in the constitutive equation when grains are added to a clay-water suspension is an experiment conceptually similar to the ones which study the viscosity change caused by adding non-colloidal particles to a Newtonian fluid (see 3.1.2). There the effect is a viscosity increase. So, in the present case, one might expect that the global effect on our suspensions should also be an increase of viscous dissipation within the interstitial fluid which would yield an increase of the apparent viscosity. Coussot and Piau (1995a) did indeed show that the addition of force-free particles to a clay-water mixture gives suspensions with higher apparent viscosity. In all their results (see Figs 6.2-6.4), the flow curve level increases with added sand concentration. This result is similar to the results obtained by Major and Pierson (1992) and O'Brien and Julien (1988). Only O'Brien and Julien (1988) presented data in which they noticed a slight initial decrease in the Bingham plastic viscosity of bentonite-based materials.

Since the interstitial fluid is non-Newtonian, it appears valuable to examine changes in rheological parameters, some of them physically interpretable, corresponding to a model fitted to an experimental flow curve, rather than apparent viscosity changes. Within this framework Coussot and Piau (1995a) and Boyer (1994) found that, as long as the added solid fraction is not too large, a Herschel-Bulkley model may be fitted to data and the suspension yield stress approximately increases exponentially as ϕ_S increases. Thus, for a water-clay-sand suspension (for a given ϕ_a), we have:

$$\tau_c = c' \exp(s'\phi) \tag{6.2}$$

where ϕ is given here by equation (6.1), and c' and s' are suspension parameters. From equations (5.9) and (6.2), one obtains:

$$\tau_c = c \exp(s\phi_a + s'(1-\phi_a)\phi_S) \tag{6.3}$$

For a given solid material, when c, s and s' are known, (6.3) is a general relation which makes it possible to determine the yield stress of any mud suspension formed by mixing different amounts of grain, clay and water. Coussot and Piau (1995a) also found that K increases with higher added sand concentrations but not as rapidly as the yield stress, but there seems to be no simple relation between K and ϕ. These authors did not find significant variations of n within their range of tests.

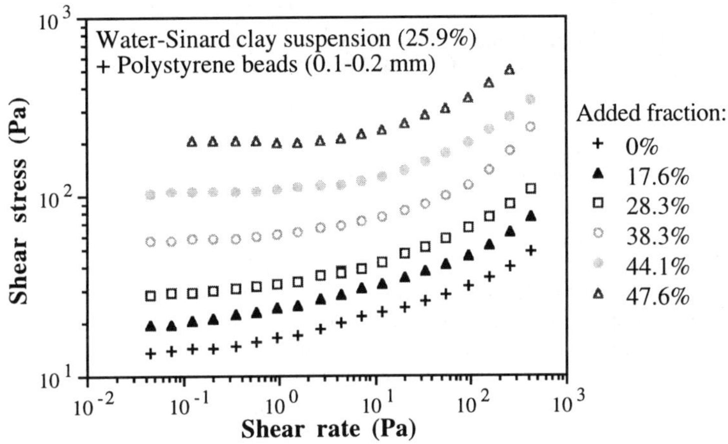

Figure 6.2: Flow curves for various added fractions of polystyrene beads in a concentrated Sinard clay-water suspension, obtained with the help of a parallel plate rheometer as in Coussot (1994). [from Boyer 1994]

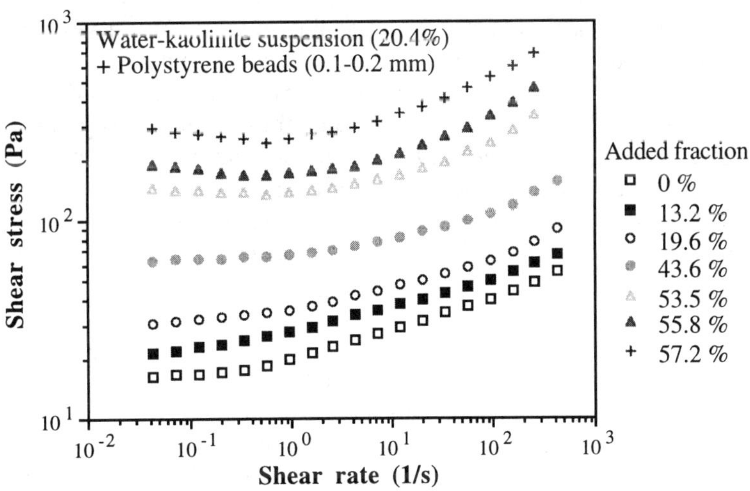

Figure 6.3: Flow curves for various added fractions of polystyrene beads in a concentrated kaolinite-water suspension. Tests were carried out in a way similar to tests by Boyer (1994) and Coussot and Piau (1995a).

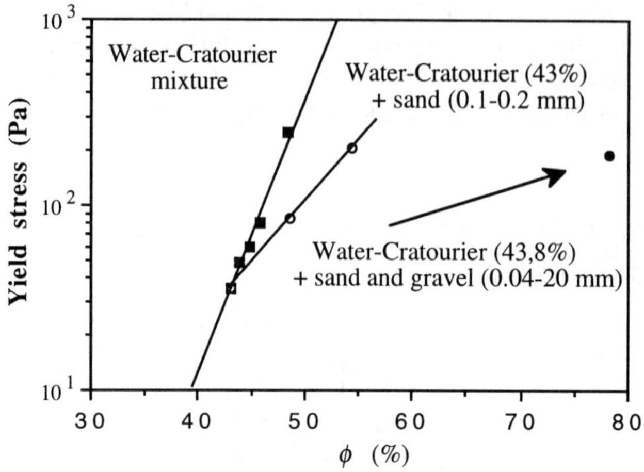

Figure 6.4: Flow curves for various added sand fractions in a concentrated Cratourier (fine debris flow sample)-water suspension [from Coussot and Piau 1995a]. One additional test was made with the large scale field rheometer (Coussot and Piau 1995b). The corresponding mixture may be seen as a clay-water suspension to which sand and pebbles have been added.

For larger solid fractions a Herschel-Bulkley model can no longer be fitted to data (see Figs 6.2-6.4). Indeed the experimental flow curve decreases for low shear rates then increases. It remains hard to conclude whether these results, similar to those obtained for very concentrated suspensions in a Newtonian fluid (see 3.1.3.3) correspond to disturbing effects or to real properties of the homogeneous mixture. Considering the large solid fraction of the corresponding mixtures, this effect might in fact be a kind of instability related to dilatant and/or frictional phenomena. Since these effects appear under various flow conditions, they should be considered as intrinsic flow properties of this type of fluid, and thus, even if they disturb rheometrical tests, must be explained by the "rheological behaviour" of the suspension. It is worth noting that, as soon as this effect appears, the corresponding yield stress, either estimated by extrapolating flow curve towards low shear rates or measured with the help of the inclined plane test, increases more rapidly than before (see Fig. 6.5) and the model (6.2) is not valid in this region.

When non-colloidal particles are added to a clay-water mixture, the fraction of interstitial fluid in a given total volume decreases. Thus, everything suggests that grains take the place of some regions of the clay particle network at the origin of the yield stress of the interstitial fluid. These grains may consolidate the new interacting particle network to a greater or lesser degree depending on the conditions of stress transmission (Batchelor 1970) between the clay-water system and the particles. This problem has not been solved in the ideal case of spheres in a perfectly homogeneous yield stress fluid either.

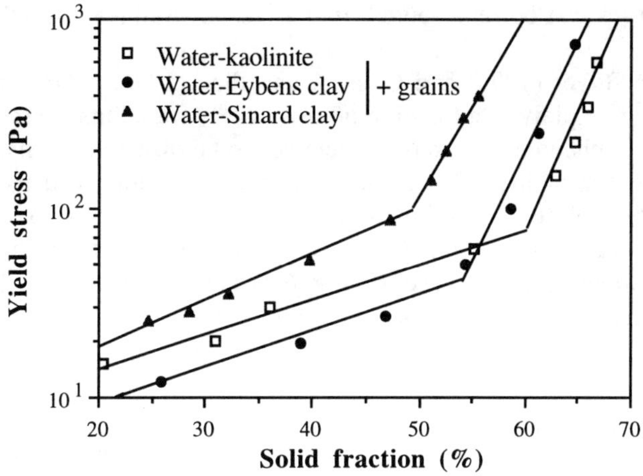

Figure 6.5: Clay-water suspensions + sand or polystyrene beads: yield stress as a function of total solid fraction. [data from Coussot and Piau (1995a), Boyer (1994) and Fig. 6.3]

In our case, it is worth noticing that the yield stress increase when adding sand is formally similar (exponential) to the yield stress increase observed when adding clay to water, though these two particle types interact quite differently with water. When grains are added to a Newtonian fluid the suspension remains Newtonian as the interstitial fluid has a larger viscosity. Here, in a similar way, the exponential yield stress increase suggests that, for all these fluids, the interstitial clay-water system, via the clay particle network, remains the prime factor affecting the suspension behaviour. This is also supported by the fact that, for low and moderate fractions of added grains, the number of direct contacts between these grains is negligible. However, when a critical, total, solid fraction is reached, a new network of more or less direct contacts between all solid particles (clay and grains) could form, which would explain the change in the yield stress curve as a function of solid fraction.

6.2.1.2 *Addition of an expanded grain size distribution to a clay-water mixture*

The behaviour of mixtures formed by suspending an expanded grain size distribution in a fine particle-water suspension has rarely been studied. Only a few recent works have presented tests of natural, coarse suspensions in large-scale prototype rheometers (Major and Pierson 1992, Coussot and Piau 1995b, Phillips and Davies 1991). All experimental problems mentioned in Chapter 4 can occur with these materials but, with regard to the scale of the tests, they are more difficult to control or avoid than with conventional laboratory rheometers. Thus existing data must be considered as giving only a rough idea of the behaviour of moderately coarse suspensions within a very limited shear rate range. Within this framework it appears that many flow curves can be represented by a Herschel-Bulkley model* (see Fig. 6.6).

Additionally, data due to Boyer (1994) and Coussot and Piau (1995a) showed that yield stress increases more rapidly as a function of how poorly sorted the coarse material is (Cf Fig. 6.4). This phenomenon must be related to the observation that the maximum packing concentration increases as the grain size distribution widens. It should be noted that when plotted as a function of the ratio of added solid fraction to the maximum packing fraction for the added granular phase alone, the yield stress approximately follows a single curve (Cf Fig. 6.7) as for the viscosity of force-free particle suspensions.

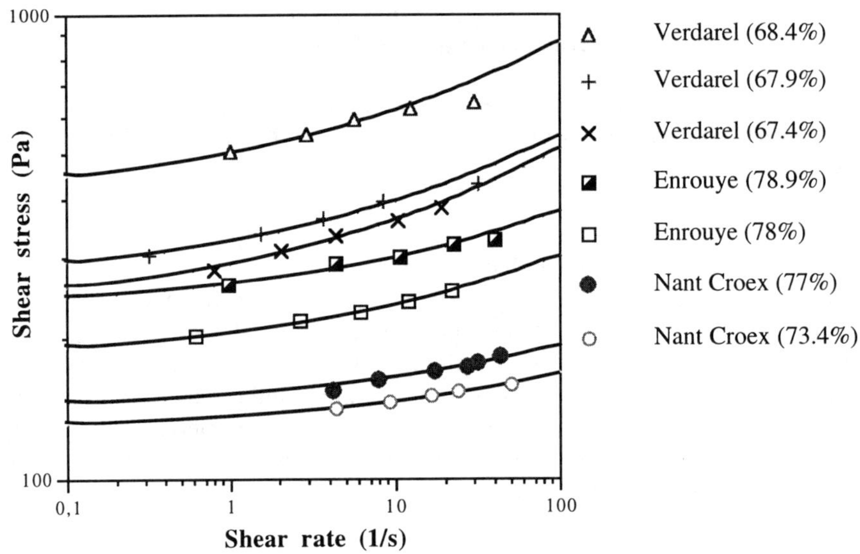

Figure 6.6: Coarse mud suspensions: flow curve. Data from Coussot and Piau (1995b). Herschel-Bulkley model fitted on data with n=1/3. Enrouye: (78%) τ_c=179 Pa, K=26.3; (78.9%) τ_c=232 Pa, K=31; Verdarel: (67.4%) τ_c=230 Pa, K=60; (67.9%) τ_c=265 Pa, K=60; (68.4%) τ_c=400 Pa, K=100; Nant Croex: (73.4%) τ_c=127 Pa, K=8.7; (77%) τ_c=140 Pa, K=11.4. (K in Pa.s$^{1/3}$).

Since he determined the yield stress of coarse suspensions from inclined plane tests, Boyer (1994) was able to obtain a good estimate of the true yield stress of suspensions even for large solid fractions, that is for suspensions that we suspect exhibit a minimum in flow curve. His results showed the same trends as for mud suspensions with smaller grain size distributions tested with a rotational rheometer (Cf Fig. 6.7): an initial exponential increase in yield stress then a sudden change in the rate of increase. With regards to rheometrical tests with finer suspensions (see

* Major and Pierson (1992) nevertheless concluded that a Bingham model represents quite well the behavior for shear rates above 5s^{-1}.

6.2.1.2), we think that, in Boyer's tests (1994), the sudden change in the slope for the poorly sorted materials might also reflect the appearance of a minimum.

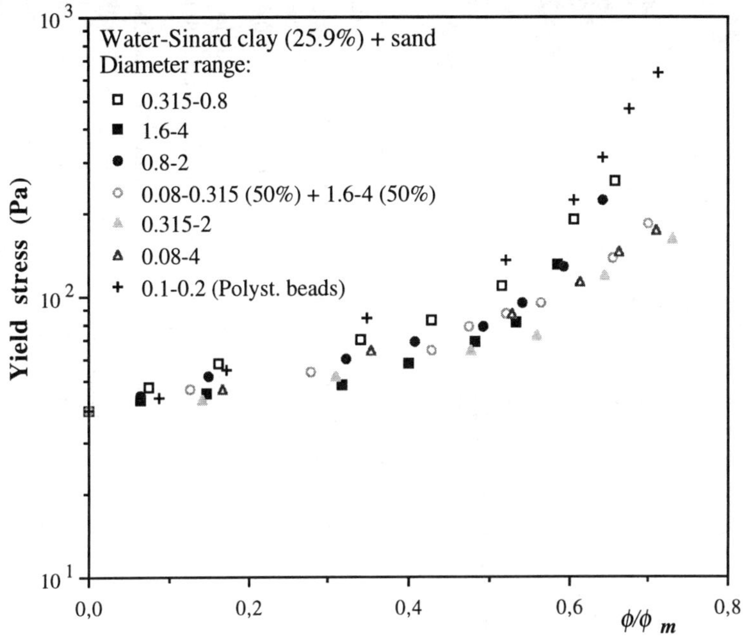

Figure 6.7: Water-Sinard clay suspension + sand of various grain size distributions: Yield stress as a function of the ratio of the solid fraction to the maximum packing solid fraction (for sand alone). The yield stress has been estimated from inclined plane tests (Cf Coussot and Boyer (1995)). [from Boyer 1994]

6.2.2 *Rheological classification*

6.2.2.1 *Experimental observations*

In Figure 6.8 many results presented in the literature were gathered in a "solid fraction vs fine to solid particle ratio" diagram similar to 6.1.2. We may easily distinguish two regions of clearly different behaviour types. The first one (on the right) includes points corresponding to mixtures whose simple shear flow curve can be represented from a general point of view by a Herschel-Bulkley model. Depending on the authors, the power parameter was either set to 1 (Bingham model) or fitted to the data. When the results included data in a wide shear rate range, the power parameter was found to be around 1/3. The second zone includes points corresponding to mixtures whose simple shear flow curve seemed to exhibit a minimum. It should be noted that one point corresponding to a flow curve with a minimum is situated in the first zone. This might be explained by the fact that the exact extent of each zone could slightly fluctuate as a function of material characteristics (grain size distribution, clay type, etc.).

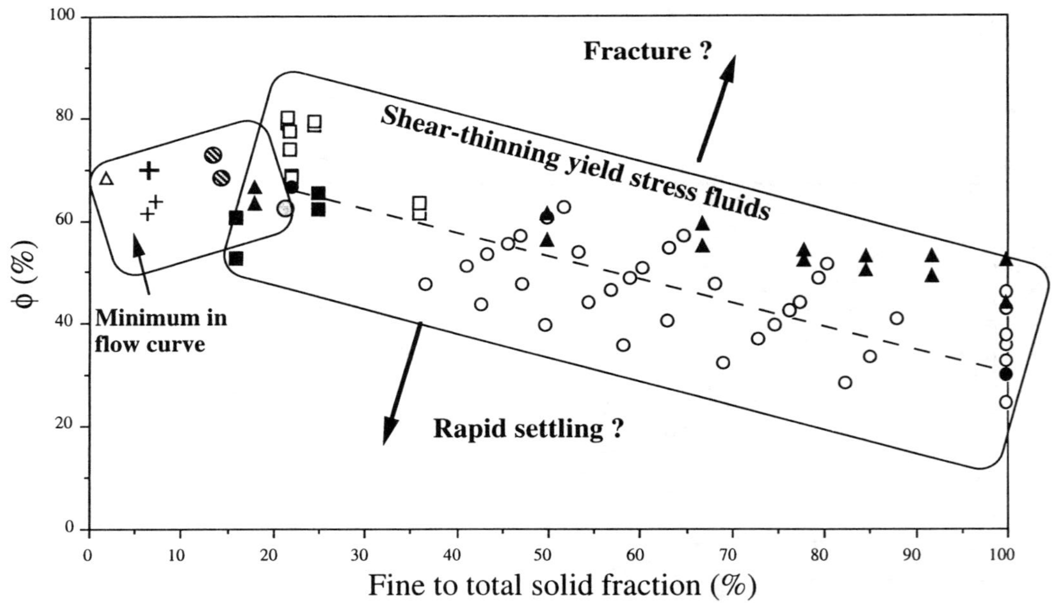

Shear-thinning yield stress fluids

- ○ Debris flow fine fraction (< 0.04 mm) + sand (0.1-0.2 mm) (Coussot 1992)
- □ Debris flow fraction < 20 mm (Coussot and Piau 1995)
- ■ Debris flow fraction < 35 mm (Phillips and Davies 1991)
- ● Debris flow fine fraction (< 0.074 mm) + sand (O'Brien and Julien 1988)
- ▲ Debris flow fine fraction (< 0.074 mm) + sand (Major and Pierson 1992)

Fluids (seemingly) with a minimum in flow curve

- △ Debris flow fraction < 20 mm (Coussot and Piau 1995)
- + Dilute clay-water system+ polystyrene beads (0.1-0.2 mm) (Coussot 1992)
- **+** Debris flow fraction < 120 mm (Phillips and Davies 1991)
- ◯ Concentrated Sinard clay-water mixture + polystyrene beads (Boyer 1994)
- ⊘ Concentrated Kaolinite-water mixture+polystyrene beads (Cf Fig.6.3)

Figure 6.8: Behaviour of some different water-debris mixtures as a function of fine to solid fraction ratio and total solid fraction.

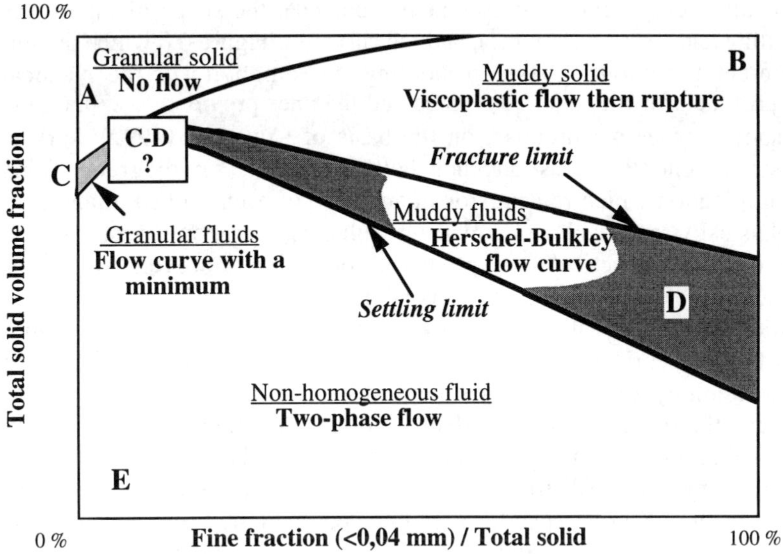

Figure 6.9: Conceptual rheological classification of water-debris mixtures.

With respect to our physical explanations of the behaviour of clay-water systems + sand (see 6.2.1.1), we suggest the following, more general, explanations for the behaviour of natural mud suspensions: the region of Herschel-Bulkley behaviour might correspond to mixtures for which the fine fraction is large enough for the interstitial fine particles-water mixture to lubricate most of the relative movements of the grains and impose its behaviour type on the whole mixture, in a similar way to a dilute or moderately concentrated suspension of grains in a Newtonian fluid. The zone of "unstable" behaviour (minimum in flow curve?) corresponds to mixtures for which the fine particle fraction is very low. In this case the interstitial fine particle-water mixture can still lubricate the relative movements of the grains at high shear rates and then direct contacts are seldom on average, because grains tend to disperse slightly. However at low shear rates direct frictional contacts between grains are likely to be predominant, as in the case of a large concentration of grains in a water-clay mixture. This could explain the possible minimum in flow curve: the energy necessary to induce flow at low shear rates, when a direct contact network exists, is higher than the energy necessary to maintain flow at moderate shear rates, when particle motions are lubricated by fine layers of interstitial fluid.

Between the two zones of distinct behaviour there is an area for which we have no data. We cannot actually conclude as to whether natural mixtures corresponding to such characteristics are difficult to test with a rheometer or whether they did not appear in these results by chance.

6.2.2.2 *Generalization*

All the above rheometrical results correspond to water-debris mixtures with solid concentrations for which neither size-preferential settling nor fracture occurs during tests. Additionally there is good correspondence between the shape of the curves for critical settling and fracture (Fig. 6.1), and the curves on Figure 6.8 which delimit zones of different behaviour types. Considering the simplicity of the physical explanations provided for the ϕ_i and ϕ_s curves and the ones provided to explain the different behaviour types we propose, on the basis of Coussot's (1992) work, a reasonable, strict general correspondence between these two diagrams. This provides a classification which remains rough and has not been proved in any case but appears physically reasonable, in agreement with existing rheometrical data and conceptually useful for practical flow prediction or future improvements to our rheological knowledge of natural water-debris mixtures.

This classification is obviously simplified and aims at representing the main types of behaviour of water-debris mixtures. Figure 6.9 illustrates the resulting generalized model of the rheology of water-debris mixtures as a function of both solid concentration and the fine content of the debris. Figure 6.9 is intended as a general, conceptual model and has been drawn without considering the effects of differing clay mineralogies or grain size distribution (e.g., unimodal versus bimodal, degree of sorting, mean size, etc.). Curves specific to certain debris mixtures could be derived from appropriate laboratory experiments as outlined above, but would probably conform qualitatively to the curves in Figure 6.9. On the basis of these generalized curves, we have identified 5 zones (A-E) associated with different types of flow behaviour (Fig. 6.9):

1- *Zone A: Low fine to solid particle ratio and high solid concentration:* Because of the high solid grain concentration no flow is possible without grain breakage.

2- *Zone B: High fine to solid particle ratio and high solid concentration:* The grain and clay concentration is too high for the material to flow rapidly without fracture. However no grain packing network exists and a very slow flow can thus take place, induced by the homogeneous deformation of the interstitial fine particle-water mixture.

The limit between zones A and B cannot be deduced from our previous reasonings. However we presume that, when the fine particle-water matrix occupies most of the volume, even if the solid concentration is close to 100%, at least a very slow flow can occur. This is why the border line was drawn as joining the upper horizontal axis at a point between 0 and 100%. The upper horizontal line (ϕ=100%) is not a part of zone B.

3- *Zone C: Low fine to solid particle ratio and water saturation.* In this zone, though grain concentration is high and close to the maximum packing concentration, the corresponding water-debris mixtures can flow thanks to slight dilatancy or a configuration change. This leads to a phenomenon which, as a first approximate, might be represented by a minimum in flow curve.

4- *Zone C-D: Transitional zone:* In this zone the maximum grain packing concentration is not reached but a slow, moderate or rapid flow cannot take place without fracture because of the high concentration of the interstitial fine particle-water mixture.

5- *Zone D: Moderate fine to solid particle ratio and high or moderate solid concentration:* In this zone the mass behaviour is essentially controlled by the interstitial fine particle-water mixture behaviour. The flow curve can be well represented by a Herschel-Bulkley model with parameters higher than the corresponding parameters of the interstitial fine particle-water mixture.

6- *Zone E: Moderate fine to solid particle ratio and low solid concentration:* In this zone an immediate settling of the coarsest particles takes place at rest so that a slow homogeneous flow is not possible.

6.3 RHEOLOGIES OF OTHER SLURRIES

In industry and nature various slurries which are in fact concentrated suspensions containing either fine or coarse particles, or both, may be encountered: mineral waste products of mining or refining (iron-ore, phosphate, bauxite), fly ash, coal slurries, foodstuffs, drilling muds, fresh cement or concrete, sludges, etc. Debris flows, muds and mudflows have been considered in previous sections. Foodstuffs, which involve a great variety of components will not be considered here.

From a general point of view, for all the above materials there exists a critical solid concentration beyond which the coarsest particles do not settle under usual flow conditions. These materials may therefore be considered as homogeneous non-Newtonian fluids. Otherwise material motions must be considered within the framework of two-phase flows. In the following, we shall present some elements concerning the rheological properties of more or less natural slurries and attempt to relate them to their microstructure. It is worth noting that, from a general point of view, only limited amounts of rheological data on these materials are available. Furthermore experimental problems commonly encountered with concentrated suspensions (Cf Chapter 4) were often not controlled and the experimental procedures for determining the constitutive equation or yield stress were often unappropriate (stress ramp, yield stress extrapolation from data corresponding to high shear rates, smooth tool surfaces, uncontrolled edge effect changes, etc.). Finally most of these slurries appear to exhibit viscoplastic behaviour so that their free surface flows may be described with the help of tools developed in the following chapters. However, in some cases, they exhibit strong thixotropic properties which require further development.

6.3.1 *Coal slurries*

Industrial activity in coal conversion processes has stimulated some research on coal suspensions. Indeed, suspending coal appears to be the best means (economical and with limited chemical transformations) of transporting and storing this material. Coal suspensions must be made so as to optimize their transport properties and their ability to be stored for long durations. The material is crushed so that its grain size distribution falls within the range [1; 100 μm]. These particles are dispersed in a gas or liquid (water, fuel, alcohol, etc.). Research aims to find the suspension with the lowest viscosity along with the largest solid fraction. This means that the grain size distribution and the interstitial fluid must be well chosen. A surfactant may be added in order to improve solid particle wetting (François and Antonini 1988).

Few rheological studies have been devoted to this type of suspension. Wildemuth and Williams (1985) proposed the far most extensive study. They tested three coal grain size distributions suspended in three interstitial fluids of different viscosities. It was demonstrated that these suspensions exhibit viscoplastic behaviour for sufficiently high solid fractions. These authors also proposed a viscosity model taking into account a possible change in the maximum packing fraction with flow intensity (more precisely shear stress).

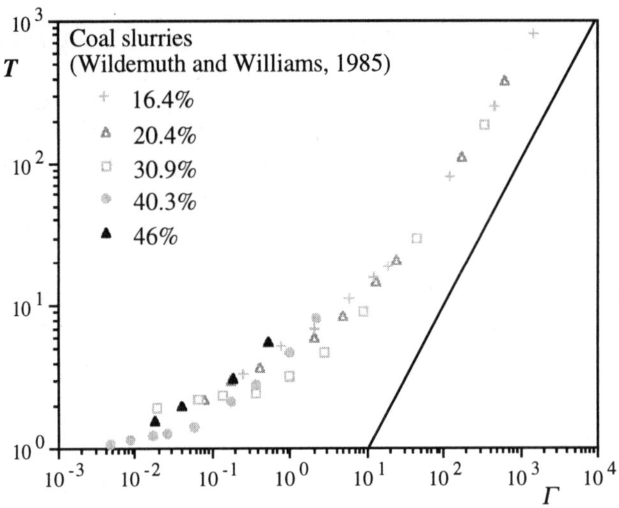

Figure 6.10: Flow curve data for coal slurry at various solid fractions in a dimensionless shear rate-shear stress diagram as in Figure 5.6. The transition towards Newtonian behaviour for increasing shear rate and decreasing solid fraction is apparent. [from data of Wildemuth and Williams (1985)]

The experimental curves presented by Wildemuth and Williams (1985) are quite similar to those already shown in Chapter 5 for fine mud suspensions. Indeed, within their experimental range, the suspensions have a Newtonian behaviour for sufficiently low solid fractions or high shear rates whereas for sufficiently large solid fractions or low shear rates their behaviour is clearly viscoplastic. Thus there is a generalized transition to Newtonian behaviour for increasing values of Γ, as with fine natural mud suspensions (see Fig. 6.10). This might be due to the fact that the suspended particles develop relatively weak colloidal interactions between each other. Then a continuous interaction network may exist at low shear rates but its strength is lower than that of the corresponding mud suspension with the same solid fraction. At higher shear rates this network is broken and hydrodynamic interactions become predominant.

Usui (1995) also presented some results concerning the thixotropic properties and the influence of the grain size distribution of coal-water mixtures. He essentially presented a theoretical model rather similar to those proposed for clay-water mixtures (see 5.2). This model takes into account particle size distribution and surface conditions but was shown to be incapable of quantitatively predicting some flow properties. The authors concluded that better knowledge of the break-up process of clusters is needed.

6.3.2 *Fresh concrete*

Fresh concrete conists of water, cement, which is a chemical product made of clay and crushed rocks, and grains (sand and pebbles). The size of cement particles ranges from a few to one hundred microns. Additives are often added to this mixture in order to change its solidification time or viscosity. The microstructure of this mixture type is quite similar to that of a coarse mud suspension in the sense that this is a suspension of non-interacting particles in a fine suspension (water + cement + additives). However the behaviour of the fine matrix may differ considerably from a fine mud suspension.

A large variety of types of behaviour may finally be obtained by changing additives, fractions of the different materials and grain size distribution. In any case, if, as a first approximation, we neglect thixotropic properties, the resulting mixtures are essentially viscoplastic fluids (Murata 1984). This explains the development of various tests which mainly determine the spreading limit of fresh concrete under different conditions (see Chapter 12). Since these techniques are essentially empirical, comparative tests they do not make it possible to predict flow properties of fresh concrete under various conditions. This has motivated viscometric studies of fresh concrete (Ukraincik 1980, Tattersall and Banfill 1983, Tanigawa et al. 1987, de Larrard et al. 1993, Hu 1995). The behaviour of fresh concrete slurries has, in general, been represented by a Bingham model mainly because, as for mud suspensions, experiments were carried out within limited shear rate ranges. Thixotropy and dilatancy have also been mentioned. In parallel various experimental problems (in particular those resulting from particle size) have been mentioned so that Tattersall and Banfill (1983) finally proposed using a special device to determine a "constitutive equation" by tossing a small volume of mixture with a rotating asymmetrical vane. Unfortunately, though in this case the material should be subjected to a shear more developed than in other tests, this test does not involve a controlled viscometric flow. A recent study tried to overcome experimental difficulties and proposed a special large-scale parallel plate viscometer with lateral walls (Hu 1995). In addition an extensive study of the influence of various parameters on the behaviour of fresh concrete was carried out by Hu (1995) using this viscometer.

Without going further into the details of all possible behaviour types of fresh cement and concrete (see Banfill (1990)) we may at least note that two boundary behaviour types may easily be predicted, in the light of results obtained with natural mud suspensions. When the coarse particle fraction is sufficiently low the form of the constitutive equation should be governed by that of the interstitial matrix, as suspended coarse particles tend to increase the overall stress level. When the coarse particle fraction reaches a critical value, a granular network may form which will to a larger extent modify the constitutive equation: it may considerably increase the yield stress or/and induce a minimum in the flow curve. Since most data were obtained at moderate or high shear rates no clear evidence of the last effect may be found in the literature except in the work of Ukraincik (1980).

6.3.3 *Sewage sludges*

Sewage sludges are suspensions of essentially organic matter in water. Their solid concentration varies from a few percent to several dozen percent during their treatment. They are first stored in primary sedimentation tanks. Then they are removed to a solid bowl decanter centrifuge to produce a cake of approximately 35% solids by weight (almost saturated). Water removal is enhanced by the addition of polyelectrolyte prior to centrifugating. The sludges are then dispatched to incineration. Their behaviour mainly depends on component types and concentrations.

Few rheological studies of such materials have been carried out. Colin (1970) and Dick and Ewing (1967) have shown that low concentrated suspensions, corresponding to those found in aeration tanks, (say between 0.5 and 10% of weight concentration) are generally yield stress fluids. More precisely Chilton et al. (1996) have successfully used a Herschel-Bulkley model to calculate head loss of sewage sludges in pumping systems. The yield stress was found to increase exponentially (Dick and Ewing 1967) or as a power law (Colin 1970) with solid fraction. Thixotropic effects were also observed (Dick and Ewing 1967, Colin 1980). A more complete survey of the literature is proposed in Chavarria (1980). For higher solid fractions, corresponding to dewatered sewage sludges, the suspensions appear to exhibit very complex behaviour types which were summed up with the term "granuloviscous material" by Kirby (1988). This term was first proposed by Cheng and Richmond (1978) to describe materials with rheological properties that are both granular and viscous. It is likely that this in fact refers to materials for which relevant rheometrical tests can hardly be obtained because of numerous and drastic disturbing effects caused by high solid fraction and high yield stress.

6.3.4 *Red muds*

Red mud is the waste material from the bauxite refining industry, it is the solid remaining after the aluminium oxide has been dissolved in hot concentrated sodium hydroxide. It consists mainly of silica, iron oxide and hydrated alumina, along with other components, suspended in a strongly alkaline solution. This waste product from the refining process is first washed to remove the bulk of the sodium hydroxide which is recycled into the extraction process and then dewatered in a gravity thickener from which the underflow is pumped to the settling ponds. As in the case of coal slurries cost effectiveness requires the highest density and the lowest viscosity.

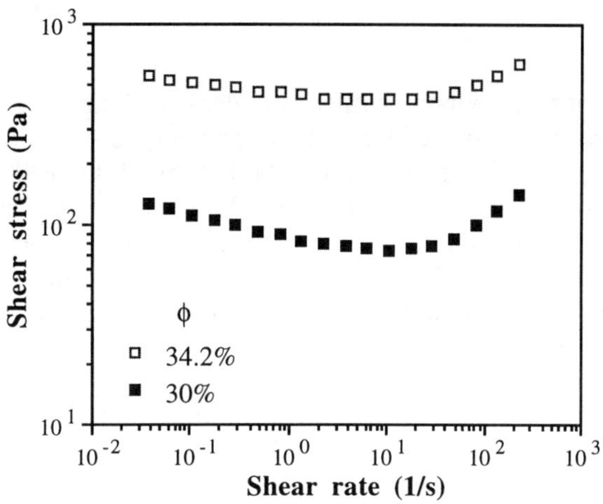

Figure 6.11: Flow curve for a red mud suspension for two solid fractions. The steady state curve was determined using a special procedure involving successive increases and decreases of the set shear rate. The fluid was first sheared at a high level until it reached an approximately steady state. Then the set shear rate was suddenly decreased to the lowest possible level (in the decreasing part of the flow curve). Under these conditions the shear stress first increased to a plateau and subsequently decreased because of the instability of this flow situation (Cf Chapter 9). We assume that the level of this plateau corresponds to the "steady state" situation. With this procedure the results were quite reproducible.

Measurements at high shear rates on various red mud suspensions led Jones et al. (1994) to consider that they were dealing with Bingham fluids whose properties change significantly with pH. It was nevertheless demonstrated by Nguyen and Boger (1983) that a Herschel-Bulkley or Casson model is more appropriate for representing red mud behaviour within a wide shear rate range. These suspensions were also shown to exhibit strong thixotropic properties (Nguyen and Boger 1983). All the above data were obtained at relatively high shear rates. We were able to make some tests on a red mud from the south of France within a large shear rate range. The experimental procedure involved all the recommendations made in Chapter 4. In Figure 6.11 the steady state flow curve is presented: these suspensions exhibit a minimum in the flow curve similar to some clay-water suspensions (see 5.1.1.2.4). In addition it is worth noticing that strong thixotropic effects were observed. Finally the behaviour obtained is similar to that of some bentonite-water mixtures as described in Coussot et al. (1993).

6.3.5 *Drilling fluids*

Drilling muds have several functions. A wellbore is filled with dense drilling fluid so that the downhole hydrostatic pressure is sufficiently high to prevent closure of the hole. The fluid also lifts rock cuttings from the drill bit at the bottom of the well to the surface. Bentonite-water suspensions are able to fulfil these roles. They are pumped down the drillstring and through the nozzles in the drill bit forming high-speed jets which dislodge rock cuttings. The mud then flows up the annulus between the drillstring and the rock. When the flow becomes laminar (close to the drill bit), because of the high viscosity, these fluids are able to support rock cuttings which are thus lifted to the surface. Because of their high yield stress they are also capable of maintaining the rock cuttings immobile. Some additives are often added to drilling fluids, for example barite in order to increase mud density and polymers to change their rheological properties. Because the wellbore is kept at a higher pressure than the pores of the surrounding rock the water tends to penetrate the rock while clay particles are too large to enter the pores. A filter cake therefore forms at the rock surface. The mechanical properties of this filter cake are of interest since one needs to know under what conditions it could be removed. Additional drilling fluid problems concern the interaction of the fluid with the shale which may lead to shale swelling or an increase in drilling mud viscosity because of shale disintegration. Finally a cement slurry is pumped down the casing and up the annular space between casing and hole, where it displaces drilling mud. This ensures the mechanical stability of the casing. The success of this operation depends on how well mud has been displaced by cement (see Cartalos et al. (1994)).

The rheology of drilling fluids is closely linked to that of bentonite-water mixtures which are rather complex fluids, it has been demonstrated that their behaviour depends a great deal on pH and electrolyte concentration (Heath and Tadros 1983). The essential rheological properties of bentonite-water mixtures are yield stress (API 1982, Alderman et al. (1991)) and thixotropy (Cf 5.1.1.2.3). Recent data shows that they might also exhibit a shear stress minimum in the flow curve (Cf 5.1.1.2.4). Reviews of current knowledge may be found in Sherwood (1994) and Baylocq et al. (1994). When polymer molecules are added to the water-bentonite mixtures they are adsorbed on clay particles which induce steric interactions and changes in the viscosity and yield stress of the initial suspension. The filter cake, which has a larger solid fraction, is very strong so that it resembles a paste. This has motivated quasi-static studies (squeeze-film rheometry (see Sherwood et al. (1991)) of mudcakes.

6.3.6 *Volcanic lava*

Lava flowing during volcanic eruptions results from the liquefaction, transformation and upflow of magma through the earth's crust. Their main components are liquid magma, crystals resulting from solid-liquid transition because of pressure or temperature changes, and gas bubbles. Pure liquid magmas at high temperature are known to be Newtonian fluids whose viscosity decreases with temperature (Shaw 1972) and increases with silica content. However, for temperature closer to the crystallisation range, the viscosity of magmas was shown to be shear-dependent (Spera et al. 1988). In addition viscoelasticity and normal stress effects were observed (Spera et al. 1998).

The temperature of lavas when they flow out of the crater is generally close to the crystallisation range of the liquid magma and continuously decreases during subsequent flow. Then crystals form and grow all along the flow so that a solid crust finally forms at the lava surface which might have a significant effect on the overall flow. Though they have long been considered as Newtonian fluids (Tanguy and Biquand 1967, Walker 1967, 1973), magma or lava material were shown at least to exhibit yield stress (Robson 1967, Shaw et al. 1968, Gauthier 1973, Pinkerton and Sparks 1978) at sufficiently low temperatures, probably because of the formation of a percolating network of contacts between crystals. This yield stress rapidly increases with temperature or with crystallisation (McBirney and Murase 1984) which complicates the treatment of flow dynamics. In addition little rheological data on natural flowing material is available and existing data have been obtained using fairly imprecise techniques (Shaw et al. 1968, Pinkerton and Sparks 1978, Fink and Zimbelman 1986, Wadge and Lopes 1991). As a first approximation, the material making up lava resembles mudflows since we are dealing with concentrated suspensions which can be considered as yield stress fluids. However, because the parameters of their constitutive equation change significantly during flow due to thermal exchanges (Danes 1972, Dragoni 1989, Crisp and Baloga 1990, Worster and Huppert 1993, Fink and Griffiths 1990), desciptions of flow dynamics remain speculative. It should also be noted that other phenomena such as degassing could also play a significant role in flow (Sparks and Pinkerton 1978, Jaupart 1991).

6.4 REFERENCES

Alderman, N.J., Meeten, G.H., and Sherwood, J.D. 1991. Vane rheometry of bentonite gels. *J. Non-Newt. Fluid Mech.*, 39, 291-310.

American Petroleum Institute (A.P.I.) 1982. Standard Procedure for testing drilling fluid. *A.P.I. Bull.* 13B, 9th Ed., Dallas.

Banfill, P.F.G. 1990. *Rheology of Fresh Cement and Concrete*. Chapman and Hall, London.

Batchelor, G.K. 1970. The stress system in a suspension of force free particles. *J. Fluid Mech.*, 41, 545-570.

Baylocq, P., Cartalos, U., and Piau, J.M. 1994. Rheological characterisation and structural modelling of clay polymer systems. in *Progress and Trends in Rheology IV*, C. Gallegos (Ed.), Steinkopf, Darmstadt, 618-620.

Boyer, S. 1994. *Détermination des propriétés rhéologiques des laves torrentielles à l'aide d'expériences d'écoulements sur un plan incliné*. DEA MMGE, INPG, Grenoble.

Cartalos, U., Rivereau, A., Haciislamoglu, M., and Lecourtier, J. 1994. Defining the rheological properties of cement slurries for the successful cementation of an oil well. in *Progress and Trends in Rheology IV*, C. Gallegos (Ed.), Steinkopf, Darmstadt, 275-277.

Chavarria, F.E. 1980. *Etude rhéologique des suspensions de flocs biologiques*. Ph.D. thesis, USTL, Montpellier.

Cheng, D.C.-H., and Richmond, R.A. 1978. Some observations on the rheological behaviour of dense suspensions. *Rheol. Acta*, 17, 446-453.

Chilton, R., Stainby, R., and Thompson, S. 1996. The design of sewage sludge pumping systems. *J. Hydr. Res.*, 34, 395-408.

Colin, F. 1970. Application de techniques rhéologiques à l'étude des boues résiduaires. *Centre Belge d'Etude et de Documentation des Eaux*, 317, 178-187.

Coussot, P. 1992. *Rheology of debris flows - Study of concentrated suspensions.* Ph.D. thesis, INPG, Grenoble, France. (in French)

Coussot, P. 1994. Steady, laminar, flow of concentrated mud suspensions in open channel. *J. Hydraul. Res.*, 32, 535-559.

Coussot, P., and Boyer, S. 1995. Determination of yield stress fluid behaviour from inclined plane test. *Rheol. Acta*, 34, 534-543.

Coussot, P., Leonov, A.I. and Piau, J.M. 1993. Rheology of concentrated dispersed systems in a low molecular weight matrix. *J. Non-Newt. Fluid Mech.*, 46, 179-217.

Coussot, P., and Piau, J.-M. 1995a. Effect of an addition of force-free particles on the rheological properties of fine suspensions. *Can. Geotechn. J.*, 32, 263-270.

Coussot, P. and Piau, J.M. 1995b. A large-scale field coaxial cylinders rheometer to study the rheology of natural coarse suspensions. *J. Rheol.*, 39, 105-124.

Crisp, J., and Baloga, S. 1990. A model for lava flows with two thermal components. *J. Geophys. Res.*, 95, 1255-1270.

Danes, Z.F. 1972. Dynamics of lava flows. *J. Geophys. Res.*, 77, 1430-1432.

de Larrard, F., Szitzkar, J.C., and Hu, C. 1993. A rheometer for fresh concrete. *Bull. Liaison Lab. Centr. Ponts Chaussées*, 186, 55-59. (in French)

Dick, R.I., and Ewing, B.B. 1967. The rheology of activated sludge. *J. WPCF*, 39, 543-560.

Dragoni, M. 1989. A dynamical model of lava flows cooling by radiation. *Bull. Volcanol.*, 51, 88-95.

Engelund, R. and Wan, Z. 1984. Instability of Hyperconcentrated flow. *J. Hydr. Eng.*, 110, 219-233.

Fink, J.H., and Zimbelman, J.R. 1986. Rheology of the 1983 Royal Gardens basalt flows, Kilauea Volcano, Hawaii. *Bull. Volcanol.*, 48, 87-96.

Fink, J., and Griffiths, R. 1990. Radial spreading of viscous-gravity currents with solidifying crust. *J. Fluid Mech.*, 221, 485-509.

François, O., and Antonini, G. 1988. Le charbon liquide. *La Recherche*, 202, 1038-1048.

Gauthier, F. 1973. Field and laboratory studies of the rheology of Mount Etna lava. *Phil. Trans. R. Soc. Lond.*, A274, 83-98.

Heath, D., and Tadros, T.F. 1983. Influence of pH, electrolyte, and poly-vinyl-alcohol addition on the rheological characteristics of aqueous dispersions of sodium montmorillonite. *J. Colloid Interface Sci.*, 93, 307-319.

Hu, C., 1995. *Rheology of fresh concrete.* Ph.D. thesis, Ecole Nationale des Ponts et Chaussées, Paris. (in French)

Jaupart, C. 1991. Effects of compressibility on the flow of lava. *Bull. Volcanol.*, 54, 1-9.

Jones, R.L., Marwood, T.M., and Horsley, R.R. 1994. Red mud rheology - The behaviour of iron(III) oxide suspensions in strongly alkaline solutions. in *Progress and Trends in Rheology IV*, C. Gallegos (Ed.), Steinkopf, Darmstadt, 660-662.

Kirby, J.M. 1988. Rheological characteristics of sewage sludge: a granuloviscous material. *Rheol. Acta*, 27, 326-334.

McBirney, A.R., and Murase, T. 1984. Rheological properties of magmas. *Ann. Rev. Earth Planet. Sci.*, 12, 337-357.

Major, J.J., and Pierson, T.C. 1992. Debris flow rheology : Experimental analysis of fine-grained slurries. *Water Resour. Res.*, 28, 841-857.

Murata, J. 1984. Flow and deformation of fresh concrete. *Matériaux et Constructions* (Paris), 17, 117-129.

Nguyen, Q.D., and Boger, D.V. 1983. Yield stress measurement for concentrated suspensions. *J. Rheol.*, 27, 321-349.

O'Brien J.S. and Julien, P.Y. 1988. Laboratory analysis of mudflows properties. *J. Hydraul. Eng.*, 114, 877-887.

Phillips, C.J., and Davies, T.R.H. 1991. Determining rheological parameters of debris flow material. *Geomorphology*, 4, 101-110.

Pinkerton, H., and Sparks, R.S.J. 1978. Field measurements of the rheology of lava. *Nature*, 276, 383-384.

Robson, G.R. 1967. Thickness of Etnean lavas. *Nature*, 216, 251-252.

Shaw, H.R. 1972. Viscosities of magmatic silicate liquids: an empirical method of prediction. *Am. J. Sci.*, 272, 870-893.

Shaw, H.R., Wright, T.L., Peck, D.L., and Okamura, R. 1968. The viscosity of basaltic magma: an analysis of field measurements in Makaopuhi lava lake, Hawaii. *Am. J. Sci.*, 266, 225-264.

Sherwood, J.D. 1994. Thixotropic mud, compacted cake and swelling shale. *Les Cahiers de Rhéologie* (Groupe Français de Rhéologie), XIII, 111-119.

Sherwood, J.D., Meeten, G.H., Farrow, C.A., and Alderman, N.J. 1991. Squeeze-film rheometry of non-uniform mudcakes. *J. Non-Newt. Fluid Mech.*, 39, 311-334.

Sparks, R.S.J., and Pinkerton, H. 1978. Effect of degassing on rheology of basaltic magma. *Nature*, 276, 385-386.

Spera, F.J., Borgia, A., and Strimple, J. 1988. Rheology of melts and magmatic suspensions: 1. Design and calibration of concentric cylinder viscometer with application to rhyolitic magma. *J. Geophys. Res.*, 93, 10,273-10,294.

Tanguy, J.C., et Biquand, D. 1967. Quelques propriétés physiques du magma actuel de l'Etna. *C. R. Acad. Sci.*, Paris, 264, 699-701.

Tanigawa, Y., Mori, H., Tsutsui, K., and Kurokawa, Y. 1987. Constitutive law and yield condition of fresh concrete. *Trans. Japan Concrete Inst.*, 9, 47-54.

Tattersall, G.H., and Banfill, P.F.G. 1983. *The rheology of Fresh concrete*, Pitman, Boston.

Ukraincik, V.. 1980. Study on fresh concrete and masonry structures. *Cement Concrete Res.*, 10, 203-212.

Usui, H. 1995. A thixotropy model for coal-water mixtures. *J. Non-Newt. Fluid Mech.*, 60, 259-275.

Wadge, G., and Lopes, R.M.C. 1991. The lobes of lava flows on Earth and Olympus Mons, Mars. *Bull. Volcanol.*, 54, 10-24.

Walker, G.P.L. 1967. Thickness and viscosity of Etnean lavas. *Nature*, 213, 484-485.

Walker, G.P.L. 1973. Lengths of lava flows. *Phil. Trans. R. Soc. London*, A274, 107-118.

Wildemuth, C.R., and Williams, M.C. 1985. A new interpretation of viscosity and yield stress in dense slurries: coal and other irregular particles. *Rheol. Acta*, 24, 75-91.

Worster, G., and Huppert, H.E. 1993. The crystallization of lava lakes. *J. Geophys. Res.*, 98, 15,891-15,901.

CHAPTER 7

Fundamental equations for viscoplastic flows

In this chapter we shall establish the detailed form of flow equations for viscoplastic fluids and in particular Herschel-Bulkley fluids. For this purpose, the three-dimensional form of the constitutive equation to be introduced in motion equations is of prime necessity. In the second part of this chapter the flow equations are written and a dimensionless form is deduced. In the last part we shall discuss the physical meaning of the obtained dimensionless numbers.

7.1 A 3D FORMULATION OF THE CONSTITUTIVE EQUATION FOR VISCOPLASTIC FLUIDS

Most natural and industrial slurries or muds obtained by suspending a large number of particles interacting via colloidal forces or direct contact in water are non-Newtonian fluids exhibiting a yield stress (τ_c), which needs to be overcome for flow to take place (see Chapter 6). The main characteristic of yield stress fluids is that they give rise to thick deposits (stationary volumes) on steep slopes whereas unyielding fluids go on flowing as long as surface tension effects remain negligible (see Chapter 12). The real, physical existence of yield stress has been the subject of numerous discussions (Barnes and Walters 1985, Hartnett and Hu 1989, Astarita 1990, Schurz 1992, Evans 1992, De Kee and Chan Man Fong 1993, Spaans and Williams 1995). In particular the discussion concerned the question of whether so-called viscoplastic fluids have a real yield stress below which they can be considered as solids or simply exhibit a very high viscosity at low shear stress levels. At the very least, it was concluded that the yield stress has a practical, engineering reality. We shall simply consider that, for such a fluid, under common conditions of observation, there generally exists an abrupt change in behaviour around a given shear stress value, that we can call the yield stress. Below this critical stress value the fluid is deformed in an essentially elastic manner. Above this critical value the fluid flows. The elastic behaviour may be taken into acount in the constitutive equation (Doraiswamy et al. 1991). However, since the critical strain before flow is in general small, elastic properties may reasonably be neglected in a first approximation. The following equation therefore makes it possible to represent the behaviour in steady simple shear of most viscoplastic fluids:

$$\gamma = 0 \Leftrightarrow \tau < \tau_c \; ; \; \gamma \neq 0 \Leftrightarrow \tau = \tau_c + f(\gamma) \tag{7.1}$$

where τ is the shear stress magnitude, γ the shear rate magnitude and f a positive continuous function of γ which depends on each material and is such that $f(0)=0$. Eq.(7.1) expresses both the yielding behaviour and the increase in viscous dissipation with shear rate.

The above model (eq. (7.1)), as determined from simple shear tests, does not account for flow situations that are more complex than simple shear. In order to carry out a systematic analysis of the complete system of flow equations under various flow conditions it is necessary to have a general, tensorial expression of this constitutive equation. A solution to this problem has been proposed for Bingham fluids whose simple shear behaviour corresponds to (2.24). In that case the von Mises criterion used in plasticity was assumed to be valid for the unsheared regions within the fluid. The final expression (Oldroyd 1947, Prager 1961, Fredrickson 1964, Piau 1979) therefore takes the form:

$$\boldsymbol{D} = 0 \text{ when } \sqrt{-T_{II}} < \tau_c \text{ and the stress tensor is indeterminate;} \tag{7.2a}$$

$$\boldsymbol{T} = \tau_c \frac{\boldsymbol{D}}{\sqrt{-D_{II}}} + 2\mu_B \boldsymbol{D} \text{ otherwise.} \tag{7.2b}$$

where τ_c and μ_B are the fluid yield stress and plastic viscosity, and D_{II} the second invariant of $\boldsymbol{D}$ which writes:

$$D_{II} = \frac{1}{2}((\mathrm{tr}\boldsymbol{D})^2 - \mathrm{tr}(\boldsymbol{D}^2)) \tag{7.3}$$

In order to represent the general behaviour described by equation (7.1) in simple shear, this approach may be generalized, and the three-dimensional constitutive equation for a viscoplastic fluid becomes:

$$\boldsymbol{D} = 0 \text{ when } \sqrt{-T_{II}} < \tau_c \text{ and the stress tensor is indeterminate;} \tag{7.3a}$$

$$\boldsymbol{T} = \tau_c \frac{\boldsymbol{D}}{\sqrt{-D_{II}}} + \mathcal{F}(D_{II})\,\boldsymbol{D} \text{ otherwise.} \tag{7.3b}$$

where $\mathcal{F}$ is a positive continuous function of D_{II} which is such that $\mathcal{F}(D_{II})\sqrt{-D_{II}}$ tends to 0 when D_{II} tends to 0. For a Bingham fluid, $\mathcal{F}$ is the constant "plastic viscosity", μ_B. As (7.2) expression (7.3) is justified physically, since it gives (7.1) in simple shear, and mathematically, since rigid regions are again found from the isotropic von Mises criterion independently of the exact form of $\mathcal{F}$.

It has been shown in Chapters 5 and 6 that the constitutive equation in simple shear of most natural mud suspensions can be represented by a Herschel-Bulkley model within a relatively wide shear rate range. We thus exclude materials such as very concentrated force-free particle suspensions or some pure clay-water mixtures which exhibit thixotropy under common experimental conditions. Flow problems with the latter suspension types have not yet received sufficient attention to be consistently presented in this book. For natural mud suspensions it was also shown in Chapter 5 that the exponent of the empirical Herschel-Bulkley model is generally close to 1/3. In the following Chapters we develop as far as possible the general case corresponding to any n value and in some cases we give more details concerning the case n=1/3. The three-dimensional expression of the constitutive equation of a Herschel-Bulkley fluid is directly given by (7.3) where (Schowalter 1978, Chen and Ling 1996):

$$\mathcal{F}(D_{II}) = \frac{2^n K}{(\sqrt{-D_{II}})^{1-n}} \tag{7.4}$$

Under simple shear conditions Equation (5.3) derives from (7.3b)-(7.4).

7.2 CONSERVATION EQUATIONS

Let us consider the free surface flow of a mud suspension under the action of gravity over a surface whose steepest slope to the horizontal is i. The whole set of equations describing this flow is obtained from the general motion and continuity equations (2.4) and (2.7) expressed for example in an orthonormal frame $(\boldsymbol{x},\boldsymbol{y},\boldsymbol{z})$ such that $\boldsymbol{g}=g(\sin i)\boldsymbol{x} - g(\cos i)\boldsymbol{y}$, in which we introduce the constitutive equation (7.3)-(7.4), taking into account the incompressibility of the fluid. If $\sqrt{-T_{II}} < \tau_c$ all the velocity components are equal to zero. Otherwise:

$$\frac{\partial u}{\partial x} + \frac{\partial v}{\partial y} + \frac{\partial w}{\partial z} = 0 \tag{7.5}$$

$$\begin{aligned}\rho\frac{du}{dt} = &-\frac{\partial p}{\partial x} + K\Big[\frac{\partial}{\partial x}\Big(\frac{2}{\alpha^{1-n}}\frac{\partial u}{\partial x}\Big) + \frac{\partial}{\partial y}\Big(\frac{1}{\alpha^{1-n}}\Big(\frac{\partial u}{\partial y}+\frac{\partial v}{\partial x}\Big)\Big) + \frac{\partial}{\partial z}\Big(\frac{1}{\alpha^{1-n}}\Big(\frac{\partial u}{\partial z}+\frac{\partial w}{\partial x}\Big)\Big)\Big]\\ &+ \tau_c\Big[\frac{\partial}{\partial x}\Big(\frac{2}{\alpha}\frac{\partial u}{\partial x}\Big) + \frac{\partial}{\partial y}\Big(\frac{1}{\alpha}\Big(\frac{\partial u}{\partial y}+\frac{\partial v}{\partial x}\Big)\Big) + \frac{\partial}{\partial z}\Big(\frac{1}{\alpha}\Big(\frac{\partial u}{\partial z}+\frac{\partial w}{\partial x}\Big)\Big)\Big] + \rho g(\sin i)\end{aligned} \tag{7.6a}$$

$$\begin{aligned}\rho\frac{dv}{dt} = &-\frac{\partial p}{\partial y} + K\Big[\frac{\partial}{\partial x}\Big(\frac{1}{\alpha^{1-n}}\Big(\frac{\partial u}{\partial y}+\frac{\partial v}{\partial x}\Big)\Big) + \frac{\partial}{\partial y}\Big(\frac{2}{\alpha^{1-n}}\frac{\partial v}{\partial y}\Big) + \frac{\partial}{\partial z}\Big(\frac{1}{\alpha^{1-n}}\Big(\frac{\partial v}{\partial z}+\frac{\partial w}{\partial y}\Big)\Big)\Big]\\ &+ \tau_c\Big[\frac{\partial}{\partial x}\Big(\frac{1}{\alpha}\Big(\frac{\partial u}{\partial y}+\frac{\partial v}{\partial x}\Big)\Big) + \frac{\partial}{\partial y}\Big(\frac{2}{\alpha}\frac{\partial v}{\partial y}\Big) + \frac{\partial}{\partial z}\Big(\frac{1}{\alpha}\Big(\frac{\partial v}{\partial z}+\frac{\partial w}{\partial y}\Big)\Big)\Big] - \rho g(\cos i)\end{aligned} \tag{7.6b}$$

$$\begin{aligned}\rho\frac{dw}{dt} = &-\frac{\partial p}{\partial z} + K\Big[\frac{\partial}{\partial x}\Big(\frac{1}{\alpha^{1-n}}\Big(\frac{\partial u}{\partial z}+\frac{\partial w}{\partial x}\Big)\Big) + \frac{\partial}{\partial y}\Big(\frac{1}{\alpha^{1-n}}\Big(\frac{\partial v}{\partial z}+\frac{\partial w}{\partial y}\Big)\Big) + \frac{\partial}{\partial z}\Big(\frac{2}{\alpha^{1-n}}\frac{\partial w}{\partial z}\Big)\Big]\\ &+ \tau_c\Big[\frac{\partial}{\partial x}\Big(\frac{1}{\alpha}\Big(\frac{\partial u}{\partial z}+\frac{\partial w}{\partial x}\Big)\Big) + \frac{\partial}{\partial y}\Big(\frac{1}{\alpha}\Big(\frac{\partial v}{\partial z}+\frac{\partial w}{\partial y}\Big)\Big) + \frac{\partial}{\partial z}\Big(\frac{2}{\alpha}\frac{\partial w}{\partial z}\Big)\Big]\end{aligned} \tag{7.6c}$$

where α is defined as:

$$\alpha = 2\sqrt{-D_{II}}$$

$$= 2\sqrt{\frac{1}{2}(\frac{\partial u}{\partial x})^2+\frac{1}{2}(\frac{\partial v}{\partial y})^2+\frac{1}{2}(\frac{\partial w}{\partial z})^2+\frac{1}{4}(\frac{\partial u}{\partial y}+\frac{\partial v}{\partial x})^2+\frac{1}{4}(\frac{\partial u}{\partial z}+\frac{\partial w}{\partial x})^2+\frac{1}{4}(\frac{\partial v}{\partial z}+\frac{\partial w}{\partial y})^2} \tag{7.7}$$

7.3 DYNAMIC SIMILARITY

Two physical systems have similar evolutions in time and space when the ratio of two corresponding variables of these systems, that is variables describing corresponding (analogous) physical phenomena, is constant and only depends on the dimension of these variables. Similarity considerations are essential because they make it possible to determine the conditions under which two apparently different flows are physically equivalent (or similar). Then the conditions to respect for consistently studying a given flow at another scale can be derived. Moreover this approach will reveal the dimensionless numbers which govern a given flow and should be preferentially used for a dimensionless analysis of any flow characteristics. The following approach is rather different from so-called dimensional analysis. Indeed, here the starting point of this similarity analysis is not a more or less complete set of dimensional variables but the complete set of equations determining the system evolution.

We now use the following dimensionless variables:

$$x_+ = \frac{x}{L} \; ; \; y_+ = \frac{y}{L} \; ; \; z_+ = \frac{z}{L} \tag{7.8}$$

$$u_+ = \frac{u}{V} \; ; \; v_+ = \frac{v}{V} \; ; \; w_+ = \frac{w}{V} \; ; t_+ = \frac{t}{\frac{L}{V}} \; ; p_+ = \frac{p}{\rho V^2} \tag{7.9}$$

where L and V are the reference length and velocity. Equations (7.5)-(7.6) may therefore be rewritten as:

$$\frac{\partial u_+}{\partial x_+} + \frac{\partial v_+}{\partial y_+} + \frac{\partial w_+}{\partial z_+} = 0 \tag{7.10}$$

$$\frac{\mathrm{d}u_+}{\mathrm{d}t_+} = -\frac{\partial p_+}{\partial x_+} + \frac{\tan i}{F_r^2 G}\,\phi_1(u_+,v_+,w_+) + \frac{\tan i}{F_r^2 H_b G} f_1(u_+,v_+,w_+) + \frac{\tan i}{F_r^2} \tag{7.11a}$$

$$\frac{\mathrm{d}v_+}{\mathrm{d}t_+} = -\frac{\partial p_+}{\partial y_+} + \frac{\tan i}{F_r^2 G}\,\phi_2(u_+,v_+,w_+) + \frac{\tan i}{F_r^2 H_b G} f_2(u_+,v_+,w_+) - \frac{1}{F_r^2} \tag{7.11b}$$

$$\frac{\mathrm{d}w_+}{\mathrm{d}t_+} = -\frac{\partial p_+}{\partial z_+} + \frac{\tan i}{F_r^2 G}\,\phi_3(u_+,v_+,w_+) + \frac{\tan i}{F_r^2 H_b G} f_3(u_+,v_+,w_+) \tag{7.11c}$$

where $\phi_{i=1,2,3}$ and $f_{i=1,2,3}$ are homogeneous dimensionless functions and Fr, G and H_b are dimensionless numbers defined as:

$$Fr = \frac{V}{\sqrt{gL\cos i}} \; ; \; G = \frac{\rho g L \sin i}{\tau_c} \; ; \; H_b = \frac{\tau_c}{K}\left(\frac{L}{V}\right)^n \qquad (7.12)$$

Two systems with the same dimensionless equations and with similar boundary and initial conditions will behave in similarity. Indeed, in that case, the system changes are determined by the identical set (7.10)-(7.11) and the dimensional variables of both systems may be related to the dimensionless variables with the help of equations (7.8) and (7.9). Then, at any time, the ratio of two corresponding variables from the two systems is equal to a dimensionless ratio of the reference variables and thus is constant.

The similarity between two channelled mudflows (k=1,2) requires the similarity between boundary and initial conditions (as in hydraulics, reference pressures must be chosen appropriately (Cf Comolet (1982))), which, in particular, should ensure that channel slopes are equal. Additionally the equality of the set of dimensionless motion equations (7.11) requires the equality of the dimensionless numbers (defined by (7.12)) of the two flows, which is equivalent to the following conditions:

$$n = n_1 = n_2 \; ; \; \frac{V_2}{V_1} = \sqrt{\lambda} \; ; \; \tau_{c2} = \tau_{c1}\frac{\rho_2}{\rho_1}\lambda \; ; \; K_2 = K_1\frac{\rho_2}{\rho_1}\lambda^{(1+n/2)} \qquad (7.13)$$

where ρ_k, K_k, n_k, and τ_{ck} (k=1,2) are respectively the density and the three Herschel-Bulkley parameters of fluid k, and L_k and V_k are the reference length and velocity of flow k. Here, the scale ratio is λ, which is equal to L_2/L_1. It should also be noted that:

$$Q_2 = Q_1\,\lambda^{5/2} \; ; \; T_2 = \sqrt{\lambda}\,T_1 \qquad (7.14)$$

where Q_k and T_k (k=1,2) are the flow rate and time at each scale.

Thus, here, unlike in conventional hydraulics, if a natural flow is to be modelled at a small-scale, the similarity conditions require the use of a second fluid with smaller K and τ_c parameters as defined by (7.13). It should be noted that K and τ_c are not reduced proportionally. In general it is rather difficult to respect completely and precisely the similarity conditions. For concentrated mud suspensions K and τ_c increase rapidly with solid concentration [see Chapters 5 and 6]. Then it is relatively easy to follow a global similarity for example by using the same fluid type at a lower solid concentration (but without coarse particles). In that case we will for example choose the solid fraction giving the exact value of τ_c for the model fluid as required by (7.13). Then only the order of magnitude of the resulting value K will be the one required by (7.13).

Obviously these results only concern homogeneous flows. Additionally it is worth noticing that all these methods are valid only when, despite scale change, the flow is still governed by the basic equations (7.5)-(7.6). When any previously negligible effect (surface tension, etc.) becomes significant when changing scale, the above similarity considerations fail.

Three dimensionless parameters (given in (7.12)) appeared in the dimensionless motion equations (7.10)-(7.11). First it should be noted that other dimensionless parameters could have been chosen, which might be obtained by combinations of the former. For example a dimensionless number, Re_b, with the same physical meaning as the Reynolds number may have been used (Piau, 1996):

$$Re_b = \frac{F_r^2 H_b G}{\tan i} = \frac{\rho V^{2-n} L^n}{K} \tag{7.15}$$

The first number (F_r) is the usual Froude number which is commonly used in hydraulics. There the flow depth is used as reference length. Here it has been modified in order to take into account the cosinus of the channel angle, which may differ significantly from 1 considering the steep slopes on which mudflows take place. Nevertheless it still expresses the ratio of a reference kinetic energy to a potential energy due to gravity.

The second number (G) ("gravity number") expresses the ratio of the potential energy (due to gravity) of a fluid portion to viscous dissipations related to the yield stress term in Herschel-Bulkley model. Taking the channel hydraulic radius as the reference length, G is the ratio of the wall shear stress to the fluid yield stress. Then, in practice, flow only starts when G is greater than 1 (see Chapter 8). Roughly speaking, when G increases, the flow velocity increases, since the external force (due to gravity) increases.

The third dimensionless number (H_b) ("Herschel-Bulkley number") reflects the ratio of the flow-independent part of viscous dissipations (related to yield stress) to the shear-dependent one (related to $\mathcal{F}$). In simple shear the order of magnitude of H_b is that of the ratio of the two terms appearing in the constitutive equation (5.3). Roughly speaking, when H_b decreases, flow velocity increases. It should be borne in mind that H_b has also been called Od by Piau (1996) thus referring to Oldroyd's work (1947). For n=1, H_b is the Bingham number :

$$B = \frac{\tau_c L}{\mu_B V} \tag{7.16}$$

7.4 REFERENCES

Astarita, G. 1990. Letter to the editor: The engineering reality of the yield stress. *J. Rheol.*, 34, 275-277.

Barnes, H.A., and Walters, K. 1985. The yield stress myth?. *Rheol. Acta*, 24, 323-326.

Chen, C.L., and Ling C.H. 1996. Granular-flow rheology: role of shear-rate number in transition regime. *J. Eng. Mech.*, 122, 469-480.

Comolet, R. 1982. *Mécanique expérimentale des fluides*. Masson, Paris.

De Kee, D., and Chan Man Fong, C.F. 1993. Letter to the editor: A true yield stress?. *J. Rheol.*, 37, 775-776.

Doraiswamy, D., A.N. Mujumdar, I. Tsao, A.N. Beris, S.C. Danforth, and Metzner, A.B. 1991. The Cox-Merz rule extended: a rheological model for concentrated suspensions and other materials with a yield stress. *J. Rheol.*, 35, 647-685.

Evans, I.D. 1992. Letter to the editor: On the nature of the yield stress. *J. Rheol.*, 36, 1313-1316.

Fredrickson, A.G. 1964. *Principles and Applications of Rheology*. Prentice-Hall, Englewood Cliffs.

Hartnett, J.P., and Hu, R.Y.Z. 1989. The yield stress - An engineering reality. *J. Rheol.*, 33, 671-679.
Oldroyd, J.G. 1947. A rational formulation of the equations of plastic flow for a Bingham solid. *Proc. Cambridge Phil. Soc.*, 43, 100-105.
Piau, J.M. 1996. Flow of a yield stress fluid in a long domain. Application to flow on an inclined plane. *J. Rheol.*, 40, 711-723.
Piau, J.M. 1979. Non-Newtonian fluids. *Techniques de l'Ingénieur*, A 710-711. (in French)
Prager, W., 1961. *Introduction to Mechanics of Continua*. Ginn and Co, Boston.
Schowalter, W.R. 1978. *Mechanics of non-Newtonian fluids*. Pergamon Press, Oxford.
Schurz, J. 1992. Letter to the editor: A yield value in a true solution. *J. Rheol.*, 36, 1319-1321.
Spaans, R.D., and Williams, M.C. 1995. Letter to the Editor: At last, a true liquid-phase yield stress. *J. Rheol.*, 39, 241-246.

CHAPTER 8

Uniform flows

We shall start by considering the simplest flow case, in other words flow corresponding to uniform conditions, for which the flow characteristics do not change along a given axis. Obviously this ideal situation is not often encountered in practice. However, as in other viscous flows or in hydraulics, this is the basis for the subsequent study of more difficult flow problems (gradually varied flows, transient flows, flow stability, etc.). In many cases, the uniform flow approximation makes it possible to have an overview of the main flow characteristics and an idea of the effects of changes in boundary conditions on flow characteristics (geometry, flow rate, etc.).

Most natural or industrial flows involve water or air. Because these fluids have low viscosity most of these flows are turbulent. It follows that, unlike laminar flows, they cannot be directly described mathematically with the help of their constitutive equation (by solving Navier-Stokes equations). In this case, one can approach the problem numerically or use empirical resistance formulae giving the friction factor as a function of channel roughness, Reynolds number and other possible dimensionless parameters. Thus, before starting to study any flow, it is of prime importance to determine whether turbulence is negligible or not. This is the purpose of the first section. The following sections consider uniform flow problems under laminar conditions since this is thought to be the most common case for mudflows. In the second part we shall treat the general case of Herschel-Bulkley fluid flows and in the third part we will propose further developments for the case n=1/3.

8.1 CRITERIA FOR THE ONSET OF TURBULENCE IN VISCOPLASTIC FLOWS

Transition to turbulence occurs when internal inertia effects (i.e. stress originating in local velocity fluctuations) become significant compared to viscous effects. From this definition it results that turbulence may be observed with any kind of fluid flow, in particular with viscoplastic flows, as soon as these inertia effects are sufficiently large. Though it might appear somewhat difficult to imagine turbulence developing when a (rigid) plug (see 8.2.1) is theoretically predicted by the laminar hypothesis, this can nevertheless occur. In fact, for a fully developed turbulent flow of a viscoplastic fluid, a genuinely rigid part cannot exist because fluid elements would significantly diffuse through it. Moreover when turbulence occurs, local fluid behaviour is different from behaviour under laminar conditions.

From the above definition it also results that a relevant criterion for the onset of turbulence must take into account the exact form of the constitutive equation. A review of the errors which may be committed when this principle is not taken into account was proposed by Hanks (1986). Well-known criteria for the onset of turbulence and frictions formulae have been established (see for example Chow (1959) or Munson et al (1994)) for Newtonian fluids but very few studies have concerned viscoplastic fluid flows (Naik 1983, Hanks 1986, Rickenmann 1990). Essentially the transition to turbulence in the case of confined viscoplastic (using the Bingham model) fluid flows has been considered in literature (Hanks 1963b, Hanks and Dadia 1971, Hanks and Ricks 1974, Le Fur and Martin 1967, Thomas 1963, Heywood and Richardson 1978, Metzner 1980, Park et al. 1989, Mannheimer 1991) but the proposed criteria cannot yet be considered as fully validated by experiments. It remains that turbulence is predicted to occur for larger Reynolds numbers as the Non-newtonian character of the fluid is more pronounced. For example Frigaard et al. (1994) found that, for Bingham fluid flows in pipes, the minimum Reynolds number (as defined by eq. (7.15)) for linear instability increases almost linearly with increasing Bingham number (as defined by eq. (7.16)). The only avalaible work concerning turbulent free surface flows was proposed by Naik (1983) for Bingham fluids. Using Hanks' approach (1963) Naik obtained a criterion for the onset of turbulence in a channelled flow and proposed a formula for the wall resistance of fully turbulent flows in a rectangular open channel.

Here we need a criterion for the onset of turbulence in Herschel-Bulkley fluids. The method of Hanks (1963a,b) is partly empirical and thus is less sound than the rigorous approach of Frigaard et al. (1994), but it nevertheless provides a straightforward, physically reasonable approach to turbulence, which, in addition, appears to be in relatively good agreement with various types of experimental data (Hanks 1963b, 1986). It involves first considering that turbulence develops when a slender rotating rod of fluid can experience resonance frequencies at which rotational motion couples with transverse vibrational motion in a catastrophic manner. Then it is postulated that this resonance phenomenon occurs when the magnitude of the angular acceleration term reaches a scalar multiple of the diffusional momentum flux vector (Hanks 1963a). The coefficient of proportionality is then inferred from the one observed for Newtonian fluids.

For a uniform, free surface flow over an infinitely wide, inclined plane this theory suggets that turbulence occurs when the velocity reaches a value such that $404 g \sin i = du^2/dy$ where u is evaluated at a critical value of y, y_c*, corresponding to the maximum of the right-hand side of this equation. Using the developments of Section 8.2 for Herschel-Bulkley fluids and in particular the expression for the velocity distribution (8.11), we deduce that:

$$y_c* = (1-(\frac{m}{2m+1})^{1/(1+m)})y_0 \tag{8.1}$$

and that the flow becomes turbulent when the flow depth becomes larger than

$$h = \frac{1}{\rho g \sin i}\left[\tau_c + K\left(\frac{404(m+1)\rho(g\sin i)^2}{K\nu}\right)^{1/(2m+1)}\right] \quad (8.2)$$

where $\nu=\left[\left(\frac{m}{2m+1}\right)^{m/(m+1)} - \left(\frac{m}{2m+1}\right)^{(2m+1)/(m+1)}\right]$

Note that (8.2) can also be formulated so as to give the critical wall shear stress ($\rho gh\sin i$) for the onset of turbulence. The corresponding critical mean velocity may be deduced by introducing the value of h given by (8.2) in the expression (8.13) where q is replaced by Uh. The above criterion cannot be expressed in a simple way as a function of the dimensionless numbers derived in 7.3.

Further theoretical and experimental works on turbulence in viscoplastic fluid flows is needed. However, for our purposes, this will not hinder the following developments. Indeed, even if it is only a rough approximation to reality, the above criterion (8.2) makes it possible to demonstrate that many natural or industrial mudflows are laminar. Let us consider the case of a Herschel-Bulkley fluid (τ_c=13.7 Pa, K=4.5, ρ=1350 kg/m^3) flowing at a rate of 5.3 l/s in a rectangular channel whose width is 17 cm and whose slope is 12%. [This flow corresponds to the more "liquid" and "rapid" laboratory test carried out by Coussot (1994).]. The remaining fluid depth on an infinitely wide inclined plane should be 8.6 mm (see Chapter 12). According to (8.2) the critical fluid depth for the onset of turbulence is 3.2 cm whereas the uniform fluid depth predicted from the laminar assumption (8.13) is 2.8 cm. The corresponding critical velocity is 2.8 m/s whereas the mean velocity here is 1.1 m/s. This means that, in this case, we should be close but still before turbulence transition. For this flow, Coussot (1994) found good agreement between measurements and theoretical flow characteristics computed from the laminar constitutive equation and motion equations, which confirms that turbulence was negligible. It is worth noticing that, according to the results of Chapter 9, the flow should become "hydraulically" unstable, that is roll waves should form, long before it could become turbulent since the critical Froude number corresponding to these flows is respectively 5 and 2.1 while roll waves are predicted to develop in mudflows when the Froude number is larger than 0.25. Roll waves were not always observed during the corresponding experiments of Coussot (1994) probably because the channel was not long enough for the instability to develop in some cases but the qualitative theoretical result remains. In fact, fundamentally, mudflows in very long channels mainly take three forms when the slope is progressively increased: below the asymptotic slope there is no flow (see 8.2); above this slope a uniform laminar flow can be obtained; then, above a critical slope not far from the asymptotic one, roll waves occur. In most cases turbulence can hardly develop between the last two slopes. Mudflows in channels which do not remain uniform along a sufficiently long portion can remain stable in a wider slope range and can even become turbulent when the slope is sufficiently steep. The question of the length necessary for turbulence development (in comparison with length for roll wave development) nevertheless remains open.

In practice industrial flows that are slower or more viscous than the above flow should be laminar. Additionally, with the help of similarity considerations (see Chapter), we deduce that the dimensionless parameters (H_b, Fr, G, etc.) of all flows similar to this particular flow have the same values and thus should be in laminar regime. The flow described above is for example similar to a natural, *a fortiori* laminar, mudflow with τ_c=1348 Pa, through a channel whose width and slope are 10.2 m and 12%, flowing at a velocity of 7.8 m/s and a discharge of 148 m^3/s. It is also similar to a flow with τ_c=674 Pa in a channel whose width is 5.1 m with a discharge of 26 m^3/s and a velocity of 5.5 m/s. Many other natural mudflows or debris flows generally involve more viscous material or are less rapid. Thus it is expected that the following theory under laminar conditions will make it possible to describe natural and industrial flows to a large extent.

8.2 UNIFORM LAMINAR FLOWS OF HERSCHEL-BULKLEY FLUIDS

8.2.1 *Flow on an infinitely wide inclined plane*

Here we shall consider the laminar, uniform flow of an incompressible, Herschel-Bulkley fluid in the absence of wall slip on an inclined plane with a slope i relative to the horizontal. This flow type has been studied by Howard (1963) and Paslay and Slibar (1958) for a Bingham fluid. Lateral walls are assumed to be at an infinite distance so that they do not affect the flow. As elsewhere in this book we use the frame of reference of Cartesian coordinates (O,x,y,z) such that $\mathbf{y}$ is perpendicular to the plane and $\mathbf{x}$ directed downwards (Cf Fig. 8.1).

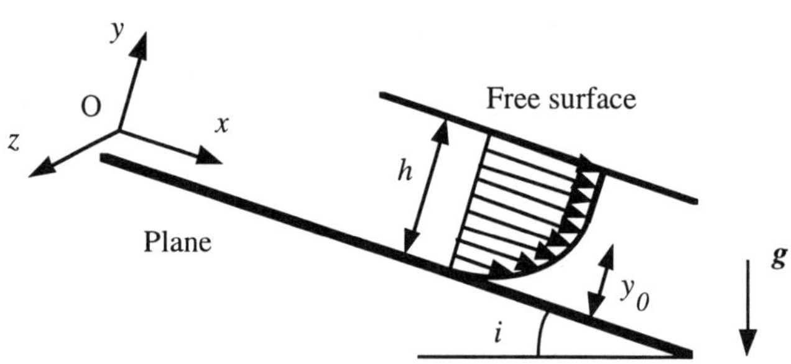

Figure 8.1: Free surface flow on an infinitely wide inclined plane: frame of reference and aspect of the velocity profile within the fluid.

We assume a steady, uniform flow is established so that $\mathbf{u}$ does not depend on x and z. Because of the symmetry of the problem regarding each (x,y) plane, w is equal to zero. From the mass equation (2.4), one deduces that v does not depend on y. Since v is equal to zero on the channel bottom, it is equal to zero elsewhere. The only non zero velocity component is u which only depends on y. Remark that this velocity distribution corresponds to a simple shear (Cf 2.3.2). Then, from the three-dimensional form of the constitutive equation (7.3)-(7.4), we deduce:

$$\tau_{xx}=\tau_{yy}=\tau_{zz}=0 \; ; \; \tau_{xz}=\tau_{yz}=0 \; ; \; \tau_{xy}=\tau_{yx}=\tau(y) \; ; \; p=p(y) \tag{8.3}$$

with:

$$\tau = \varepsilon(\tau_c + K|\gamma|^n) \text{ when } \gamma = \frac{du}{dy} \neq 0 \quad (\varepsilon = \frac{\gamma}{|\gamma|}) \tag{8.4a}$$

$$|\tau| \leq \tau_c \text{ when } \gamma = 0 \tag{8.4b}$$

In steady state the momentum equation (2.7) writes:

$$\frac{d(\tau)}{dy} + \rho g \sin i = 0 \tag{8.5}$$

$$-\frac{dp}{dy} - \rho g \cos i = 0 \tag{8.6}$$

from which we deduce:

$$\tau = -\rho g(\sin i)y + A \; ; \quad p = -\rho g(\cos i)y + B \tag{8.7}$$

A and B are two constants which can be determined from the boundary condition at the free surface:

$$\Sigma . \boldsymbol{n} = p_0 \tag{8.8}$$

where $\boldsymbol{n}$ is the outer unit normal vector and p_0 the atmospheric pressure. One may note that the pressure distribution as given by (8.7) is of hydrostatic type. From (8.7) and (8.8) it results that the free surface is necessarily parallel to the inclined plane. We finally obtain:

$$\tau = \rho g(\sin i)(h-y) \; ; \quad p = \rho g(\cos i)(h-y) + p_0 \tag{8.9}$$

The wall shear stress (τ_p) (at y=0) is therefore equal to $\rho g(\sin i)h$. For $\tau_p < \tau_c$ there can be no flow. The maximum possible fluid depth in that case, which is also the (asymptotic) minimum fluid depth of a uniform flow, is:

$$h_0 = \frac{\tau_c}{\rho g \sin i} \tag{8.10}$$

Assuming that the fluid flows downwards (ε=1) and using (8.4) and (8.9) in conjonction with the condition of no wall slip ($u(0)$=0), we obtain, by integration, the velocity distribution within the fluid (de Kee et al. 1990):

$$u(y) = \frac{\alpha}{1+m} [y_0^{(1+m)} - (y_0 - y)^{(1+m)}] \text{ when } y \leq y_0 \tag{8.11 a}$$

$$u(y) = u(y_0) = \frac{\alpha}{1+m} y_0^{(1+m)} \text{ when } h \geq y \geq y_0 \tag{8.11 b}$$

with $m=\frac{1}{n}$, $y_0 = h - \frac{\tau_c}{\rho g \sin i}$; $\alpha = (\frac{\rho g \sin i}{K})^m$

From (8.11) it appears that there is a region close to the free surface ($h \geq y \geq y_0$) where the fluid is not sheared. This region is commonly called a "plug". The velocity distribution (8.11) may also be expressed in dimensionless variables:

$$V(Y) = (1 - (1 - Y)^{1+m}) \text{ when } Y<1 \qquad (8.12\text{ a})$$

$$V(Y) = 1 \text{ when } 1<Y<\frac{G}{G\text{-}1} \qquad (8.12\text{ b})$$

where $V = \frac{u(y)}{u(y_0)}$, $Y = \frac{y}{y_0}$ and $G = \frac{\rho g h(\sin i)}{\tau_c}$.

Equations (8.11) and (8.12) are relevant only when $G>1$, i.e. when there is a flow. It is worth noticing that, within the frame of this dimensionless representation:
- the plug begins at $Y=1$ and its thickness is $1/(G-1)$.
- for a given fluid, from one flow to another only the plug thickness varies.

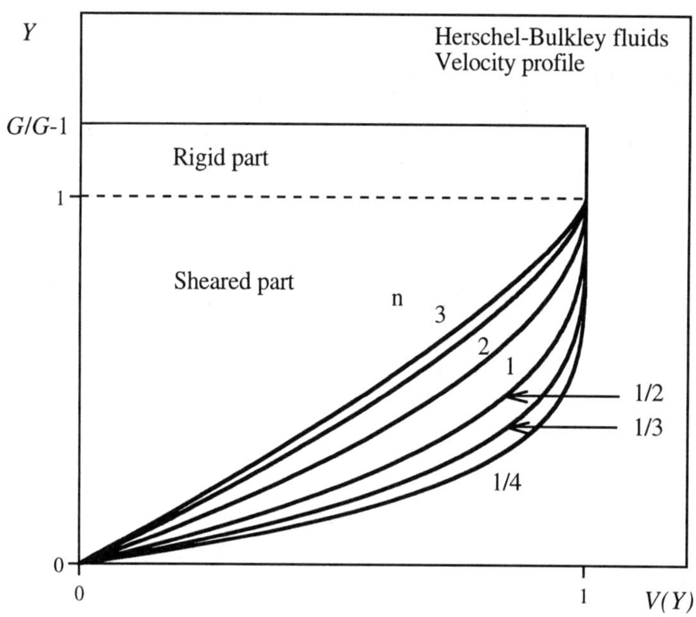

Figure 8.2: Velocity distribution over the cross-section for a uniform flow on an inclined plane, for different *n* values.

The form of velocity distribution for different values of n is shown in Figure 8.2. It is of interest that when n is small it would be possible to approximate this distribution by a straight line with a slope of $(1+m)$ and a thicker, unsheared region. When n is greater than 1 it would be possible to approximate the velocity distribution by a straight line starting from zero and extending as far as the base of the plug. All other things being equal, the fluid shear near the wall grows more intense as n decreases. The aspect of the vertical velocity distribution for this plane flow resembles what is observed at the free surface for a rectangular channel flow (Fig. 8.3) because there, in the absence of wall slip, the side walls play a role similar to that of the channel bottom ($\boldsymbol{u}$=0 along channel walls).

Figure 8.3: Velocity profile at the free surface for steady uniform flow of a mud suspension (τ_c=25 Pa) within a rectangular channel (i=10%; L=20 cm). At a given instant some black powder was dropped along a line on the fluid surface perpendicular to the flow direction. The picture shows the shape of the deformed line after a certain duration. The central undeformed region corresponds to the plug.

The discharge by unit length through a vertical cross-section is deduced from (8.11):

$$q = \int_0^h u(y)\mathrm{d}y = \frac{\alpha}{1+m} y_0^{(1+m)} \left[h - \frac{1}{2+m} y_0\right] \tag{8.13}$$

which may also be written in the dimensionless form:

$$H_b^{-m} = \frac{1}{(m+1)(m+2)} G^{-2} (G-1)^{(m+1)} ((m+1)G+1) \tag{8.14}$$

with $H_b = \frac{\tau_c}{K} (\frac{h}{U})^n$

where U is the mean flow velocity. It is interesting to compare the mean velocity of water and mudflow under the same initial and boundary conditions. The main practical difference is that water can reach a high velocity for small depths while mud suspension remains static (Cf Fig. 8.4).

The practical validity of the above equations was checked by Coussot (1994) in the case of fine mud suspensions. Komatina (1996) found slightly less good agreement between theory and experiments probably because he used a Bingham or pseudoplastic (power law) model fitted to rheological data. It must be emphasized that, for comparisons of this sort between theory and channel flow tests, the accuracy and validity of the rheological model within a sufficiently large shear rate range is of prime importance.

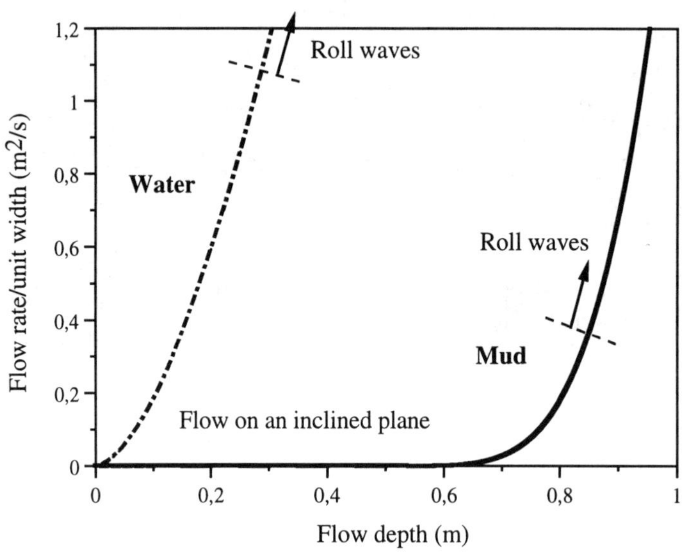

Figure 8.4: Flow rate by unit channel width as a function of flow depth for a mud suspension (τ_c=1400 Pa, K=500 Pa.s$^{1/3}$, ρ=2300 kg/m^3) and pure water in uniform flow on a wide, inclined plane (i=12%), as computed respectively from present theory and conventional hydraulics methods (Manning friction equation (see Chow 1959), we have taken 25 for the value of the Manning coefficient). The final parts of these curves, corresponding to large mean velocities, are not realistic since instabilities (roll waves) for water and mud (see Chapter 9) or turbulence for mud may appear. The limits beyond which roll waves should appear according to theory (see Chapter 9) have been drawn.

8.2.2 *Flow in a semi-cylindrical open channel*

Here we shall only consider the case when the semi-cylindrical channel is filled with fluid. The components of velocity $\boldsymbol{u}$ are taken in a cylindrical coordinate frame of reference (z,r,θ) where Oz is the central downward axis of the cylinder. The exact solution of mass and momentum equations (see Appendix I) can be found by assuming:

$$u_r = u_\theta = 0\ ;\ u = u_z = u_z(r) \tag{8.15}$$

Then, from the constitutive equation (7.3)-(7.4) we find that the only non-zero component of $\boldsymbol{T}$ is $\tau_{rz}=\tau(r)$ with:

$$\tau = \varepsilon(\tau_c + K|\gamma|^n) \text{ when } \gamma = \frac{\mathrm{d}u}{\mathrm{d}r} \neq 0 \quad (\varepsilon = \frac{\gamma}{|\gamma|}) \tag{8.16 a}$$

$$|\tau| \leq \tau_c \text{ when } \gamma=0 \tag{8.16 b}$$

In uniform flow p does not depend on z and the momentum equation is written:

$$-\frac{\mathrm{d}p}{\mathrm{d}r} + \rho g(\cos i)(\cos\theta) = 0 \tag{8.17}$$

$$-\frac{\mathrm{d}p}{r\mathrm{d}\theta} - \rho g(\cos i)(\sin\theta) = 0 \tag{8.18}$$

$$\frac{\mathrm{d}(\tau)}{\mathrm{d}r} + \frac{\tau}{r} + \rho g(\sin i) = 0 \tag{8.19}$$

which, making allowance for boundary conditions (8.8), may be integrated as:

$$p = \rho g(\cos i)r(\cos\theta) + p_0\ ; \qquad \tau = -\frac{1}{2}\rho g(\sin i)r \tag{8.20}$$

Here the wall shear stress is:

$$\tau_p = \frac{1}{2}\rho g(\sin i)R \tag{8.21}$$

It is worth noticing that the only solution making it possible to respect boundary conditions for stress tensor and velocity is such that the free surface is the longitudinal cross-section of the cylinder, which corresponds to $\theta=\pm\pi/2$. The free surface can be curved only if the fluid exhibits finite normal stress differences. Indeed, in that case, Equations (8.17)-(8.19) are modified by these additional stress terms.

Assuming that the fluid flows downwards ($\varepsilon = -1$) and using (8.16) and (8.20), the velocity distribution can be deduced:

$$u(r) = \frac{\alpha}{1+m}\,[(R-r_0)^{(1+m)} - (r-r_0)^{(1+m)}] \quad \text{when } R \geq r \geq r_0 \tag{8.22 a}$$

$$u(r) = u(r_0) = \frac{\alpha}{1+m}\,(R-r_0)^{(1+m)} \quad \text{when } r \leq r_0 \tag{8.22 b}$$

with $r_0 = \dfrac{2\tau_c}{\rho g(\sin i)}$ and $\alpha = (\dfrac{\rho g(\sin i)}{2K})^m$.

The dimensionless expression for this velocity distribution is again given by equation (8.12) with now:

$$V = \frac{2^m(m+1)}{\alpha\, r_0^{(1+m)}}\,u(R-r),\ \ Y = \frac{R-r}{r_0}\ ,\ r_0 = R - \frac{2\tau_c}{\rho g(\sin i)} \ \text{and}\ G' = \frac{\rho g R(\sin i)}{2\tau_c}$$

The dimensionless discharge equation is written:

$$H_b'^{-m} = \frac{1}{m+1}\,G'^{-3}\,(G'-1)^{(m+1)}\,[G'^2 - \frac{2}{m+3}(G'-1)^2 - \frac{2(G'-1)}{m+2}] \tag{8.23}$$

with $H_b' = \dfrac{\tau_c}{K}\,(\dfrac{R}{U})^n$

8.2.3 *Confined flows*

Bird et al [1982] reviewed a number of analytical solutions for Bingham fluid flows in closed conduits with cylindrical or planar symmetry. For most of these geometries, the case of a Herschel-Bulkley fluid flow could be treated in an analogous way because this model is in some respects similar to the Bingham one and it is consequently easy to integrate the equations. It is not our purpose to review these solutions. It should simply be noted that in the basic case of a flow through a cylindrical conduit the velocity distribution is similar to (8.22) where $\rho g(\sin i)/2$ is replaced by the driving pressure loss by unit length. For other geometries without any cylindrical or planar symmetry no analytical solution may be found (Hanks 1986). Taylor and Wilson (1997) provided a detailed numerical analysis of the flow of a Bingham fluid through a rectangular channel. In particular they found that stationary (rigid) regions of varying size remain in the corners, in all cases.

8.2.4 *Uniform flow in an open channel of any cross section*

8.2.4.1 *Analytical or numerical approach*

When the channel cross-section has no particular property of symmetry the momentum equation can no longer be solved analytically. In particular, for the inclined plane flow, when the lateral walls are too close to each other, equation

(8.11) fails to predict experiments (Coussot 1994). In addition the analytical solution proposed by Johnson (1970) in the case of a Bingham fluid flowing through a semi-elliptical channel is wrong (Johnson and Rodine 1984, Whipple 1994, Taylor and Wilson 1997). Thus, in these cases, numerical or empirical solutions must be found. From a general point of view, the main problem solving the complete system of flow equations is therefore to determine the boundaries of the rigid domains.

In order to solve specific problems with the boundaries of complicated shapes, one can use variational principles similar to those developed by Oldroyd (1947), Prager (1954), Yoshioka and Adachi (1974) or Duvault and Lions (1976) for laminar Bingham fluid flows. However these approaches have scarcely been used for solving any practical problems. This is probably due to the fact that in these variational formulations, one has to find the extremum of non-differentiable functionals (Leonov 1988). On the basis of previous Russian mathematical works, Leonov (1988) proposed an alternate variational approach where the Bingham model is replaced by a model approximately similar (without yield) making it possible to obtain exact two-side bounds for a potential function and dissipation. One can also use finite-element numerical methods (Bercovier and Engelman 1980, Beris et al 1985) to solve directly the complete system of flow equations along with boundary and initial conditions. It is always more practical to use a bi-viscous model rather than the Bingham model in finite-element formulations (Beverly and Tanner 1992; O'Donovan and Tanner 1984). There the fluid is assumed to behave as a Newtonian fluid (whose viscosity is μ') up to a critical stress or strain rate and as a Bingham fluid beyond (with a plastic viscosity μ'' much smaller than μ'). However this kind of simplification may lead to large errors especially when it is used to predict slow and rapid flows. In any case the characteristics of viscoplastic fluid flows through complex geometries can be very complex. For example Szabo and Hassager (1992) have shown, on the basis of simple arguments, that viscoplastic flows in eccentric annular geometries may exhibit a singularity in shear stress. As far as we know, all these methods or results have not been yet applied to Herschel-Bulkley fluids for any actual flow type.

8.2.4.2 *Empirical approach*

For use in practical applications we intend here to determine a flow equation capable of predicting real conditions for a large range of depth to width ratios (at least up to 1) as in conventional hydraulics. For a rectangular or a trapezoidal channel no such analytical expression can be found directly from momentum and mass equations. To solve this problem a semi-empirical approach was proposed by Kozicki and Tiu (1967). Their starting point is the Rabinowitsch (1929)-Mooney (1931) equation demonstrated to be exact for flow in semi-cylindrical or infinitely wide channels:

$$\gamma_w = a\tau_p \frac{d[\frac{2U}{R_H}]}{d\tau_p} + b[\frac{2U}{R_H}] \tag{8.24}$$

where γ_w is the shear rate at the wall. For an infinitely wide inclined plane we have $\gamma_w=\gamma(y=0)$, a=1/2 and b=1. For a half-cylinder we have $\gamma_w=\gamma(r=R)$, a=1/4 and b=3/4.

They therefore assume that (8.24) is also formally valid for mean variables (mean wall shear stress) for any fluid flowing steadily in a channel of any cross-section. In the resulting equation, they suggest using the values of the two parameters (*a* and *b*) determined (analytically) for simple (Newtonian) fluids. For a fluid following a Herschel-Bulkley model with n=1/3 and flowing in an open channel without wall slip, the Kozicki and Tiu (1967) approach leads to the following implicit equation for the mean wall shear stress τ_p:

$$\frac{2U}{R_H} = \frac{\tau_p^{-b/a}}{K^3}\left[\frac{\tau_p^{b/a+3}-\tau_c^{b/a+3}}{b+3a} - \frac{3\tau_c(\tau_p^{b/a+2}-\tau_c^{b/a+2})}{b+2a} + \frac{3\tau_c^2(\tau_p^{b/a+1}-\tau_c^{b/a+1})}{b+a} - \frac{\tau_c^3(\tau_p^{b/a}-\tau_c^{b/a})}{b}\right] \quad (8.25)$$

For a rectangular channel the appropriate values of *a* and *b* as a function of the channel section may be found for example in Kozicki and Tiu (1986). With this model the difference with the experimental results of Coussot (1994) was found to be much larger than 30% for aspect ratios (depth to width ratio) greater than 0.4. Thus, in our case, this approach appeared to be of little value since its predictions are in poor agreement with experiments for large aspect ratios, whereas it was expected to improve predictions of the simplified theory of 8.2. In fact this is not really surprising. Parameters *a* and *b* of the Rabinovitsch-Mooney equation should depend on the form of the velocity distribution which depends to a large extent on fluid behaviour (see Hanks (1986)) especially when there is no particular channel symmetry.

In order to provide an approximate, analytical formula for the mean velocity through a channel of any cross-section, Coussot (1994) proposed the following alternate approach. It may be noted that for both an infinitely wide plane and a semi-cylindrical channel the dimensionless discharge equations (8.14) and (8.23) take the form of relations between H_b and G, where G is in each case the ratio of wall shear stress to yield stress and where H_b is written using the maximum flow depth as the characteristic length. It should be noted that, with the help of the momentum balance over a fluid portion, this generalized form of G may be written as a function of flow characteristics:

$$G = \frac{\tau_p}{\tau_c} = \frac{\rho g (\sin i) R_H}{\tau_c} \quad (8.26)$$

As a consequence, for an open channel of any cross-section, the flow rate equation could be expressed in terms of an equation linking H_b and G:

$$H_b = f(G, a_1, a_2, \ldots) \quad (8.27)$$

where the coefficients (a_1, a_2,...) would only depend on the geometrical characteristics of the wetted section. These coefficients are constant for the infinitely wide plane or the semi-cylindrical channel (eqs. (8.14) and (8.23)) because in each case there is only one characteristic length (h or R), and yet for any other channel cross-section they will depend on aspect parameters since the form and size of the wetted area depend on flow depth. A roughly similar approach was proposed by Whipple (1994) for Bingham fluids. This type of approach has no solid theoretical ground but, as we shall see below, makes it possible to represent real conditions reasonably well and thus provides some practical formulae at least for mudflows.

8.3 FURTHER DEVELOPMENTS FOR n=1/3

In this particular case which is of considerable practical interest for mud suspensions (Cf Chapter 5 and 6) one easily obtains from (8.14) and (8.23) the following approximated formula for G as a function of H_b:

$$G = 1 + a\,(H_b)^{-0,9} \tag{8.28}$$

where in both cases H_b is written using the maximum flow depth (h or R) and G is the ratio of wall shear stress to yield stress. For an infinitely wide inclined plane we have a=1.93 and for a semi-cylindrical channel we have a=1.98. Equation (8.28) is an approximation of (8.14) and (8.23) within 5% for (G-1) in the range [0.2; 20] which covers a wide range of laboratory, industrial or natural flows.

8.3.1 *Rectangular channel*

Generalizing equation (8.28) to obtain an equation of the type of (8.27), Coussot (1994) proposed the following empirical formula for the flow of a Herschel-Bulkley fluid (with n=1/3) in a rectangular channel:

$$a = 1.93\text{-}0.43\,\text{atan}\left[\left(\frac{10h}{L}\right)^{20}\right] \qquad \left(\frac{h}{L} < 1\right) \tag{8.29}$$

where h is the flow depth and L the channel width. For a low enough aspect ratio (h/L<0.1) this gives the approximate discharge equation (8.28) valid for an infinitely wide channel. Equation (8.29) was able to predict all the experimental results of Coussot (1994) within an acceptable range of error (30%), considering the various possible experimental errors made during the tests needed for this comparison (separate channel and rheometrical tests).

8.3.2 *Trapezoidal channel (basis b, edge slope: 45°)*

In this case a similar analysis led Coussot (1994) to propose:

$$a = 1.93\text{-}0.6\,\text{atan}\left[\left(\frac{0,4h}{b}\right)^{20}\right] \qquad \left(\frac{h}{b} < 4\right) \tag{8.30}$$

where h is the maximum flow depth. This equation was able to predict (but slightly overestimates) all the experimental results of Coussot (1994) to within 35%.

8.3.3 *Channel of any cross-section*

We propose a practical, general method for establishing the discharge equation in dimensionless form for a channel with a given, relatively convex, cross-section ($\mathcal{S}$). Let us choose, for a rectangular, trapezoidal (edge slope: 45°) or semi-cylindrical form, a cross-section resembling $\mathcal{S}$. We then fit it to $\mathcal{S}$ as well as possible so as to determine the characteristic lengths of the chosen cross-section form, taking into account the fact that the surfaces of the two sections must be equal. Then the discharge equation through $\mathcal{S}$ can be estimated by using the equation corresponding to the fitted cross-section, out of (8.23), (8.28)-(8.29) or (8.28)-(8.30), along with the appropriate expression for G as given by (8.26).

It is worth noting that the resulting discharge expression may suffer from the particular semi-empirical expressions for a in (8.29) or (8.30): the mean velocity may suddenly increase extremely rapidly around a critical value of the aspect ratio. This characteristic is certainly not realistic even though the mean velocity varies relatively rapidly with flow depth because we are dealing with shear-thinning yield stress fluids.

8.3.4 *Optimum cross-section*

A common practical problem is to determine the best hydraulic cross-section defined as the cross-section of the minimum area for a given flow rate. Here we shall consider the equivalent problem of determining the cross section characteristics allowing the maximum flow rate to be conveyed for a given area. We assume that fluid characteristics (ρ, τ_c, K) are given and that the slope (i) and the cross-section area (S) of the channel are fixed.

For a rectangular channel we can use (8.28) and (8.29) which give an approximate, simple formula for the flow rate (Q). Then we must find the value of L for which Q reaches its maximum. Numerically it may be shown that the minimum value for $[\frac{S}{h_0{}^2}]$ making it possible to have a flow is 8. A study of the variations of Q as a function of L demonstrates that, for $[\frac{S}{h_0{}^2}]$ is less than 100, the maximum flow rate is reached for the approximate value (within 3%):

$$L = 1.77\, h_0 \left[\frac{S}{h_0{}^2}\right]^{0.4} \tag{8.31}$$

Then:

$$h = \frac{S}{L} = 0.565\, h_0 \left[\frac{S}{h_0{}^2}\right]^{0.6} = 0.32\, L \left[\frac{S}{h_0{}^2}\right]^{0.2} \tag{8.32}$$

and the corresponding maximum flow rate is given as:

$$Q = 0.27\, h_0^3 \left(\frac{S}{h_0^2}\right)^{1.6} \left(\frac{\tau_c}{K}\right)^3 \left[\frac{0.56(S/h_0^2)^{0.4}}{0.64+(S/h_0^2)^{-0.2}} - 1\right]^{3.33} \tag{8.33}$$

This result relies on equation (8.29) which is assumed to be valid for $h<L$. The expression for the mean velocity for $h>L$ is currently unknown but it is likely that, in this case, wall effects are larger so that the obtained flow rate, keeping the same cross-section, decreases. As a consequence equations (8.31)-(8.33) certainly provide a good approximation for the optimum cross-section and flow rate.

Moreover it is worth noting that, here, the discharge varies significantly as we approach the optimum cross-section, unlike in water hydraulics (Munson et al. 1994), all the more when (S/h_0^2) is small. In particular when this ratio is close to its minimum possible value, i.e. 8, a narrow range of cross-sections makes it possible to have a flow. Thus, in practice, it is of great interest to devise a channel with a cross section as close as possible to the ideal values to avoid flow stoppage or significant flow depth increase because of rapid changes in boundary conditions. Finally it should be noted that, within our range of S, the ratio L/h, corresponding to the optimum, ranges from 1.12 to 2 depending on fluid characteristics, which means that the best cross-section form is more compact than in hydraulics where the best cross-section is reached for $L/h = 2$ (using Manning's equation for wall friction) (Munson et al. 1994). Additionally, all other things being equal, the ratio L/h corresponding to the optimum cross-section decreases with slope.

Equation (8.31)-(8.33) make it possible to devise a ideal channel for mudflows. In practice, for a fixed cross-section (S), one has to estimate i, ρ and τ_c from which to deduce h_0 and then compute L from (8.31). Then the maximum flow rate which can go through the channel is obtained from (8.33). For trapezoidal channels, a similar expression cannot yet be obtained because of the limited range of validity of equation (8.30).

8.3.5 *Degree of reliability of the above formulae*

We aim here to estimate the uncertainty on flow prediction for the above theoretical equations. As an example let us take the particular case of the flow on an infinitely wide inclined plane. We can differentiate equation (8.14) to obtain:

$$\frac{\Delta h}{h} = \frac{1}{20}\left(1-\frac{1}{G}\right)\left(4+\frac{1}{G}\right)\frac{\Delta q}{q} + \frac{3}{20}\left(1-\frac{1}{G}\right)\left(4+\frac{1}{G}\right)\frac{\Delta K}{K} + \frac{1}{4}\left(3+\frac{1}{G}\right)\frac{1}{G}\frac{\Delta \tau_c}{\tau_c} - \frac{i}{\tan i}\frac{1}{10}\left(\frac{1}{G^2}+\frac{3}{G}+6\right)\frac{\Delta i}{i} \tag{8.34}$$

It is worth noticing that the uncertainty originating in errors on q and K takes its maximum value for $\frac{1}{G}=0$ whereas the uncertainty originating in errors on τ_c and i is obtained for $G=1$. If, for example, we assume the following reasonable values (for laboratory tests) (Coussot 1994) for the respective maximum relative errors on q, K, τ_c and i: 10%, 10%, 20% and 5%; we find that the maximum total relative error on the prediction of h is, in percent (assuming i=tan$i$):

$$\frac{1}{2}\left[\frac{7}{G^2}+\frac{21}{G}+22\right] \tag{8.35}$$

which takes its maximum value for G=1. The maximum relative error on h is therefore 25%. Uncertainties for the prediction of natural flows are generally larger because the uncertainty on each variable is larger.

8.4 REFERENCES

Bercovier, M., and Engleman, M. 1980. A finite-element method for the incompressible non-Newtonian flows. *J. Comp. Phys.*, 36, 313-326.

Beris, A.N., Tsamopoulos, J.A., Armstrong, R.C., and Brown, R.A. 1985. Creeping motion of a sphere through a Bingham plastic. *J. Fluid Mech.*, 158, 219-244.

Beverly, C.R., and Tanner, R.I. 1992. Numerical analysis of three-dimensional Bingham plastic flow. *J. Non-Newt. Fluid Mech.*, 42, 85-115.

Bird, R.B., Dai, G.C., and Yarusso, B.J. 1982. The rheology and flow of viscoplastic materials. *Rev. Chem. Eng.*, 1, 1-70.

Chow, V.T. 1959. *Open-channel hydraulics.* Mc Graw-Hill, Civil Engineering Series, New York.

Coussot, P. 1994. Steady, laminar, flow of concentrated mud suspensions in open channel. *J. Hydr. Res.*, 32, 535-539.

De Kee, D., Chhabra, R.P., Powley, M.B., and Roy, S. 1990. Flow of viscoplastic fluids on an inclined plane: Evaluation of yield stress. *Chem. Eng. Comm.*, 96, 229-239.

Duvaut, G., and Lions, J.L. 1976. *Inequalities in mechanics and physics.* Springer Verlag, Berlin.

Frigaard, I.A., Howison, S.D., and Sobey, I.J. 1994. On the stability of Poiseuille flow of a Bingham fluid. *J. Fluid Mech.*, 263, 133-150.

Hanks, R.W. 1963a. The laminar-turbulent transition for flow in pipes, concentric annuli, and parallel plates. *AIChE J.*, 9, 45-48.

Hanks, R.W. 1963b. The laminar-turbulent transition for fluids with a yield stress. *AIChE J.*, 9, 306-309.

Hanks, R.W. 1986. Principles of slurry pipeline hydraulics. in *Slurry and suspension flow properties*, Encyclopedia of Fluid Mechanics, N.P. Cherimissinoff (Ed.), Gulf Publishing Company, London, Chapter 6, 213-276.

Hanks, R.W., and Dadia, B.H. 1971. Theoretical analysis of the turbulent flow of non-Newtonian slurries in pipes. *AIChE J.*, 17, 554-557.

Hanks, R.W., and Ricks, B.L. 1974. Laminar turbulent transition in flow of pseudoplastic fluids with yield stress. *J. Hydronautics*, 8, 163-166.

Heywood, N.I., and Richardson, J.F. 1978. Rheological behavior of flocculated and dispersed aqueous kaolin suspensions in pipe flow. *J. Rheol.*, 22, 599-613.

Howard, C.D.D. 1963. Flow of clay-water suspensions. *J. Hydraul. Div.*, ASCE, HY5, 89-97.

Johnson, A.M. 1970. *Physical processes in Geology.* Freeman Cooper and Co, San Francisco.

Johnson, A.M., and Rodine, J.R., 1984. Debris flow. in *Slope instability*, D. Brunsden and D.B. Prior (Ed.), John Wiley and Sons, New-York, Chapter 8.

Komatina, D. 1996. Estimation of frictional resistance to a non-Newtonian open-channel flow by application of Bingham and pseudoplastic rheological models. *J. Theor. Appl. Mech. (Yugloslav Society of Mechanics)*, 22, 21-40.

Kozicki, W., and Tiu, C. 1986. Parametric modeling of flow geometries in non-Newtonian flows. in *Flow dynamic and transport phenomena, Encyclopedia of Fluid Mechanics*, N.P. Cherimissinoff (Ed.), Gulf Publishing Company, London, Chapter 8, 199-252.

Kozicki, W., and Tiu, C. 1967. Non-Newtonian flow through open channels. *Can. J. Chem. Eng.*, 45, 127-134.

Le Fur, B., and Martin, M. 1967. Laminar and transitional flow of drilling muds and various suspensions in circular tubes. *J. Fluid Mech.*, 30, 449-464.

Leonov, A.I. 1988. Extremum principles and exact two-side bounds of potential: functional and dissipation for slow motions of nonlinear viscoplastic media. *J. Non-Newt. Fluid Mech.*, 28, 1-28.

Mannheimer, R.J. 1991. Laminar and turbulent flow of cement slurries in large diameter pipe: a comparison with laboratory viscometers. *J. Rheol.*, 35, 113-133.

Metzner, A.B. 1980. Criticism of "Rheological behavior of flocculated and dispersed aqueous kaolin suspensions in pipe flow" by Heywood and Richardson. *J. Rheol.*, 24, 115-121.

Mooney, M. 1931. Explicit formulas for slip and fluidity. *J. Rheol.*, 2, 210.

Munson, B.R., Young, D.F., and Okiishi, T.H. 1994. *Fundamentals of Fluid Mechanics*. John Wiley & Sons, New York.

Naik, B. 1983. *Mechanics of mudflow treated as the flow of a Bingham fluid.* Ph.D. thesis, Washington State University, Washington.

O'Donovan, E.J., and Tanner, R.I. 1984. Numerical study of the Bingham squeeze film problem. *J. Non-Newt. Fluid Mech.*, 15, 75-83.

Oldroyd, J.G. 1947. A rational formulation of the equations of plastic flow for a Bingham solid. *Proc. Cambridge Phil. Soc.*, 43, 100-105.

Park, J.T., Mannheimer, R.J., Grimley, T.A., and Morrow, T.B. 1989. Pipe flow measurements of a transparent non-Newtonian slurry. *J. Fluids Eng.*, 111, 331-336.

Paslay, P.R., and Slibar, A. 1958. Flow of an incompressible viscoplastic layer on an inclined plane. *Trans. Soc. Rheol.*, II, 255-262.

Prager, W. 1954. On slow viscoplastic flow. in *Studies in Mathematics and Mechanics: R von Mises Presentation Volume*, Academic Press, New York, 208-216.

Rabinowitsch, B. 1929. Uber die viskosität und elastizität von solen. *Zeit. Physik. Chem.*, A145, 1-26.

Rickenmann, D. 1990. *Bedload transport capacity of slurry flows at steep slopes*. D. Vischer (Ed.), Mitteilungen, VAW, Zürich.

Szabo, P., and Hassager, O. 1992. Flow of viscoplastic fluids in eccentric annular geometries. *J. Non-Newt. Fluid Mech.*, 45, 149-169.

Taylor, A.J., and Wilson, D.R. 1997. Conduit flow of an incompressible, yield-stress fluid. *J. Rheol.*, 41, 93-101.

Thomas, D.G. 1963. Non-Newtonian suspensions. *Ind. Eng. Chem.*, 55, 18-29.

Whipple, K.X. 1994. Personal communication.

Yoshioka, N., and Adachi, K. 1971. On variational principles for a non-Newtonian fluid. *J. Chem. Eng. Japan*, 4, 217-220.

CHAPTER 9

Roll waves

On steep slopes water flows may evolve in almost periodic surges whose front velocity and depth are greater than those of the corresponding uniform flow. A detailed description of this phenomenon is given in Figures 9.1 and 9.2. This may be observed for example in steep rivers (Cornish 1934) or in shallow-water flows (around 1 mm deep) due to rain on steep roads. This phenomenon, referred to as roll waves, may occur in other fluid flows under certain specific conditions. Roll waves in mudflows assume a special importance. Indeed, because of the shear-thinning yield stress behaviour of mud suspensions, when increasing the slope, for a given flow rate, the mean flow velocity increases rapidly from zero beyond a critical slope which is generally already rather steep (Cf Chapter 8). Beyond a critical velocity roll waves occur. Thus the range of slopes for which stable flows are obtained may be quite narrow. It is worth noticing that this macroscopic hydraulic instability may in general be disconnected from internal, local instability due to inertia (turbulence) (see Section 8.1) and from flow instability resulting from peculiar rheological behaviour (minimum in flow curve (see 5.1.1.2.4)).

Roll waves have been frequently observed in mudflows in nature (Wan and Wang 1994, Li et al. 1983, Khang 1990, Davies 1990, Delahaye and Coussot 1996) and in the laboratory (Engelund and Wan 1984, Wan and Wang 1994, Coussot 1994, Figure 9.3). More generally debris flows are also often described as periodic surges (see for example the review of Coussot and Meunier (1996)). However, the time separating each surge appears to be relatively long, which means that, in this case, other phenomena could have been superimposed on simple common roll waves.

In the following we first review the possible, theoretical approaches for the mathematical prediction of roll waves. Then a simple, general, theoretical analysis along with a physical interpretation of this phenomenon are proposed. Finally we apply these results to mudflows considered as Herschel-Bulkley fluids.

9.1 ROLL WAVE PREDICTION

9.1.1 *General considerations*

9.1.1.1 *Stability criterion*

Most works have concerned the stability criterion for simple fluid flows. Two approaches (3D and 1D) must be distinguished. The first one consists in examining the stability of the general expressions for mass and momentum conservations in laminar regimes (Yih 1963, Benjamin 1957, Lin 1967). Then the so-called Orr-Sommerfeld equation is deduced. The stability of the system including this equation and boundary conditions is then analysed with the help of various techniques (Drazin and Reid 1981, Joseph 1976, Stuart 1971). The exact stability criterion may thus be obtained after some rather complex calculations.

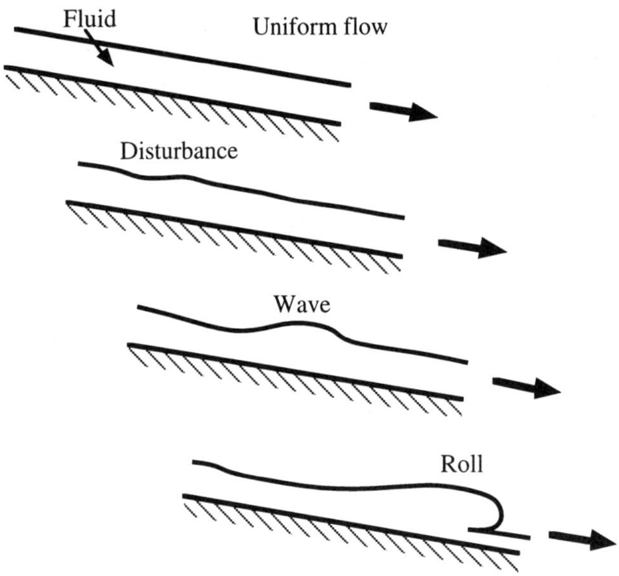

Figure 9.1: Longitudinal aspect of roll wave formation. Free surface at different instants and positions following a wave. A small elevation of the free surface appears somewhere and propagates downstream as a wave whose velocity is higher than the fluid velocity. Its depth increases simultaneously until it reaches a limiting value, after which it develops into a roll at the front of which fluid essentially moves downwards.

The second approach is basically similar but more approximate. It consists in studying directly the stability of the system obtained by integrating mass and momentum equations over a cross-section. The result highly depends on the choice of the wall shear stress expression. Jeffreys (1925) first used this technique with Chezy's formula, followed by Dressler and Pohle (1953) with a generalized power-law resistance expression. Related, slightly different approaches were developed by

Vedernikov (1946), Craya (1951), Ishihara et al. (1954) and Mayer (1959). In these works, approximations concerned especially the wall shear stress expression and the momentum coefficient. Some improvements were proposed by Chen (1993) who, on the basis of a more complete calculation, derived a corrected expression whose predictions are in agreement with those of the first (complete) approach. A valuable, concise, review of stability criteria for turbulent and laminar flows was proposed by Trowbridge (1987).

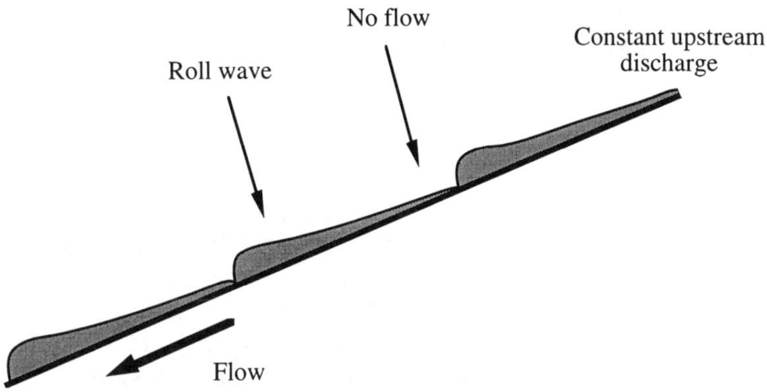

Figure 9.2: Longitudinal aspect of the free surface of an unstable flow. Successive waves are separated by shallow depth regions. Since we are dealing with yield stress fluids which give rise to finite static layers on steep slopes (see Chapter 12), almost no flow is observed in the separating regions.

Figure 9.3: Roll waves in laboratory mudflows (water-kaolinite mixture, τ_p =25 Pa).

9.1.1.2 *Wave profile*

Dressler (1949) provided a mathematical theory for the exact profile of roll waves by fitting shock waves between periodic stretches of smooth profiles. However, this theory remains undeterministic unless the wave speed and length are given. Other authors also provided periodic solutions of the unstable equations with some improvements (Hwang and Chang 1987, Merkin and Needham 1986). From a general point of view this problem is difficult and has not yet been completely solved. It has also been studied in the case of power-law fluids (Ng and Mei 1994) and initiated in the case of a Bingham fluid (Liu and Mei 1994). Further work is needed for yield stress fluids.

9.1.2 *Stability analysis for yield stress fluids*

The problem of the stability of the free surface flow of a yield stress fluid is basically the same as that of simple viscous fluids, since, as soon as it flows, the yield stress fluid has similar overall properties. However an additional obstacle for a general, either numerical or analytical, three-dimensional treatment of the instability is the existence of an unsheared region. Studies which concerned yield stress fluid instability in free surface flows made approximations. Ng and Mei (1994) used a power-law model instead of a yield stress constitutive equation. Hjorth (1990) assumed the yielding part to be of constant depth. Liu and Mei (1994) assumed a velocity profile within the fluid layer. This last approach provides so far the deepest, theoretical insight into free surface flow instability of yield stress fluids. However, apart from the fact that they assumed a Bingham model, their results are hard to handle from a practical point of view.

Since we aim at providing practical formulae and an easily understandable approach from a physical point of view we will finally prefer the simple, approximate, generalized approach of Trowbridge (1987) which relies on a one-dimensional treatment and which, for simple fluids, provides results that agree with those based on more accurate analyses. The slightly different approach of Wang et al. (1993) is more approximate and rather partial. Hjorth (1990) argued that Trowbridge's analysis did not take into account the fact that for viscous fluid flows the growth of the unstable mode is related to shear at the surface, and he therefore claimed that this method was unable to predict flow instability for yield stress fluids. Hjorth proposed an analysis based on a bi-viscous model in two dimensions, which led him to conclude that free surface flows of yield stress fluids are always stable when the free surface is subjected to oscillations. Unfortunately, to simplify the problem, he was forced to consider plug thickness as a constant. This hypothesis is obviously an approximation but its consequences on his reasoning are enormous. Indeed, if this hypothesis were satisfied, instability would never develop, as the fluid would be forced to flow with a completely constant and rigid plug above it. In these conditions it is not surprising that, following Hjorth's reasonings, unlike various field and laboratory observations, there seems to be no way for any instability to develop.

9.2 APPROXIMATE STABILITY ANALYSIS

In the following we shall use a simple one-dimensional linear stability analysis which leads to a general approximate criterion from which one can derive the corresponding criterion for mudflows. The weaknesses and validity limits of this theory will be discussed as we go along.

9.2.1 *One-dimensional equations in transient regime*

Here we consider an unsteady flow through an open channel of any cross-section . We assume that, within the fluid, the pressure distribution is of hydrostatic type so that p is given by equation (8.9). The mass and momentum equations in integral form (2.5)-(2.8) may be applied to a fluid portion comprised between two wetted sections which are perpendicular to the channel and separated by a small length dx. We assume longitudinal motion far outweighs vertical motion. In addition we neglect the effect of normal stresses (see Chapter 10). All this means that we are close to the situation of uniform flow. Then, as a projection along the x direction, when dx tends towards zero, one obtains the following one-dimensional momentum equation:

$$\rho\frac{\partial(SU)}{\partial t}+\rho\frac{\partial(\beta U^2 S)}{\partial x}=\rho g(\sin i)S-\tau_p P-\rho g(\cos i)\frac{\partial(y_G S)}{\partial x} \tag{9.1}$$

where y_G is the depth of the gravity center of the cross-section, U the mean velocity through the cross-section, τ_p the mean wall shear stress, S and P the wetted area and perimeter of the cross-section and β the momentum coefficient defined by equation (10.10). One also obtains the one-dimensional mass equation:

$$\frac{\partial S}{\partial t}+\frac{\partial(SU)}{\partial x}=0 \tag{9.2}$$

For flow in prismatic channels of arbitrary cross-section, it is easy to demonstrate that:

$$\frac{\partial(y_G S)}{\partial x}=S\frac{\partial h}{\partial x} \tag{9.3}$$

Then, with the help of (9.3), the set of equations (9.1) and (9.2) can be rewritten as:

$$\frac{\partial h}{\partial t}+U\frac{\partial h}{\partial x}+\frac{S}{\mathcal{L}}\frac{\partial U}{\partial x}=0 \tag{9.4}$$

$$\frac{\partial U}{\partial t}+\beta U\frac{\partial U}{\partial x}-(\beta\text{-}1)\frac{U\mathcal{L}}{S}\frac{\partial h}{\partial t}+U^2\frac{\partial\beta}{\partial x}=g(\sin i)-\frac{\tau_p}{\rho R_H}-g(\cos i)\frac{\partial h}{\partial x} \tag{9.5}$$

where $\mathcal{L}$ is the free surface width, defined as:

$$\mathcal{L} = \frac{\partial S}{\partial h} \tag{9.6}$$

For steady uniform flows of Herschel-Bulkley fluids β is equal to 10/9 for an infinitely wide inclined plane while it is equal to 6/5 for a semi-cylindrical channel. Moreover the variations of β with flow depth should be much smaller than the variations of other dimensionless variables. Thus it appears reasonable to neglect terms proportional to (β-1) or $\partial\beta/\partial x$ in the left-hand term of (9.5). The aim of this approximation is to deal with a relatively simple approach which may be more easily understood both mathematically and physically. More accurate results can be obtained when taking into account the momentum coefficient at the different stages of the calculations (Chen 1993) but the difference with results obtained otherwise remains small.

Now we assume that the mean wall shear stress is a differentiable single-valued function of the local, instantaneous flow depth and mean velocity:

$$\tau_p = \tau_p(U,h) \tag{9.7}$$

This function also involves some parameters which depend on channel roughness and mechanical properties of the flowing material. For uniform flow on an infinitely wide inclined plane or through a semi-cylinder, expression (9.7) can be deduced from (8.14) and (8.23) respectively. In the general case, in order to express (9.7) in a more explicit way, we shall make a similar assumption to that made for gradually varied water or mud flows (see Chapter 10). When flow depth and velocity vary locally and slightly around their theoretical steady values corresponding to uniform flow, we assume that a correct estimation of the shear stress at the wall, at a specific point, is given by the value of the uniform flow of comparable discharge and height. This assumption has also been made in stability analyses for many other types of fluids (Trowbridge 1987). It should be noted that the two-dimensional approach (Liu and Mei 1994) which consists in assuming a similar velocity profile to that of the uniform flow of same flow depth and mean velocity in each cross-section, is formally similar to the present approach.

It should also be noted that, for a steady, uniform flow, U and h do not depend on t and x and thus (9.5) simply gives the balance between gravity and the resistance to flow due to boundary friction (already contained in (8.26)):

$$g(\sin i) = \frac{\tau_p}{\rho R_H} \tag{9.8}$$

9.2.2 *Linearized equations*

We now assume that, starting from this uniform situation with contant U and h, the flow is slightly disturbed so that the new mean velocity over a cross-section is $(U + v(x,t))$ and the new flow depth is $(h + \eta(x,t))$ with v and η being small compared to U and h. If we introduce these new values in (9.4) and (9.5), assuming β to be constant and equal to 1 and neglecting second-order terms, we obtain:

$$\frac{\partial \eta}{\partial t} + U\frac{\partial \eta}{\partial x} + \frac{S}{\mathcal{L}}\frac{\partial v}{\partial x} = 0 \tag{9.9}$$

$$\frac{\partial v}{\partial t} + U\frac{\partial v}{\partial x} = \frac{\eta}{R_H{}^2}\tau_p\frac{\mathrm{d}R_H}{\mathrm{d}h} - \frac{\eta}{R_H}\frac{\mathrm{d}R_H}{\mathrm{d}h}\frac{\partial \tau_p}{\partial R_H} - \frac{v}{R_H}\frac{\partial \tau_p}{\partial U} - g(\cos i)\frac{\partial \eta}{\partial x} \tag{9.10}$$

where we have taken into account equation (9.7) and the following development at the first order:

$$\frac{\tau_p(h+\eta, U+v)}{R_H(h+\eta)} = \frac{\tau_p}{R_H} - \frac{\eta}{R_H{}^2}\tau_p\frac{\mathrm{d}R_H}{\mathrm{d}h} + \frac{\eta}{R_H}\frac{\mathrm{d}R_H}{\mathrm{d}h}\frac{\partial \tau_p}{\partial R_H} + \frac{v}{R_H}\frac{\partial \tau_p}{\partial U} \tag{9.11}$$

The derivatives of wall shear stress and hydraulic radius in (9.11) are understood to be evaluated in uniform conditions and at the specific values U and h. We now consider a new set of variables (X,T) moving with velocity U relative to the initial frame of coordinates:

$$X = x - Ut \quad \text{and} \quad T = t \tag{9.12}$$

so that for any variable $f(x,t)$ we have:

$$\frac{\partial f}{\partial T} = \frac{\partial f}{\partial t} + U\frac{\partial f}{\partial x} \quad \text{and} \quad \frac{\partial f}{\partial X} = \frac{\partial f}{\partial x} \tag{9.13}$$

Using (9.13) in (9.9) and (9.10) and eliminating v leads to:

$$-\frac{\partial^2 \eta}{\partial T^2} + \frac{S}{\mathcal{L}}g(\cos i)\frac{\partial^2 \eta}{\partial X^2} = \frac{S}{\rho \mathcal{L} R_H}\frac{\mathrm{d}R_H}{\mathrm{d}h}\left(\frac{\tau_p}{R_H} - \frac{\partial \tau_p}{\partial R_H}\right)\frac{\partial \eta}{\partial X} + \frac{1}{\rho R_H}\frac{\partial \tau_p}{\partial U}\frac{\partial \eta}{\partial T} \tag{9.14}$$

9.2.3 *Stability analysis*

The perturbation η is searched out as the sum of independent (normal) modes, each with the form:

$$\eta(X,T) = \mathrm{Re}\,\{a \exp[\mathrm{j}\chi(X - cT)]\} \tag{9.15}$$

where j is the imaginary unit, a is a complex amplitude, χ is real wave number which can take on any positive value, and c is a complex constant which must be determined. The real part of c is the propagation velocity of the spatially periodic disturbance defined by (9.15). Introducing (9.15) into (9.14) one obtains, after simplification, the following eigenvalue relation:

$$\chi c^2 + \frac{1}{\rho R_H}\frac{\partial \tau_p}{\partial U}\mathrm{j}c - \frac{S}{\mathcal{L}}g\chi(\cos i) - \frac{S}{\rho \mathcal{L} R_H}\frac{\mathrm{d}R_H}{\mathrm{d}h}\left(\frac{\tau_p}{R_H} - \frac{\partial \tau_p}{\partial R_H}\right)\mathrm{j} = 0 \tag{9.16}$$

In this linear stability analysis (Cf Joseph (1976)) the flow is considered as unstable as soon as it is possible to find one root of (9.16) whose imaginary part is positive so that the disturbance amplitude (defined by (9.15)) will grow with time.

Consider first the case when $\partial\tau_p/\partial U$ is negative, i.e. the flow curve of the fluid decreases with shear stress at least for shear rate around a value which makes it possible to have a mean velocity equal to U. Then one of the solutions for c has a positive imaginary part so that the flow is always unstable. This demonstrates the instability of free surface flows of some clay-water suspensions (see 5.1.1.2.4) whose flow curve exhibits a minimum (Engelund and Wan 1984, Trowbridge 1987).

When $\partial\tau_p/\partial U$ is positive, one root of (9.16) has a positive imaginary part and thus the flow is unstable when:

$$\left(\frac{\tau_p}{R_H} - \frac{\partial\tau_p}{\partial R_H}\right)\frac{\mathrm{d}R_H}{\mathrm{d}h}\sqrt{\frac{S}{\mathcal{L}}} > \frac{\partial\tau_p}{\partial U}\sqrt{g(\cos i)} \qquad (9.17)$$

For flow on an infinitely wide inclined plane, $R_H = S/\mathcal{L} = h$, so that (9.17) becomes:

$$\frac{\left(\tau_p - h\frac{\partial\tau_p}{\partial h}\right)}{\frac{\partial\tau_p}{\partial U}} > \sqrt{gh(\cos i)} \qquad (9.18)$$

It should be noted that this linear stability analysis only provides a sufficient condition for instability. Flow might be unstable when criterion (9.18) is not respected. More details about this problem from a general point of view can be found in Joseph (1976). The above straightforward approach gives a simple, relevant, approximate instability criterion in the case of free surface flow in an open channel whose width to depth ratio is large. For example, for the laminar flow of a Newtonian fluid the present criterion is $Fr > 0.5$, while the asymptotic criterion for very long waves obtained by Benjamin (1957) and Yih (1963) based on a rigorous analysis of the complete, linearized, two-dimensional equations of motion is $Fr > 0.527$. For other channel types, the discrepancy between theory and reality might be larger but this problem has not been studied systematically. The present result could also be improved at large wave numbers by including surface tension and vertical accelerations (Chin et al. 1986) or longitudinal transport of momentum due to turbulent velocity fluctuations (Needham and Merkin 1984).

9.2.4 *Wave length*

When there exists a dominant mode (χ_0) for which the temporal growth rate (χc) is larger than all the others, one can predict that, at least during the first instants of roll wave formation, the length of the periodic waves will be equal to $1/\chi_0$. However, in our case, it is easily found from equation (9.16) that χc increases monotically from a minimum for $\chi=0$ to a maximum constant value when χ increases from zero to infinity. Thus this theory does not predict the existence of a finite χ value corresponding to a maximum of χc, and consequently no dominant mode can be computed theoretically. In practice, it is likely that the temporal growth is in fact limited by phenomena which have not been taken into account in the above approach, such as surface tension or vertical accelerations.

9.2.5 *Physical interpretation*

The following remarks are essentially inspired from Trowbridge's work (1987). In the absence of wall resistance, equation (9.14) may be solved exactly and the corresponding solutions, in the mobile frame, are permanent waves with velocity:

$$c_1 = \pm\sqrt{\frac{S}{\mathcal{L}} g(\cos i)} \tag{9.19}$$

If the effect of friction dominates, the solution to equation (9.14) in the mobile frame is permanent kinematic waves with absolute velocity:

$$c_2 = \frac{S}{\mathcal{L}} \frac{\mathrm{d}R_H}{\mathrm{d}h} \left(\frac{\tau_p}{R_H} - \frac{\partial \tau_p}{\partial R_H}\right) / \frac{\partial \tau_p}{\partial U} \tag{9.20}$$

The instability criterion (9.17) then corresponds to $|c_1| < c_2$, that is a situation for which the kinematic wave speed exceeds the linear frictionless wave speed. Furthermore, it may easily be seen from (9.16) that, for very large wave numbers (χ), the disturbance speed tends towards the frictionless wave speed whereas for very small wave numbers, it tends towards the kinematic wave speed.

Trowbridge (1987) went on to consider the different forms of energy dissipation due to the disturbance. His conclusion is that the flow is linearly unstable if the rate of working by gravity exceeds the rate of energy dissipation by boundary friction. To complete these physical comments we propose a slightly different, simpler approach. Let us consider a disturbance in the form of a periodic linear frictionless wave initiated within the flow. It is the solution to equation (9.14), in which the right-hand side is taken to be zero, and thus η may be expressed in the form of equation (9.15) with $c=c_1$. Then, from the continuity equation in the mobile frame and assuming no phase shift between v and η, v may be expressed as:

$$v = \eta\sqrt{\frac{\mathcal{L}}{S} g(\cos i)} \tag{9.21}$$

Consider now the difference (Δ) of gravity force minus resistant force due to wall shear stress acting on a portion of fluid (dx) crossed by this disturbance. From (9.11) and (9.21) we easily obtain:

$$\Delta = \frac{\eta}{R}\sqrt{\mathcal{L}S}dx \left[\sqrt{\frac{S}{\mathcal{L}}} \frac{\mathrm{d}R_H}{\mathrm{d}h}\left(\frac{\tau_p}{R_H} - \frac{\partial \tau_p}{\partial R_H}\right) - \frac{\partial \tau_p}{\partial U}\sqrt{g(\cos i)}\right] \tag{9.22}$$

Equation (9.22) shows that when the instability criterion (9.17) is respected, Δ is positive. Then, as a first approximation forgetting pressure differences and momentum flux through this portion of fluid, the tendency of the disturbance will be to grow in velocity and thus in depth. This provides a physical explanation for instability: the initial frictionless wave is not damped by friction and gravity effects, but, on the contrary, tends to develop, thus finally affecting flow significantly.

9.3 MUDFLOW INSTABILITY

The discharge equation corresponding to a steady, uniform free surface flow on an infinitely wide inclined plane was derived (see 8.2.1) for a Herschel-Bulkley fluid with a constitutive equation given in the general form by (7.3)-(7.4). Using ξ (equal to 1-1/G) in (8.14), we obtain:

$$(\frac{\partial\xi}{\partial U}) = \frac{1}{U}\,\xi(1-\xi)(1-\frac{\xi}{2+m}) \quad \text{and} \quad (\frac{\partial\xi}{\partial h}) = \frac{\xi(\frac{\xi}{2+m}-1)(1-\xi)}{h(\frac{2}{2+m}\xi^2-2\xi+(m+1))} \tag{9.23}$$

Then, from (9.23) and (9.18) along with the expression of G, and thus of ξ, as a function of τ_p (8.26), we obtain the following criterion for the development of an instability in a steady uniform flow on an inclined plane:

$$Fr = \frac{U}{\sqrt{gh(\cos i)}} > \frac{(m+1)G^2-mG-1}{(m+1)^2G^2+mG+1} \tag{9.24}$$

Note that this result is quite different from the criterion obtained for water flows. Using the Manning-Strickler friction formula for a turbulent water flow on an inclined plane, the instability criterion is Fr>2.25. The present result for mud suspensions predicts that no stable flow should be obtained for Fr values greater than 1/(m+1), whose value is generally close to 0.25 (Cf Chapters 5 and 6). Additionally the critical Froude number increases monotonically when G increases from unity to infinity.

The validity of equation (9.24) was checked experimentally by Coussot (1994). His results concerning flows in a rectangular channel (with a ratio of fluid depth to channel length smaller than 0.1) and the theoretical limit curve (equation (9.24)) are plotted in a (Fr,G) diagram in Figure 9.4. There is good overall agreement between theory and experiments, since no unstable flow occurred below the theoretical limit. However it is not possible to state that the theory is sound, since many stable flow points are situated above the theoretical limit. This discrepancy is certainly not due mainly to approximations made in the theoretical analysis but more likely to uncertainty in experimentally determining flow instability, because of the limited channel length. When the distance for wave development is much longer than channel length, rolls or discharge pulses cannot be observed. For water flows, it was proved by Julien and Hartley (1986) that this distance is roughly proportional to the ratio of flow depth to channel slope. Montuori (1961) also gave a plot for such a length (in dimensionless form) as a function of the Froude number. No similar analysis has been done for mudflows. However, if the result obtained by Julien and Hartley (1986) was valid for mudflows, this might explain the much too simplified conclusions of Wang et al. (1993) that no roll waves occur when the fluid discharge is high, the channel slope is small or the fluid yield stress is large. Indeed, everything else being equal, the latter conditions correspond to the largest flow depth to channel slope ratio and thus could correspond to distances larger than the channel length.

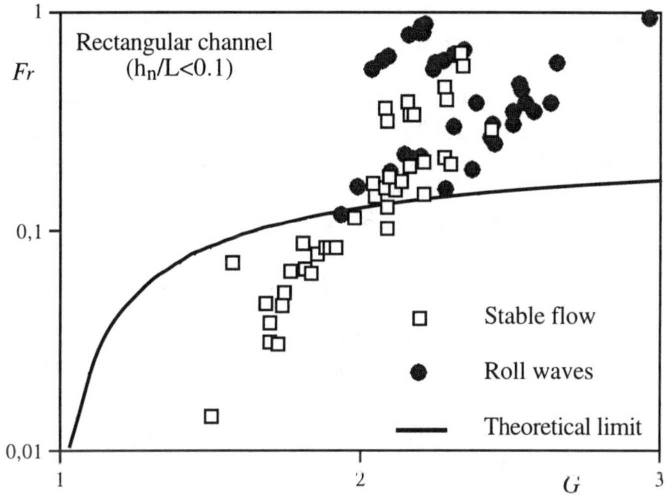

Figure 9.4: Stable and unstable mudflows (water-kaolinite suspensions) and comparison with theoretical predictions (from Coussot (1994))

From the general equation (9.17) an instability criterion can be derived for flows in rectangular channels when wall effect is not negligible. For this purpose the approximate wall shear stress expression (8.28)-(8.29) may be used. The instability criterion is then written:

$$Fr > \frac{1}{1 + \dfrac{GL}{0.3(G-1)(L+2h)}} \tag{9.25}$$

Equation (9.25) clearly shows that, all other things being equal, the greater the fluid depth to channel width ratio, the more the flow is stable, as was observed experimentally by Bossan (1993). However, it is likely that walls play some additional role that tends to damp roll waves and this effect, which is not taken into account in the above approach, should be examined by future works.

9.4 REFERENCES

Benjamin, T.B. 1957. Wave formation in laminar flow down an inclined plane. *J. Fluid Mech.*, 2, 554-574.

Bossan, E. 1993. *Hydraulique des laves torrentielles.* DEA report, INPG, Grenoble.

Chen, C.L. 1993. Unique laminar-flow stability limit based on shallow-water theory. *J. Hydr. Eng.*, 119, 816-829.

Chin, R.W., Abernathy, F.H., and Bertschy, J.R. 1986. Gravity and shear wave stability of free surface flows, 1, Numerical calculations. *J. Fluid Mech.*, 168, 501-513.

Cornish, V. 1934. *Ocean Waves and Kindred Geophysical Phenomena*, Cambridge University Press, Cambridge.

Coussot, P. 1994. Steady, laminar, flow of concentrated mud suspensions in open channel. *J. Hydr. Res.*, 32, 535-539.

Coussot, P., Leonov, A.I., and Piau, J.-M. 1993. Rheology of concentrated dispersed systems in a low molecular weight matrix. *J. Non-Newt. Fluid Mech.*, 46, 179-217.

Coussot, P., and Meunier, M. 1996. Recognition, classification and mechanical description of debris flows. *Earth-Sci. Rev.*, 40, 209-227.

Craya, A. 1951. The criterion for the possibility of roll wave formation. *Proc. Grav. Wave Symp.*, Nat. Bur. Stand.

Davies, T.R.H. 1990. Debris-flow surges - Experimental simulation. *J. Hydrology (New Zealand)*, 29, 18-46.

Delahaye, G., and Coussot, P. 1996. *Debris flows.* Video, Cemagref-Min. Agr. (in French)

Drazin, P.G., and Reid, W.H. 1981. *Hydrodynamic Stability.* Cambridge University Press, Cambridge.

Dressler, R.F. 1949. Mathematical solution of the problem of roll-waves in inclined open channels. *Comm. Pure Appl. Math.*, 2, 149-194.

Dressler, R.F., and Pohle, F.V. 1953. Resistance effects on hydraulic instability. *Comm. Pure Appl. Math.*, 6, 93-96.

Engelund, F., and Wan, Z. 1984. Instability of hyperconcentrated flow. *J. Hydraul. Eng.*, 110, 219-233.

Hjorth, P.G. 1990. Stability of free surface sediment flow. *J. Geophys. Res.*, 95, C11, 20363-20366.

Hwang, S.H., and Chang, H.C., 1987. Turbulent and inertial roll waves in inclined film flow. *Phys. Fluids*, 30, 1259-1268.

Ishihara, T., Iwagaki, Y., and Iwasa, Y. 1954. Theory of the roll-wave trains in laminar water flow on a steep slope surface - Studies on the thin sheet flow, *Trans. Japan. Soc. Civ. Eng.*, 19, 46-57. (in Japanese)

Jeffreys, H. 1925. The flow of water in an inclined channel of rectangular section. *Phil. Mag.*, 6, 49.

Joseph, D.D. 1976. *Stability of Fluid Motions.* Springer-Verlag, Berlin.

Julien, P.Y., and Hartley, D.M. 1986. Formation of roll waves in laminar sheet flow. *J. Hydr. Res.*, 24, 5-17.

Khang, Z. 1990. *Motion characteristics of debris flow at Jiangjia gully, Yunnan province.* China. Int. Res. Train. Cent. Eros. Sediment. Publ., Beijing, China.

Li, J., Yuan, J., Bi, C., Luo, D. 1983. The main features of the mudflow in Jiang-Jia Ravine. *Z. Geomorph.*, 27, 325-341.

Lin, S.P. 1967. Instability of a liquid film flowing down an inclined plane. *Phys. Fluids*, 10, 308-313.

Liu, K.F., and Mei, C.C. 1994. Roll waves on a layer of a muddy fluid flowing down a gentle slope - A Bingham model. *Phys. Fluids*, 6, 2577-2590.

Mayer, P.G. 1961. Roll waves and slug flows in inclined open channels. *Trans. ASCE*, 126, 505-534.

Merkin, J.H., and Needham, D.J. 1986. An infinite period bifurcation arising in roll waves down an open inclined channel. *Proc. Roy. Soc. Lond.*, A405, 103-116.

Montuori, *L'Energia Elettrica*, 38, 1961.

Needham, D.J., and Merkin, J.H. 1984. On roll waves down an open inclined channel. *Proc. R. Soc. Lond.*, A394, 259-278.

Ng, C.O., and Mei, C.C. 1994. Roll waves on a shallow layer of mud modelled as a power-law fluid. *J. Fluid Mech.*, 263, 151-183.

Stuart, J.T. 1971. Non-linear stability theory. *Ann. Rev. Fluid Mech.*, 3, 347-370.

Trowbridge, J.H. 1987. Instability of concentrated free surface flows. *J. Geophys. Res.*, 92, C9, 9523-9530.

Vedernikov, V.V. 1946. Characteristic features of a liquid flow in an open channel. *Proc. USSR Acad. Sci.*, 52, 207-210.

Wan, Z., and Wang, Z. 1994. *Hyperconcentrated Flow*, IAHR monograph, Balkema, Rotterdam.

Wang, Z., Lin, B., Zhang, X. 1993. Instability of non-Newtonian open channel flow. in *Contributions to non-stationary sediment transport*, W. Kron (Ed.), IHW, Universität Karlsruhe, E 1-23.

Yih, C.S. 1963. Stability of liquid flow down an inclined plane. *Phys. Fluids*, 6, 321-334.

CHAPTER 10

Gradually and rapidly varied free surface flows

In steady regime, far from boundaries, the flow in a prismatic channel is uniform. Gradually varied steady flow occurs at the approach of a boundary around which the former uniform flow conditions cannot prevail. The flow characteristics then progressively change in space, which makes it possible to ensure the transition between the two flow types. Macroscopically this is observed through the longitudinal variations of the free surface profile. Other important effects occur, such as the existence of different regimes and hydraulic jump. In some cases the transition between two flows is rapid, i.e. occurs within a distance of the order of the flow depth. Those are referred to as rapidly varied flows.

Gradually and rapidly varied flows are well known and mathematically described in standard hydraulics works (Chow 1959, Munson et al. 1994). As far as we know, with regard to viscous fluids, this problem has only been studied for confined flows through non-uniform ducts (see for example Bird et al. (1977)). With regard to confined flows of Bingham fluids few works concerning lubrication flows are available (see review by Tichy (1991)). In this particular case (gradually varied confined flows), Lipscomb and Denn (1984) demonstrated that no yielding should take place within the fluid, though some sheared parts may be approximated by plug regions. Wilson (1993) studied the problem of the squeeze flow of a Bingham fluid, and Atkinson and El-Ali (1992) the strain field for the flow near the apex of corners. In the case of free surface flows of Herschel-Bulkley fluids, Coussot (1994) provided a framework which will constitute the basis of this chapter. Liu and Mei (1989) also provided a numerical treatment of gradually varied flows of Bingham fluids assuming negligible inertia. Piau (1996) proposed a modified version of Coussot's (1994) results in which he took into account a possible effect of normal stresses.

Here we show that gradually varied mudflows considered as shear-thinning yield stress fluids are similar to water flows (supercritical and subcritical regimes, hydraulic jump, etc.). This is not really surprising since these characteristics basically derive from mass and momentum conservation principles which hold for any fluid type. However the following developments make it possible to emphasize the slight differences between gradually varied flows of water and mud suspensions.

First we establish the general flow profile equation for a channel of any cross-section and consider in more detail the case of an infinitely wide inclined plane. In the second part some energy considerations are proposed. In the last part we consider rapidly varied flows with, in particular, the problems of hydraulic jump and flow through singularities.

10.1 GRADUALLY VARIED FLOW IN A CHANNEL OF ANY CROSS-SECTION

Here we consider the case of a steady non-uniform mudflow in a prismatic open channel under the action of gravity. We assume that the flow varies gradually along the x direction (channel axis), that is $\mathrm{d}h/\mathrm{d}x \ll 1$.

We cannot find an analytical solution of the mass and momentum equations for a channel of any cross-section. Obviously for a gradually varied flow the problem is even more complex, since now lateral (along y or z axis) velocity components may be significant. We shall circumvent this problem by applying the momentum equation in integral form in conjunction with some simplifying hypotheses.

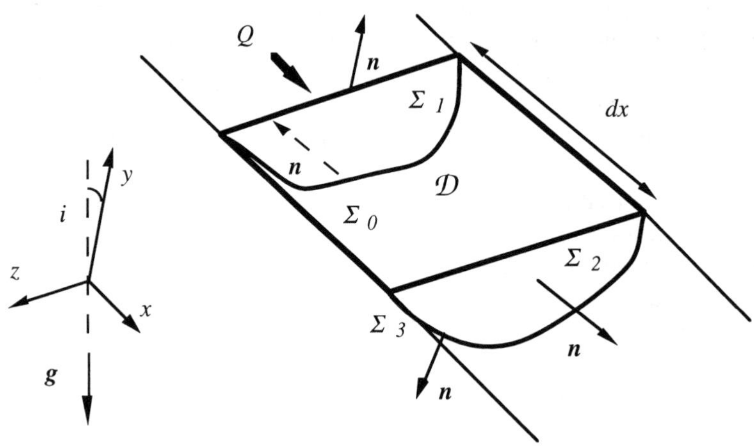

Figure 10.1: Scheme of a fluid portion $\mathcal{D}$.

Consider a fluid portion $\mathcal{D}$ (Cf Fig. 10.1) of length dx limited by two surfaces Σ_1 and Σ_2 situated in (y,z) planes. The free surface Σ_0 is fixed since the flow is steady. There is no flow through the bed surface (Σ_3). The momentum equation (2.8) applied to $\mathcal{D}$ and projected along the x direction then becomes:

$$\int_{\mathcal{D}} \rho g(\sin i)\mathrm{d}\Omega + \int_{\Sigma_0 \cup \Sigma_1 \cup \Sigma_2 \cup \Sigma_3} (\boldsymbol{x}.\Sigma.\boldsymbol{n})\mathrm{d}\sigma = \int_{\Sigma_1 \cup \Sigma_2} \rho(\boldsymbol{u}.\boldsymbol{x})(\boldsymbol{u}.\boldsymbol{n})\mathrm{d}\sigma \qquad (10.1)$$

where $\boldsymbol{n}$ is the outer unit normal to the current surface.

We now make the following assumptions:

(a) Because the flow varies gradually along the x direction the free surface is not that far from a plane parallel to an initially horizontal plane inclined so as to be

parallel to the channel axis. Thus, as for water hydraulics, it is reasonable to assume that the pressure field is of hydrostatic type as for a uniform flow (see 8.2).
(b) In each cross-section the velocity distribution (along the axis perpendicular to the plane) may be approximated by the velocity distribution of the uniform flow with the same discharge and local depth occurring on this slope with an appropriate tangential component of body force. This appropriate body force component will be written $(\sin j)g$.
(c) Surface tension and shear stress due to air resistance to flow at the free surface are negligible.

These hypotheses are quite similar to those used in water hydraulics on smooth slopes. They enable the effects of the vertical velocity component to be neglected and a reasonable estimation to be made of the velocity profile and wall shear stress in any flow section. (a) and (b) correspond in fact to the usual lubrication approximation. It is worth noting that, according to the results obtained by Lipscomb and Denn (1984) for confined flows, there should not be any yielding surface within gradually varied flows and thus assumption (b) is wrong. This result may be established for free surface flows following the demonstration of Lipscomb and Denn (1984). For example for flow over an inclined plane (Cf 8.2), according to the lubrication approximation (a), the velocity distribution in a particular section situated in x should be :

$$u(y) = \frac{\alpha}{1+m}\,[y_0^{(1+m)} - (y_0-y)^{(1+m)}] \quad \text{when } y \leq y_0 \tag{10.2}$$

$$u(y) = u(y_0) = \frac{\alpha}{1+m}\, y_0^{(1+m)} \quad \text{when } h \geq y \geq y_0 \tag{10.3}$$

with $m = \frac{1}{n}$, $y_0 = y_0(x) = h(x) - \frac{\tau_c}{\rho g(\sin j(x))}$, $\alpha = \alpha(x) = (\frac{\rho g \sin j(x)}{K})^m$

and the discharge per unit length is :

$$q = \frac{\alpha(x)}{1+m}\, y_0(x)^{(1+m)}\, [h(x) - \frac{1}{2+m}\, y_0(x)] = u(y_0)\, [h(x) - \frac{1}{2+m}\, y_0(x)] \tag{10.4}$$

If a rigid region (plug) exists within the flowing fluid, $u(y_0)$ must be constant in this region and a relation between $h(x)$ and $j(x)$ follows from (10.3). Introducing this relation in (10.4) one finds that the only solution for h and j is to be constant, which corresponds to a uniform flow situation. Thus our assumption of a true, rigid (but moving!) region is incorrect. In the case of rapidly varying channel geometries a straighter demonstration has been given by Piau (1996): uniform flows are expected upstream and downstream far from the singularity; in these regions the extent and velocity of the plugs are different; if these plugs are linked we have a single rigid body moving steadily in a non-uniform channel; this can occur only if it is strained or elongated, which is in disagreement with the hypothesis of unyielding.

Nevertheless, as recognized by Lipscomb and Denn (1984), within gradually varied flows, some sheared parts may be approximated by plug regions, which is the real meaning of hypothesis (b). More generally (a) and (b) lead to results that are similar to those which would be obtained starting from the local expression of the momentum equation on the basis of the long-wave approximation and assuming a similar velocity profile to the uniform one (as did Liu and Mei (1989) for a Bingham fluid, in addition neglecting inertia).

Piau (1996) considered that this approach might be insufficient for yield stress fluids and proposed a treatment taking into account normal stress terms originating from velocity components along the y and z directions. This effectively leads to a refined flow profile equation similar to (10.9) but with an additional term in the denominator which is related to the normal stresses. Unfortunately this term, which depends on particular, internal, flow conditions, cannot be estimated separately. In fact this approach can be useful only when it is possible to compute the whole set of velocity components numerically. Anyway it is easy to demonstrate that, within the framework of the long-wave or lubrication approximation, that is, for effectively, gradually varied flows, this term remains negligible and may become significant only when the order of magnitude of longitudinal flow depth variations is no longer much smaller than 1. This clearly appears from Piau's (1996) analysis even for a yield stress fluid. This is so because in that case simple shear prevails, which implies that normal stresses are much smaller than shear stresses (this may be checked using the 3D expression of the constitutive equation (7.3-7.4)). Thus, as expected from the above assumptions, the following approach must be seen as a rough approximation of reality as soon as the flow depth varies rapidly. A number of experiments (Coussot and Meunier 1994, Coussot and Proust 1996) have confirmed the validity of these assumptions in other cases. Though observed flow profiles were in overall agreement with theory, it was observed (Bossan 1993) that the experimental value of the critical depth measured at the downtream drop of a channel can be far from the theoretical prediction. It remains hard to determine whether this is mainly due to normal stress effects predicted by Piau (1996) in the case of rapidly varying flows or to experimental problems (influence of possible flow along the downstream side of the channel (when the fluid does not effectively falls freely), uncertainty due to the exact choice of the x-position for measurement, etc). In the following we shall mention the changes induced when taking into account normal stress effects, keeping in mind that the range of values taken by the form coefficient introduced, φ, is not yet known.

On the basis of hypotheses (a), (b) and (c), we shall use h_k and j_k (k=1,2) as flow depth and slope in Σ_k, where j_k is the slope for which a uniform flow is obtained with a discharge equal to Q and a uniform fluid depth equal to h_k (which might be defined as the maximum flow depth of the cross-section for example). The area of Σ_k is denoted S_k. From (a), the mean pressure over a cross-section Σ_k can be written:

$$p_k = \frac{1}{S_k} \int_{\Sigma_k} \rho g (\cos i)(h-y)\mathrm{d}\sigma = \rho g(\cos i) y_{Gk} \tag{10.5}$$

where y_{Gk} is the depth of the gravity center of the cross-section Σ_k. We also define τ_p the mean wall shear stress on Σ_3 as:

$$\tau_p = -\frac{1}{S_3} \int_{\Sigma_3} (\boldsymbol{x}.\boldsymbol{T}.\boldsymbol{n})\mathrm{d}\sigma \tag{10.6}$$

Then, taking into account the no-slip condition at the wall and the free surface condition, equation (10.1) becomes:

$$\rho g(\sin i)\Omega - \tau_p S_3 - [p_2 S_2 - p_1 S_1] = [\rho(\beta_2 (U_2)^2 S_2 - \beta_1 (U_1)^2 S_1)] \tag{10.7}$$

where Ω is the volume of fluid, U_k the mean fluid velocity through S_k and β_k (k=1,2) the momentum coefficient through S_k as defined by (10.13). When dx tends towards zero, we have, at the first order:

$$S_1 = S_2 = S;\ h_1 = h_2 = h;\ S_3 = P\mathrm{d}x;\ \Omega = S\mathrm{d}x = R_H P\mathrm{d}x;\ j_1 = j_2 = j;\ \beta_1 = \beta_2 = \beta \tag{10.8}$$

and, taking into account (10.8), equation (10.7) becomes:

$$\frac{\mathrm{d}h}{\mathrm{d}x} = \frac{\sin i - \sin j(h,Q)}{\cos i + \dfrac{1}{gS}\dfrac{\mathrm{d}(\beta S(U^2))}{\mathrm{d}h}} \tag{10.9}$$

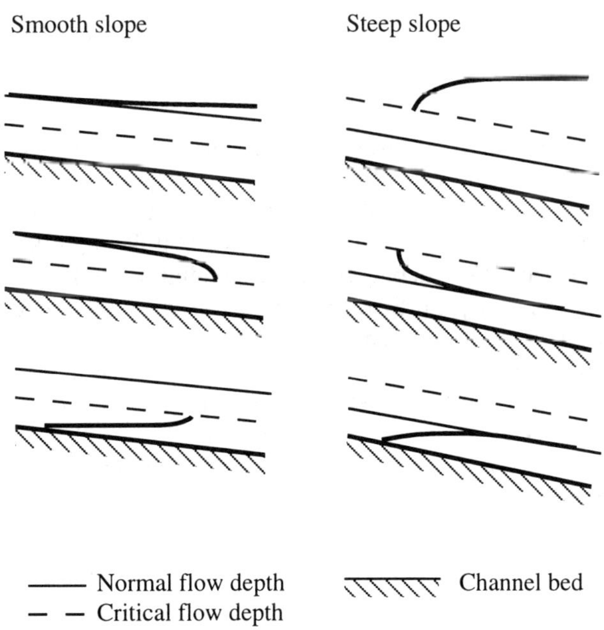

Figure 10.2: Main local flow profile types.

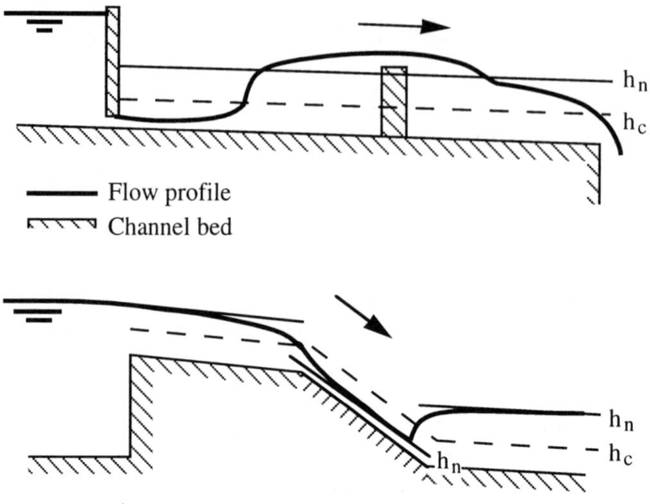

Figure 10.3: Two examples of flow profiles taking into account specific boundary conditions.

If, additionally, we neglect β variations with h, we obtain:

$$\frac{\mathrm{d}h}{\mathrm{d}x} = \frac{\sin i - \sin j(h,Q)}{\cos i - \dfrac{\beta Q^2 \mathcal{L}}{S^3 g}} \tag{10.10}$$

The corresponding formula when normal stresses effects are taken into account (Piau 1996) involves an additional term in the denominator: $[\varphi \mathcal{L} \tau_c \mathrm{sign}(\mathrm{d}h/\mathrm{d}x)]/\rho g S$.

Anyway equation (10.10) is quite similar to the flow profile equation obtained for water flows, except that here we have kept β and $\cos i$ which may be significantly different from 1. There is a normal flow depth (h_n) corresponding to uniform flow ($\mathrm{d}h/\mathrm{d}x=0$), which is obtained when $i = j$. Furthermore ($\sin i - \sin j$) generally increases monotically with h, since j decreases when h increases, and the sign of the numerator of (10.10) is that of ($h - h_n$). There is also a critical flow depth (h_c) corresponding to a change of regime ($\mathrm{d}h/\mathrm{d}x \to \infty$), which is obtained when the denominator of (10.10) equals zero. For a wide-mouthed cross-section $\mathcal{L}/S^3$ decreases monotically with h, thus h_c is unique and the sign of the denominator of (10.10) is that of ($h - h_c$). Finally we obtain:

$$\mathrm{sign}[\frac{\mathrm{d}h}{\mathrm{d}x}] = \mathrm{sign}[\frac{h-h_n}{h-h_c}] \tag{10.11}$$

which determines the form of the flow profile depending on the relative positions of h, h_n, and h_c. This result is quite similar to that obtained in usual open-channel hydraulics. As a consequence, the various flow profile types which may be obtained are formally quite similar to those encountered in open-channel hydraulics (Cf Fig. 10.2). Some examples of possible complete flow profiles with particular boundary conditions are given in Figure 10.3. Obviously a flow profile calculation using equation (10.10) is a good approximation of reality only when the flow depth is not too close to the critical depth (since in this latter zone the flow is no longer gradually varying). We may also speak of supercritical and subcritical regimes depending on whether the flow depth is below or above the critical depth, or, which is an equivalent formulation, whether the Froude number is larger or smaller than 1.

10.2 GRADUALLY VARIED MUDFLOW ON AN INCLINED PLANE

The approximation which consisted in assuming β to be constant is less justified for a mudflow than for water flows, because for a viscous flow (in laminar regime) the velocity profile is generally far from being uniform (u constant). Thus it seems interesting to study equation (10.9) more precisely when it can be computed exactly. This is possible for a flow on an infinitely wide inclined plane (see 8.2). Here the equation giving (sin$j$) is found directly from (8.13):

$$q = \frac{\alpha'}{1+m} y'_0{}^{(1+m)} \left[h - \frac{1}{2+m} y'_0\right] \tag{10.12}$$

with $y'_0 = h - \dfrac{\tau_c}{\rho g(\sin j)}$; $\alpha' = \left(\dfrac{\rho g \sin j}{K}\right)^m$

and the velocity distribution through this cross-section is given by (8.11), now using j instead of i. Under these conditions we also obtain:

$$\beta\rho(U)^2 Lh = \int_S \rho(\boldsymbol{u.n})^2 d\sigma = \rho L\left[\frac{\alpha' y'_0{}^{(1+m)}}{1+m}\right]^2 \left[h - \frac{(4+3m)}{(2+m)(3+2m)} y'_0\right] \tag{10.13}$$

Differentiating (10.12) we deduce:

$$\frac{d}{dh}(\sin j) = -\frac{\tau_c}{\rho g h^2 (1-\xi)^2} \left[\frac{(m+1)(1-\xi)}{\frac{2}{2+m}\xi^2 - 2\xi + (m+1)}\right] \tag{10.14}$$

with $\xi = y'_0/h$. Expression (10.14) is always negative since ξ is smaller than 1. Thus (sin$j$) decreases when h increases and the sign of (sin$i$ - sin$j$) is the same as that of (h-h_n). Using (10.13) and (10.14) we deduce:

$$\frac{1}{g}\frac{1}{h}\frac{d}{dh}(\beta(U)^2 h) = I(h)$$

$$= \frac{1}{g}\left[\frac{1}{1+m}\left(\frac{\tau_c}{K}\right)^m\right]^2 \left[\frac{\xi}{(1-\xi)}\right]^{(2m+1)} h \left[\frac{A\xi^3 + B\xi^2 + C\xi + D}{\frac{2}{2+m}\xi^2 - 2\xi + (m+1)}\right] \tag{10.15}$$

with:

$$A=\frac{8m^2+14m+4}{(m+2)^2(3+2m)};\qquad B=-\frac{10m^3+48m^2+68m+28}{(m+2)^2(3+2m)};$$

$$C=\frac{2m^3+16m^2+32m+18}{(m+2)(3+2m)};\qquad D=-(m+1)\qquad (10.16)$$

It can easily be demonstrated that ξ decreases when h increases. We can also show that when h tends towards 0, ξ tends towards 1 and when h tends towards infinity, ξ tends towards 0. Then it is possible to demonstrate by using (10.14) that when h tends towards infinity $I(h)$ tends towards 0 whereas when h tends towards 0, $I(h)$ tends towards minus infinity. Consequently there exists at least one critical depth. Additionally it can be demonstrated that, if $n<3$* ($m<1/3$), $dI(h)/dh$ is always negative, which shows that in this case there is only one critical depth. In the other cases theoretically there can be three critical depths. Obviously an accurate analysis of flow stability should be carried out for flow conditions corresponding to these different critical depths. It is very likely that two of them correspond to an unstable situation. Since such a study is complex and the case $n>3$ seems to be of little practical interest here we shall not consider this case in detail and assume $n<3$ in the following.

Practically speaking, as a first approximation, for determining the critical depth, we can assume that β is constant and equal to 1. For uniform flows, β can take values between 1 and 4/3 depending on the velocity distribution and m. Since for a gradually varied flow its value (which is probably close to that of the corresponding uniform flow) should change progressively and since only its cubic root plays a role in h_c, assuming β to be equal to 1 should lead to a slight error. Moreover this error should in general be slighter than that arising from the lubrication approximation in the region of critical depth (see above), where it is clear that dh/dx is not much smaller than 1. Taking $\beta=1$ we directly obtain the practical, usual (hydraulics) expression for the critical depth:

$$h_c=\sqrt[3]{\frac{q^2}{g(\cos i)}}\qquad (10.17)$$

It is worth noticing that, for a given flow rate q and slope i, this depth may be smaller than the plug height corresponding to uniform flow conditions. Moreover, though it may appear surprising, in gradually varied flows, any flow depth, and in particular the asymptotic depth corresponding to stoppage of a uniform fluid layer (Cf Chapters 8 and 12), may be reached at least locally, as soon as the flow is suitably accelerated. Under our assumptions with regard to the local velocity profile and wall shear stress, this is expressed by a fictive ($\sin j$) term which ranges from zero to infinity. However, in practice, when flow depth tends to zero, surface tension effects may become significant. At last it must be mentionned that when one takes into

* instead of 3.45 as erroneously estimated by Coussot (1994)

account normal stresses effects the formula for the critical depth (10.17) must be modified. In particular, the critical depth at the downstream end of a channel is larger and, as a first approximate (for small enough yield stress), obtained by adding $\varphi\tau_c/3\rho g\cos i$ (Piau 1996) to (10.17).

The above theoretical predictions using equation (10.9) were compared (Coussot and Meunier 1994) with experimental results corresponding to rectangular channel flows for a low aspect ratio (Cf Fig. 10.4) and showed a fairly good overall agreement.

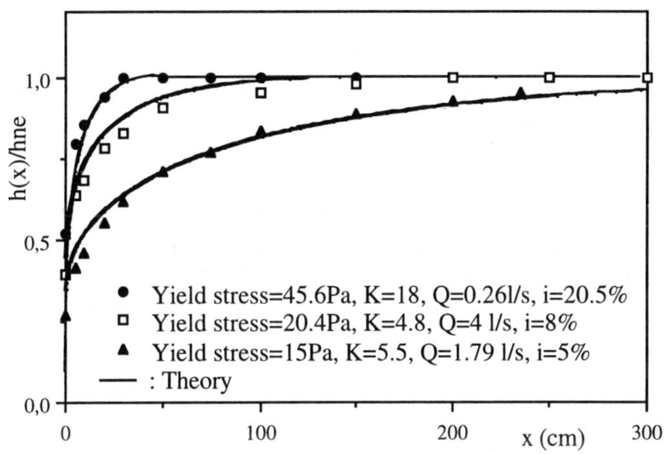

Figure 10.4: Water-Sinard clay mixtures: flow profiles corresponding respectively to a downstream channel drop for different free surface flows in a rectangular channel (channel width: 0.6 m) (aspect ratio less than 0.1). The local fluid depth was divided by the normal fluid depth. Theoretical curves were calculated using equation (10.10) neglecting side wall resistance. [from Coussot and Meunier (1994)]

10.3 ENERGY CONSIDERATIONS

With a view to calculating energy dissipation arising from rapidly varied flows, we must determine the expression for the mean specific energy (E) of the fluid in a cross-section. In steady state, the energy dissipated within a given portion of flow is equal to the difference between the specific energies of limiting (downtream and upstream) cross-sections (Bernoulli's theorem). E is expressed as follows:

$$E = \frac{1}{\rho g Q}\int_{\Sigma}\left(p + \rho g y\cos i + \frac{\rho u^2}{2}\right)\boldsymbol{u}.\boldsymbol{n}\mathrm{d}\sigma \qquad (10.18)$$

Let us consider the case of a uniform steady flow in an open channel of any cross-section. We assume that the pressure distribution is of hydrostatic type. Thus the term ($p+\rho gy\cos i$) in (10.18) is constant and equal to $\rho gh\cos i$ if the pressure reference is taken to be zero. We will use α_Σ, the kinetic energy coefficient, defined as:

$$\alpha_\Sigma = \frac{1}{\rho U^3 S} \int_\Sigma \rho u^2 \, \boldsymbol{u}.\boldsymbol{n} \, \mathrm{d}\sigma \tag{10.19}$$

The general expression for the specific energy then becomes:

$$E = h\cos i + \frac{1}{2}\frac{\alpha_\Sigma U^2}{g} \tag{10.20}$$

This expression can be used for non-uniform flows. For this purpose, it is necessary to assume that vertical motions are negligible and to use the fictive uniform velocity distribution determined as a function of the local depth and mean velocity as described in 10.1.2. The local coefficient α_Σ and consequently E are then obtained as a function of local h and U values.

From a general point of view, for a channel of any cross-section, when the velocity profile is unknown, the remaining problem for determining E is to estimate α_Σ. For flow over an infinitely wide inclined channel, the exact expression for α_Σ is:

$$(1 - \frac{67}{195}\xi)/(1 - \frac{\xi}{5})^3 \tag{10.21}$$

where ξ is equal to $(1 - 1/G)$, G being determined using the fictive slope $j(h,Q)$. (10.21) may be approximated to within 0.4% by $(1 + 0.28\,\xi)$, so that, approximately:

$$\alpha_\Sigma = 1.28 - \frac{0.28}{G} \tag{10.22}$$

It should be pointed out that, in contrast to water flows, the kinetic energy coefficient is not close to 1 and thus it may be important to use the exact expression (10.18) rather than the approximate one with α_Σ=1. Furthermore, all other things being equal, the kinetic coefficient is all the farther from 1 for a higher velocity or a smaller yield stress. These conclusions hold for other channel cross-sections. Indeed, for a semi-cylindrical channel, α_Σ is almost equal to 1 when the ratio of plug radius to channel radius is larger than 0.7, and varies from 1 to 2.46 when this ratio decreases from 0.7 to 0. For other channel cross-sections, we can at least say that, all other things being equal (mean velocity, flow depth, etc.), the kinetic coefficient will in general be all the farther from 1 the greater the wall effect. Further discussions of this problem require a systematic analysis of velocity distribution in complex cross-sections, which has not been carried out yet.

10.4 RAPIDLY VARIED FLOWS

10.4.1 *Hydraulic jump*

In a manner similar to water hydraulics, a kind of "hydraulic jump" will form as soon as a supercritical regime is imposed upstream while a subcritical regime is imposed downstream (see Figs 10.5 and 10.6). The equation for this "hydraulic jump" may be found by applying the momentum equation in integral form to a portion of fluid including the jump. Then we can use the conventional hydraulic hypotheses, which consist in assuming that, in a region not that far from the jump, flow is negligibly perturbed by this jump. If we also neglect gravity and wall resistance we obtain :

$$-\rho g\cos i[S_2 y_{G2} - S_1 y_{G1}) = \rho\,(\beta_2 (U_2)^2 S_2 - \beta_1 (U_1)^2 S_1) \qquad (10.23)$$

For a flow on an infinitely wide inclined plane it is possible to obtain for example h_1 as a function of h_2 from this equation by using equations (10.12) and (10.13) to determine β_1 and β_2. Nevertheless if one assumes that $\beta_1=\beta_2=1$ one obtains the less precise but more practical expression:

$$h_1 = h_2 \left[\frac{\sqrt{1+8Fr_2^2} - 1}{2}\right] \qquad (10.24)$$

where Fr_2 is the Froude number computed for the downward flow. (10.24) is similar to the common hydraulic jump equation for water flows and can be reversed to obtain an exactly similar expression for h_2 as a function of h_1 and Fr_1. As for the critical flow depth, one may obtain a corresponding depth from (10.23) or (10.24) which is smaller than the uniform plug depth. Lastly it may be noted that the energy loss in a hydraulic jump may be computed with the help of (10.20).

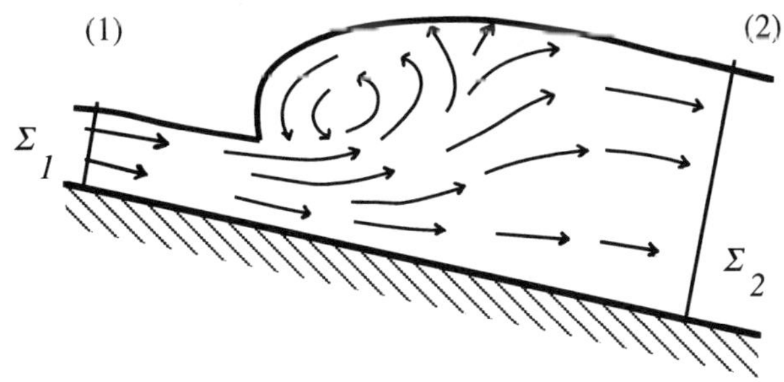

Figure 10.5: Scheme of a hydraulic jump in a longitudinal flow section. The transition between the upstream supercritical regime and the downstream subcritical regime is ensured by a roll. Observations of fluid velocity at the free surface suggest the possible internal velocity distribution as drawn by arrowed lines.

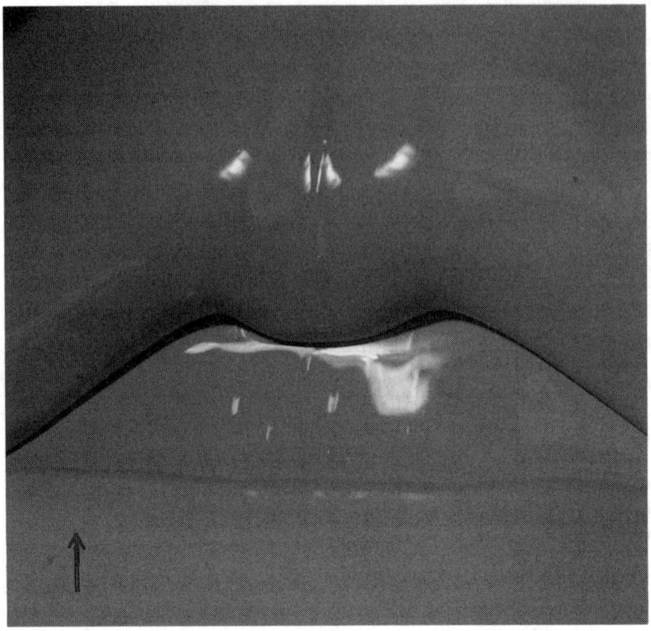

Figure 10.6: Plan view of a hydraulic jump in the laboratory with a mixture of water and kaolinite (yield stress: 20 Pa). Flow from bottom (on the picture) with small fluid depth and steep slope to top (on the picture) with high fluid depth and smooth slope. The curved form of the boundary line (hydraulic jump) between the two regimes (subcritical and supercritical) is due to lateral edge effects (channel of finite width).

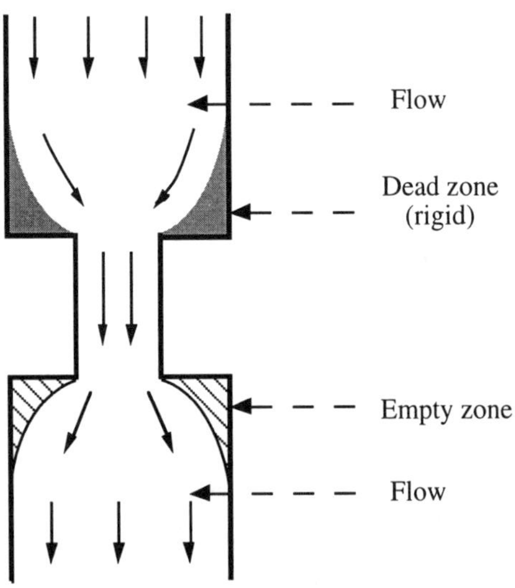

Figure 10.7: Schematic plan view of a mud flow through sudden channel narrowing and broadening.

10.4.2 *Channel singularities*

Since gradually and rapidly varied mudflows exhibit macroscopic characteristics that are similar to those of water flows and thus may be described with the help of analogous mathematical tools, mudflows through singularities will in general be macroscopically similar. In particular, energy losses may be estimated in similar ways. As a consequence we shall not review all possible cases of channel singularities but focus on the slight differences between mudflows and water flows, which arise from mud suspension yield stress. From a general point of view these differences may be expressed as follows:

* In zones where water flows would tend to recirculate, the yielding mud suspension may be subjected to insufficient stresses to remain rigid and static.
* When the fluid is not sheared everywhere it may occupy only a part of the full channel width so that the intersection between its free surface and a channel cross-section is not horizontal.
* As for a hydraulic jump, rapidly varied flows generally remain in laminar regime. Significant energy losses may then be recorded but they are due to additional laminar viscous dissipation.

These characteristics are fairly well illustrated by the following example. From tests through sudden channel narrowings and broadenings the following qualitative observations were made (Coussot and Meunier 1994, Piau 1996). Unlike water flows, mudflows do not develop large eddy zones. There are two particular fluid parts (Cf Fig. 10.7): before the narrowing, because the shear stress is small in the corners, constant rigid zones may form in each corner; after the broadening the fluid may not reach the corners and empty zones remain, due to the fact that the lateral pressure differences are incapable of overcoming the resistance that originates in fluid yield stress (Cf Chapter 12). A more quantitative analysis of these rapidly varied flows would rely on a wide range of experimental data or complete numerical solution of flow equations around the singularity. None has been provided so far in the case of free surface flows. For confined flows Wilson and Taylor (1996) have provided variational and numerical approaches to the channel entry problem using the approximate bi-viscous model.

10.5 REFERENCES

Atkinson, C., and El-Ali, K. 1992. Some boundary value problems for the Bingham model. *J. Non-Newt. Fluid Mech.*, 41, 339-363.

Bird, R.B., Armstrong, R.C., and Hassager, O. 1977. *Dynamics of Polymeric Liquids*. Wiley, New York.

Bossan, E. 1993. *Hydraulique des laves torrentielles*, DEA de Mécanique, INPG, Grenoble.

Chow, V.T. 1959. *Open-channel hydraulics*. Mc Graw-Hill, Civil Engineering Series, New York.

Coussot, P. 1994. Steady, laminar, flow of concentrated mud suspensions in open channel. *J. Hydr. Res.*, 32, 535-539.

Coussot, P., and Meunier, M. 1994. Mudflow dynamics. in *Floods and Inundations related to Large Earth Movements*, A. Armanini and G. Di Silvio (Ed.), Trent, A7.

Coussot, P., and Proust, S. 1996. Slow, unconfined spreading of a mudflow. *J. Geophys. Res.*, 101, 25,217-25,229.

Lipscomb, G.G., and Denn, M.M. 1984. Flow of Bingham fluids in complex geometries. *J. Non-Newt. Fluid Mech.*, 14, 337-346.

Liu, K.F., and Mei, C.C. 1989. Slow spreading of a sheet of Bingham fluid on an inclined plane. *J. Fluid Mech.*, 207, 505-529.

Munson, B.R., Young, D.F., and Okiishi, T.H. 1994. *Fundamentals of Fluid Mechanics.* John Wiley & Sons, New York.

Piau, J.M. 1996. Flow of a yield stress fluid in a long domain. Application to flow on an inclined plane. *J Rheol.*, 40, 711-723.

Tichy, J.A. 1991. Hydrodynamic lubrication theory for the Bingham plastic flow model. *J. Rheol.*, 35, 477-496.

Wilson, S.D.R. 1993. Squeezing flow of a Bingham material. *J. Non-Newt. Fluid Mech.*, 47, 211-219.

Wilson, S.D.R., and Taylor, A.J. 1996. The channel entry problem for a yield stress fluid. *J. Non-Newt. Fluid Mech.*, 65, 165-176.

CHAPTER 11

Mudflow force

This chapter tackles the general problem of the force exerted by a mudflow on an immersed body. This topic has various practical implications: settling threshold for a particle suspended in a mud suspension, incipient motion of a particle at rest undergoing a submerging flow, force exerted by a mud suspension flowing around an obstacle, impact force of a mudflow wave. Here the immersed bodies are assumed to be perfectly rigid and their internal response to flow force, which would require knowledge of their own behaviour, is not considered.

Once again this topic will be tackled with the help of our knowledge of mudflows considered as shear-thinning yield stress fluids. The first part of this chapter focuses on the force exerted by a steady mudflow around a totally or partially immersed obstacle. In a second part we present some theoretical considerations concerning the impact force of transient mudflows or muddy debris flows on an obstacle.

11.1 STEADY MUDFLOW AROUND AN OBSTACLE

Here we consider mudflows whose behaviour may be reasonably well described by a Herschel-Bulkley model (see Chapter 7). So far previous studies concerning the force exerted by yield stress fluids have been essentially based on the Bingham model assumption, even if it now clearly appears that the Herschel-Bulkley model is often much more appropriate to represent the behaviour within a wide shear rate range (Chhabra and Uhlherr 1988, Atapattu et al. 1995). Independently of that problem, knowledge in this field remains approximate.

First, we focus on the simplest problem of a body immersed in a flow which is uniform sufficiently far from this body. This case provides some important qualitative ideas concerning more complicated cases until further studies bring further insight on the matter. The other extreme and more complex case of a partially immersed body in a free surface flow is then considered. Finally we examine the ability of mudflows to support coarse particles and the effects of particle transport on flow characteristics.

11.1.1 *Body immersed in a uniform flow*

An excellent literature review and critical analysis of the problem of drag force exerted by yield stress fluids on a sphere was done by Chhabra and Uhlherr (1988). The consequences of steady mudflows around an obstacle may be separated into three topics: critical static force, velocity distribution within the material and drag force during flow.

11.1.1.1 *Critical static force before motion*

A common property of yield stress fluids is their ability to support finite force without flowing. This is a straightforward consequence of the yield criterion (7.3a). Thus a sufficiently small particle can be supported by a yield stress fluid at rest. In other terms, for a given particle there is a critical value of fluid yield stress making it possible to support this particle at rest. Finally, from a general point of view, the force exerted on a particle immersed in a yield stress fluid must overcome a minimum value for motion initiation. Although at first sight the problem may appear quite simple, at least for a sphere, a mathematical treatment remains difficult. The reason is that it should be based on a serious knowledge of stress distribution within yield stress fluids at rest, which is still lacking. Indeed, so far the method has been essentially to use only a simple criterion for simple shear flow initiation and to extrapolate a three-dimensional form with the help of tools developed in plasticity (see Chapter 7). However, as in plasticity, this approach only gives a criterion for flow initiation in terms of critical stresses but does not directly provide information concerning stress distribution within the material at rest. In fact, in many cases, with a yield criterion of type (7.3a), a wide range of stress distributions can occur at rest, even for a fixed value of $\mathcal{F}$.

Let us consider a sphere (diameter d) embedded in a mud suspension. The most natural and simple theoretical approach consists in assuming that, just when flow starts, the force exerted by the fluid corresponds to a shear stress equal to the yield stress all around the body and tangential to it. Then, after integration, one finds that the total resulting force in the flow direction is (du Plessis and Ansley 1967):

$$\alpha \tau_c \pi d^2 \tag{11.1}$$

where α will be referred to as the critical resistance coefficient and is here equal to $\pi/4$. This scheme is oversimplified. In reality it is very likely that the stress distribution around the body is much more complex. However this formula is dimensionally relevant, i.e. it contains all important parameters. Only the exact value of the constant α remains to be found.

Some authors have proposed refinements of the above theory. In particular, assuming that under static conditions a yield stress fluid can be considered as an ideal solid body with a yield criterion, it has been suggested that yielding might be analysed on the basis of the plasticity theory (see Hill 1950). Within this framework, the strain field is estimated or determined from maximum energy principles. In particular, the surfaces along which slip (or rupture) occurs for a solid body, or correspondingly yield stress is reached for a yield stress fluid, may be found. With the help of this technique, Johnson (1970) computed the force exerted by a yield stress fluid on a body partially immersed at its free surface. The result extrapolated to the case of submerged grains finally leads to a critical resistance coefficient of 1.4 (Hampton 1975). However a complete analysis of the problem of a sphere suspended in a yield stress fluid on the basis of plasticity theory is much more difficult and has not been done yet (Beris et al. 1985). Indeed axisymmetric problems in ideal plasticity are governed by elliptic-equation sets. In planar geometries the analogous equations are hyperbolic and slip lines now develop along the characteristic direction in which the material undergoes purely shear deformation, with the magnitude of the stress equal to the yield value (Beris et al. 1985). Though no extension of the results for planar conditions to axisymmetric problems is available, Ansley and Smith (1967) borrowed ideas from the theory of two-dimensional slip lines in a rigid plastic. They approximated the yield surface for a viscoplastic material as a toroid centred at the equator of the sphere with radius $d/2\sqrt{2}$ and thus found that α should be equal to $7\pi/8$.

A number of authors have developed variational principles which hold for the inertialess flow of a Bingham fluid (Prager 1954, Yoshioka and Adachi 1971, Duvault and Lions 1976). Glowinski et al. (1981) combined variational inequalities with either finite-element or finite-difference numerical approximations and effective minimization techniques to compute viscoplastic confined flows accurately. Beris et al. (1985) used a similar technique to determine the velocity distribution and drag force around a sphere. In their formulation the field equations and boundary conditions are first transformed to a fixed domain by mapping involving the unknown shape functions describing the locations of the yield surfaces. The new set of equations is first reduced to a non-linear algebraic equation set by a Galerkin/penalty finite-element formulation then solved by Newton's method, which iterates simultaneously for all unknowns. They found a critical value of 3.5 for α. Finally, considering the relatively good agreement of the most advanced theories of Ansley and Smith (1967) and Beris et al. (1985), we think it reasonable to use in practice a value of 3 for α.

Experimental works aiming at checking these formulae or at establishing empirical relations (Boardman and Whitmore 1961, Ansley and Smith 1967, Brookes and Whitmore 1968, Wan 1985, Uhlherr 1986, Atapattu et al. 1990) were probably considerably affected by uncertainty in yield stress determination (Chhabra and Uhlherr 1988, Atapattu et al. 1995). Indeed the yield stress value was often extrapolated from experiments carried out at high shear rate values. This generated artificial, overestimated, yield stress values (Cf 5.1.1.2.1) which cannot relevantly be used for establishing empirical formulae. At least it is worth noticing that the resulting empirical value for α is generally much lower than the theoretical ones, which is consistent with the hypothesis of yield stress overestimation from rheometry. Finally the most accurate data, obtained by Atapattu et al. (1995) who used a Herschel-Bulkley model, provide a value for α much lower than 3 (see 11.1.1.3) which is in disagreement with the above theoretical results for unexplained reasons.

11.1.1.2 *Flow field around a sphere*

Volarovitch and Gutkin (1953) were the first to realize that a sphere falling in a Bingham fluid must do so in a small envelope of fluid separating the body from rigid parts in which the stress does not exceed the yield criterion. Few attempts have been made in the literature to compute the flow field. Adashi and Yoshioka (1973) used variational principles to estimate the velocity field. However, probably because of some crude assumptions concerning stress fields, their results are in complete disagreement with those obtained by other authors. Beris et al. (1985), with the help of a finite-element method (see above), obtained the most complete set of results for the velocity field around a sphere. The general form of the sheared envelope is presented in Figure 11.1. It is composed essentially of two symmetrical lobes on each side of the sphere in the direction of the flow, while small unsheared parts remain behind and in front of the sphere. This appears to be in very good agreement with that obtained by Ansley and Smith (1967) by a slip-line theory (see above). Additionally, this qualitative result corresponds to intuitive ideas and is analogous to that obtained for free surface flow around partially submerged obstacles (see 11.1.2). These theoretical results are valid for Bingham fluids but they should obviously be qualitatively very similar for Herschel-Bulkley fluids. Nevertheless, from a quantitative point of view, the predictions of Beris et al. (1985) and Ansley and Smith (1967) are in slight disagreement with the corresponding experimental data obtained by Atapattu et al. (1995). As pointed out by Atapattu et al. (1995), the difference probably originates from the fact that the former postulated a Bingham model for the flow curve whereas, in practice, a Herschel-Bulkley model is clearly much more appropriate.

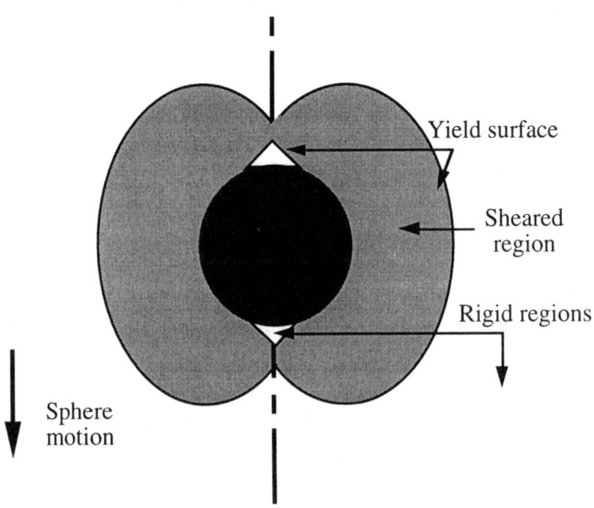

Figure 11.1: Appearance of velocity field for a sphere falling through a yield stress fluid, from Beris et al. (1985).

11.1.1.3 *Drag force*

Various experimental results have been obtained which aimed at determining the drag force (F) of a uniform flow around a sphere (Valentik and Whitmore 1965, Brookes and Whitmore 1969, Ansley and Smith 1967, Wan 1985, du Plessis and Ansley 1967, Pazwash and Robertson 1975, Hanks and Sen 1983, see also Chhabra and Uhlherr (1988) for an exhaustive review). Generally the drag coefficient defined as:

$$C_d = \frac{8F}{\rho\pi V^2 d^2} \tag{11.2}$$

where V is the relative velocity of the sphere and fluid far from the sphere, was investigated as a function of Hedstroem ($d^2\rho\tau_c/\mu_B{}^2$) or/and Reynolds (Cf (7.15)) numbers. Considering the sometimes arbitrary forms of the equations and the great uncertainty in yield stress measurements, it remains difficult to assess the validity of these results.

Ansley and Smith (1967) proposed a theoretical analysis of the problem which appears to be in relatively good agreement with Beris et al. (1985) finite-element solution and not too far from most experimental results. Even if it only provides a rough, albeit practical, estimate of the drag force, we believe it valuable to describe their analysis and use their results for further developments. For a Bingham fluid the drag force is assumed to be equal to the sum of a constant stress term corresponding to the critical stress for motion (Cf 11.1.1.1), and a velocity-dependent stress term which may be estimated by considering that the dynamic effects induced close to the sphere can be approximated as those due to a Stokes flow field (Cf 1.3.2.1). The total drag force is then given as:

$$\alpha\tau_c\pi d^2 + 3\pi\mu_B dV \tag{11.3}$$

Obviously, for incipient motion, only the first term remains and one again finds the formula (11.1). The same analysis may be done for a Herschel-Bulkley fluid with the assumption that the Stokes drag force of an Otswald (1925) fluid, which here corresponds to the velocity-dependent part of the drag force, can be estimated by the formula of Wasserman and Slattery (1964) (cited by Cho and Hartnett (1983)):

$$\alpha\tau_c\pi d^2 + 3\pi XKd^{2-n}V^n \tag{11.4}$$

where X is a correction factor with a value within the range [1.1; 1.5] when n is equal to 1/3 (Cho and Hartnett 1983). Using the value for α originating from the analysis of incipient motion (see Section 11.1.1.1) Equation (11.4) provides a rough estimate of the drag force.

Atapattu et al. (1995) carried out systematic measurements of steady velocity for spheres falling through viscoplastic fluids within a wide range of experimental conditions. Since they fitted a Herschel-Bulkley model to a fluid flow curve in a wide shear rate range, their study is the most accurate to date. By correlating data within 20%, they were able to propose an improved formula for the drag force which is expressed as equation (11.4), where α is equal to 1.7. This result is in disagreement with the values for α proposed in 11.1.1.1 but apparently provides good predictions for fall velocity. The general lack of data and theoretical analysis makes it impossible to decide in a definite way which one is the closest to reality.

11.1.2 *Free surface flow around a partially immersed obstacle*

Now we consider a problem of great practical interest for natural mudflows around dams or houses: the force exerted by a free surface flow around a partially immersed obstacle. Although some experimental studies used free surface flows to estimate the drag force on an immersed body (Brookes and Whitmore 1968, Pazwash and Roberston 1975), only one study specifically concerned the problem of the drag force exerted by a free surface flow around an obstacle (Pratx 1994).

The experiments of Pratx (1994) consisted in measuring the force exerted on a parallepiped fixed perpendicularly to the flow direction. The obstacle was immersed within the fluid up to about 1 mm above channel bottom and was fixed at its other extremity on an iron bar. The torque exerted on it was then measured. The channel width (60 cm) was large compared to flow depth so that the infinitely wide channel hypothesis was valid far from the obstacle (Cf Chapter 8). Additional specific edge effects due to the obstacle were probably negligible: the induced strain field perturbation appearing at the free surface around the obstacle did not extend closer than 15 cm to the walls. Near channel bottom the fluid moves slowly. In addition, because of its yield stress, it could hardly flow through the interval between the obstacle and channel bottom. Thus the corresponding induced force was small.

Figure 11.2: Appearance of the velocity field at the free surface of a flow around a fixed obstacle. Plan view, flow from left to right. [experiments of Pratx (1994)]

Measurements concerned only steady flows. For all tests, the appearance of the free surface profile around the obstacle was similar: downwards to the obstacle the fluid depth increases significantly while upwards to the obstacle the fluid depth decreases. Two approximately triangular dead zones form behind and before the obstacle (see Fig. 11.2). It is worth noting that these dead zones did not disappear by themselves when flow rate decreased and it was necessary to clean up the stagnant volumes after each test in order to obtain subsequent relevant results.

As a consequence, in comparison with drag force on an immersed obstacle, there is at least an additional pressure due to fluid depth difference between each side of the obstacle. Thus, in the following, we shall consider the dimensionless drag force (F^*) which is the real force (F) scaled with that due to the hydrostatic pressure of the uniform flow on a fictive obstacle (see eq. (11.5)). The experimental data of Pratx (1994) can be simply represented in a (F^*,G) diagram. Probably because the scaling of F involves L_0, (the obstacle width), for a given slope, all data appear to fall around the same curve, independently of obstacle width to fluid depth ratio (see Fig. 11.3).

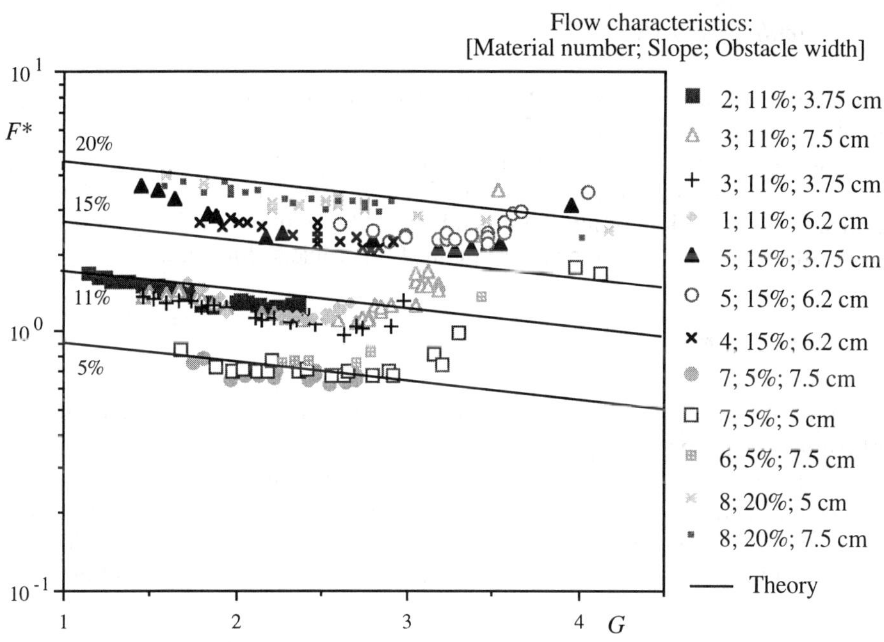

Figure 11.3: Dimensionless force (F*) exerted by a free surface flow on a fixed rectangular obstacle as a function of G. Materials: (1) τ_c=32 Pa, K=24.5, ρ=1370 kg/m^3; (2) τ_c=41 Pa, K=24, ρ=1413 kg/m^3; (3) τ_c=21 Pa, K=12.6, ρ=1350 kg/m^3; (4) τ_c=46 Pa, K=23.3, ρ=1423 kg/m^3; (5) τ_c=21 Pa, K=12, ρ=1379 kg/m^3; (6) τ_c=11.3 Pa, K=9.6, ρ=1362 kg/m^3; (7) τ_c=9.4 Pa, K=8.1, ρ=1354 kg/m^3 (K in Pa.s$^{1/3}$). Continuous lines: empirical equation (11.5). [experimental data from Pratx (1994)]

F^* first decreases slightly then suddenly increases rapidly for G larger than about 3. The similarity of the shape of these curves with those of specific energy as a function of G, which suddenly increase when the inertia term becomes significant (see Section 10.2), led Pratx (1994) to consider that the sudden F^* curve increase for $G>3$ originates in significant momentum transfer to the obstacle whereas for $G<3$ mainly "ordinary" pressure plays a significant role. Obviously further theoretical studies are needed to progress on this point. Here we only propose the following empirical formula fitted to Pratx's (1994) data within the G range [1;3] (Cf Fig. 11.3):

$$F^* = \frac{F}{0.5\rho g L_0 h^2} = \exp(10.8(\sin i) - 0.48 - 0.17\, G) \qquad (11.5)$$

Equation (11.5) makes it possible to predict Pratx's results to within 30%, which may be considered as acceptable with regard to the various sources of uncertainty (rheometry, force measurements, flow fluctuations, etc.).

11.1.3 *Settling velocity*

The above results, and in particular the value of the force (F) exerted by a sphere in motion through a yield stress fluid, may be used directly to predict the free fall velocity of a sphere in a mud suspension under the action of gravity (see Section 1.3.2.1). In particular the critical conditions for settling are obtained when the critical static force before motion given by (11.1) is equal to the buoyant force given by (1.18). Thus we obtain the critical radius for particle sedimentation:

$$R_c = \frac{3\alpha\tau_c}{g(\rho_s-\rho)} \qquad (11.6)$$

in which α is close to 3. The steady settling velocity may be found from the balance of the resistance (eq. (11.4)) with the buoyant force.

In addition, when a particle moves through a fluid not far from rigid walls, the resistance of the fluid is larger than without finite boundaries. This wall effect has been studied by Atapattu et al. (1990) in the case of a particle falling within a yield stress fluid. A critical value of sphere-to-tube diameter ratio was found to exist, below which no wall effect is discernible. This critical value depends on the properties of fluid and sphere through empirical formulae suggested by Atapattu et al. (1990) but, since the free fall velocity, which so far results from the approximate expression (11.4), is not well established, further work is needed to improve these results.

11.1.4 *Settling within a flowing mud*

Here we consider a solid particle immersed in a mudflow in motion and wish in particular to determine whether the particle will settle or not. This problem is hard to tackle from a general point of view for the same reasons as above: the stress field around the particle cannot easily be determined. Here we shall only consider particular simple situations. Let us first take a sphere immersed in an infinite volume of yield stress fluid for which boundary conditions are such that they tend to impose a homogeneous shear (γ=Cst.) in the absence of the sphere. In the frame of reference of the sphere the velocity distribution is antisymmetrical with regard to sphere centre. Thus the stress (found from eq. (7.3-7.4)) at one point of the surface is the opposite of the stress at the symmetrical point with regards to sphere centre. Hence, whatever the constitutive equation, the total stress applied to the sphere is equal to zero (there is nevertheless a finite torque). If the pressure distribution is of hydrostatic type the sphere will settle only if its density is higher than that of the fluid.

In practice mudflows often undergo heterogeneous shearing. Let us consider the case of a wide, long, solid plane immersed in a yield stress fluid. The fluid is submitted to simple shear so that fluid layers gliding over each other are parallel to the plane and perpendicular to the vertical (O*y*). We neglect the problem of plane stability. As a first approximation we can consider that the shear rate is uniform along both large plane surfaces and equal to γ_1 and γ_2. Thus the net stress applied to the plane in the *y* direction is

$$\boldsymbol{y}.(\boldsymbol{T}(\gamma_2)-\boldsymbol{T}(\gamma_1)).\boldsymbol{y} \tag{11.7}$$

For a yield stress fluid whose constitutive equation is given by (7.3-7.4) this stress is equal to zero when γ_2 and γ_1 are finite. For a more complex fluid such as those supposed to exhibit dispersive pressure (see Bagnold (1954) and Section 3.1.4.2) the net stress may be finite and thus can tend to support a plane of higher density than that of the fluid. Though the present assumption of negligible edge effects is not realistic for particles immersed in natural mudflows, this should not change the qualitative result: non neutral buoyant particles would generally tend to settle when they are embedded in the sheared part of mudflows.

11.1.5 *Support of particles in the unsheared part of a free surface mudflow*

Since the unsheared part of a free surface mudflow is rigid, the conditions (in terms of particle density, fluid yield stress, etc) under which the plug may support a particle are similar to those for an entirely static mud suspension (see 11.1.1). However the effects of this kind of support on the overall flow characteristics constitute a new problem. Davies (1986) provided an answer: a dense particle embedded in the plug should also increase the mean plug density, which induces a general settling of this plug through the sheared part of the fluid. This conclusion is erroneous because density changes must be taken into account when solving the set of motion equations.

Let us consider the case of a single particle set at the free surface of a flowing mud. We assume that the particle diameter is smaller than the plug depth (which could occur before setting the particle) and that the conditions for the support of this particle within the mud at rest are fulfilled. Thus the particle remains close to the plug surface. The stress tensor within this plug, resulting from the support of this particle, depends on plug form and channel slope and may be estimated from plasticity theory (Hill 1950). Clearly, for a particle denser than the fluid, the shear stress in a fluid region situated below the particle is larger than the shear stress obtained when there is no particle. Thus the shear stress reaches mud yield stress at a smaller fluid depth than previously (without particle) at least in a small region below the particle. We finally expect that the plug boundary is unchanged far from the particle but is slightly curved towards the free surface around and below the particle. Referring to the discussion of Section 9 concerning the non existence of plug flow within gradually varied flows, it is worth noticing that the above situation is a particular case for which a plug flow truly takes place within a not exactly uniform flow. Here this is possible because the perturbation (density heterogeneity) in boundary conditions travels with the fluid at the same velocity. Thus the plug moves but remains undeformed. This is no longer true for the initial stage during which the particle is released and slightly settles through the mud. It therefore remains an open question as to whether the particle will finally reach a definite position in the plug or will progressively settle down to the unsheared part because, during this transient stage, the local perturbation induces a gradually varied flow without a true plug. As a matter of fact the stable situation is more likely to be reached if the particle is set slowly.

A more interesting case, which may easily be quantified, is that of an isotropic distribution of coarse particles in the plug. Assume that we disperse a large number of pebbles close to the free surface so that it is as if the fluid density increased in that region but the fluid yield stress to be taken into account in the flow equations did not change. For most of the mud this particle distribution is equivalent to a uniform layer (of thickness e) of a material of density ρ_S set upon the free surface. In that case, for flow over an infinitely wide plane, we find, by integrating (8.5), that the shear stress is now written:

$$\tau = \rho g \sin i (h-y) + \rho_s g e \sin i \qquad (11.8)$$

The velocity distribution is then similar to (8.11), in which one must now use h' instead of h, with $h' = h + \rho_s e/\rho$. In particular the plug boundary is now found at a greater height (above the plane) than without the particles:

$$h + \frac{\rho_s}{\rho} e - \frac{\tau_c}{\rho g \sin i} \qquad (11.9)$$

Since everywhere the shear stress is larger and since the sheared region is thicker for a given fluid thickness, the flow rate of the fluid phase will be higher than without the solid layer set at the free surface. It is also worth noting that equation (11.9) remains true as long as e is smaller than $\tau_c/\rho g \sin i$, that is, as long as the layer of particles is smaller than the plug thickness without particles. Otherwise the fluid is sheared everywhere. In that case the particles situated along the boundary between fluid and particle layer are no longer sustained (see 11.1.4) and should start falling down to channel bottom. Then simultaneously the yield stress of the (new) suspension formed by adding coarse particles within the initial fluid increases (see below), thus the plug thickness without the particle layer increases, and the thickness of the particle layer decreases. Finally it is likely that particles will stop settling when the particle layer more or less reaches the (new) plug thickness.

It is also interesting to consider the other extreme case of small particles dispersed throughout the fluid so that it is as if a new (continuous) mud suspension were obtained. We assume that the amount of particles added is similar to the previous situation. The effect of an addition of solid (force-free) particles within a mud suspension is to increase its yield stress (see Section 6.2.1) and its mean density, which become respectively τ_c' and ρ'. For the sake of simplicity we consider the case of a clay-water mixture with a solid concentration ϕ_a in which one adds solid grains whose density is higher than that of the mixture. The solid fraction (ϕ_S) added may be written as a function of e and h:

$$\phi_S = \frac{e}{h+e} \quad (11.10)$$

Moreover the new mean density is written:

$$\rho' = \rho + (\rho_S - \rho)\phi_S \quad (11.11)$$

and, from (6.2) and (6.3) we have:

$$\tau_c' = \tau_c \exp(s'(1-\phi_a)\phi_S) \quad (11.12)$$

The shear stress at a fluid height y, is written:

$$\tau = \rho' g \sin i (h+e-y) \quad (11.13)$$

Then, for a given height y and a given total fluid (water-clay mixture) depth, it is easy to demonstrate that the stress is larger when the particles are added at the free surface than when they are mixed with the fluid. From (11.10-11.13) we deduce that, for $e \ll h$, as a first approximation, the plug boundary is now found at a height:

$$h - \frac{\tau_c}{\rho g \sin i} + \phi_S \frac{\tau_c}{\rho g \sin i}\left[G - s'(1-\phi_a) + \frac{\rho_s}{\rho} - 1\right] \quad (11.14)$$

In expression (11.14), considering the values for s' (see Coussot and Piau (1995)), the term as a factor of $s'(1-\phi_a)$ is in general predominant, which means that the plug thickness is greater than without added grains. It is worth noting that this effect is the opposite to that of adding a layer of particles at the free surface of the flow (see above). Moreover, independently of the possible value of s' it is easy to demonstrate that, all other things being equal, the plug is thinner in the first case (particles set at the free surface).

Finally, if we compare the effects of the two possibilities for adding solid particles in the mud, it appears clearly that, for a given fluid depth, the final velocity is greater when these particles are set at the free surface than when they are mixed with the fluid. Indeed the shear stress is everywhere smaller in the latter than in the former and the plug thickness is greater. Thus, when one needs to rapidly transport solid particles with the help of a yield stress mud, it will in general be more efficient to disperse the particles at the free surface than throughout the fluid.

11.2 IMPACT FORCE OF A DEBRIS FLOW ON AN OBSTACLE

Most natural mudflows or debris flows have a marked transient character (Cf Chapter 12). Here we wish to propose some practical solutions for estimating the order of magnitude of the force exerted by a debris flow wave on a fixed obstacle, since this represents the main cause of damage, either on dams built in mountain streams or on houses. For this purpose we will consider two cases: a mudflow wave made of a single material of fluid type, and a muddy debris flow including a granular front.

Very few research works have dealt with this problem. The work of Armanini and Scotton (1992), though really innovative, concerned only flows of model suspensions whose behaviour has not been determined. In addition they interpreted their results only within the framework of wave propagation and possible reflection. In practice, for dam or weir construction, civil engineers use approximate formulae to estimate the impact force of possible debris flows. In Switzerland, Austria and France, it is generally estimated that debris flow impact force is equal to n times (n ranges from 3 to 10 depending on boundary conditions) the equivalent hydrostatic force of the same amount of water and has a triangular distribution (A 1973, Deymier 1992). This result was obtained from a feedback analysis of the destruction of various dams by debris flows. In Japan, civil engineers determine an equivalent static force by considering that the kinetic energy of the flow is entirely transmitted to the dam (B 1988), which corresponds to a force equal to the second term of equation (11.15) (see also the review and work of Takahashi (1991)).

In the following we consider the impact force originating in a debris flow taken as a whole. Thus we will not consider particular local impacts due to boulders colliding with the obstacle. In practice, it is likely that this effect could provide some additional destructive force but, for the moment, this remains difficult to estimate.

11.2.1 *Impact force of a mudflow on an obstacle*

We propose to estimate this force by considering that, as a first approximation, the interface between the mudflow front and the obstacle is a plane (of width L) perpendicular to the channel axis. Then we apply the momentum equation in integral form (7.8) to a volume $\mathcal{D}$ including a small portion of fluid in contact with the obstacle at the first instant of contact (see Fig. 11.4a). The velocity of any fluid element getting out of $\mathcal{D}$ is assumed to be parallel to the obstacle. We also assume a hydrostatic pressure distribution and neglect wall shear stress and force due to gravity. We finally obtain the force per unit width exerted by the fluid perpendicular to the obstacle:

$$F_l = \frac{1}{L}\int_{\Sigma} (\rho g y \cos i + \rho u(\boldsymbol{u}.\boldsymbol{n}))\mathrm{d}\sigma = \rho g \frac{h^2}{2}\cos i + \beta\rho U^2 h \tag{11.15}$$

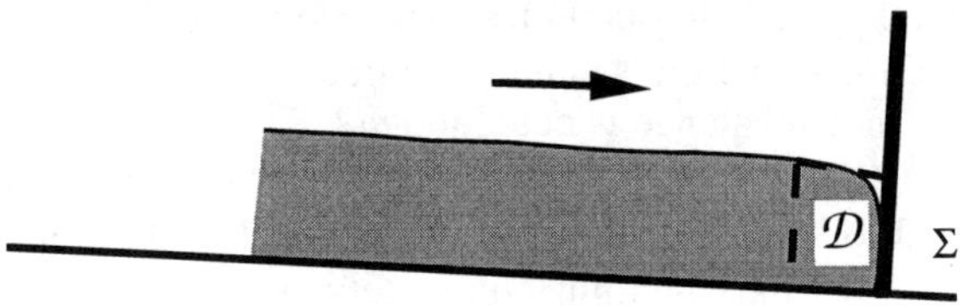

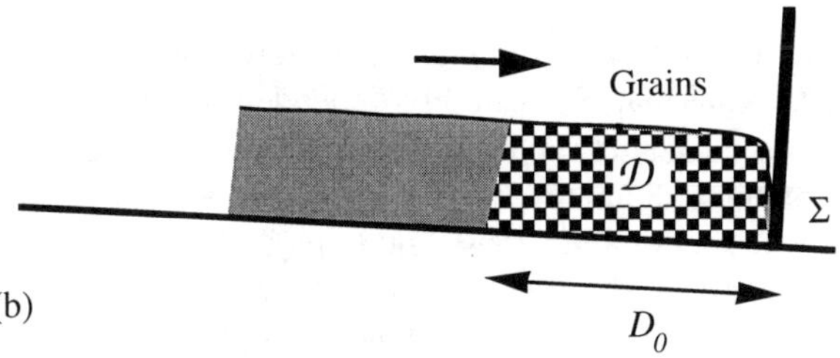

Figure 11.4: Scheme of the transient flow of a mud suspension (without (a) and with (b) granular front) onto an obstacle.

11.2.2 *Impact force of a muddy debris flow on an obstacle*

Now we consider the case of a muddy debris flow surge whose front is composed of a large number of grains immersed in a fluid whose density and viscosity are low. Under these conditions, viscous effects and momentum transfer to liquid motion are negligible. The momentum transmitted to the obstacle is mainly due to the contact of the granular medium with the obstacle. This granular medium has the following characteristics: mass m, mean particle diameter d, length D_0 along channel axis, solid volume fraction during flow ϕ (Cf Fig. 11.4b).

The exact, rheological behaviour of this granular flow remains unknown (Cf Chapter 3) but, in order to compute the impact force with the obstacle, we can take into account its general physical characteristics: grains are separated by a fluid of low viscosity so that they can easily move within this fluid in a small volume around their mean position. In this case, in agreement with Ancey's observations (1993) for sand flows (with air as interstitial fluid), we postulate that the motion of the granular matter is suddenly stopped but, unlike usual fluids, a percolating structure forms by compression due to the arrival of successive grain layers squeezed each against another. Thus the force exerted by this granular medium is due to the sudden stoppage of grains by the wall. At stoppage, the grains form a rigid structure whose downstream part is in contact with the wall. Considering that the rigid structure forms by a rearrangement of grains around their initial position within the flow, the characteristic length needed locally for that is approximately equal to the sum of the mean distances between neighbouring particles. Assuming that, during the stoppage stage, the granular medium keeps the same configuration and grains simply come closer to each other, the mean distance is equal to $\varepsilon d/2$. Thus the total free distance between the spheres is $D_0/(1+2/\varepsilon)$. Before the first instant of contact the velocity (U) of each grain is assumed to be that of the granular flow. Thus the time required to stop the motion of the granular medium completely is approximately equal to $D_0/(1+2/\varepsilon)U$. If we assume that the stoppage of each grain results in a momentum transfer to the wall corresponding to the momentum of this grain during flow, the force per unit width exerted by the material on the obstacle is:

$$F_2 = \frac{m(1+2/\varepsilon)U^2}{LD_0} = \rho h(1+2/\varepsilon)U^2 \tag{11.16}$$

where $\rho = m/hD_0L$ is the mean density of this granular flow whose (uniform) depth is h and whose width is L. Note that F_2 is highly dependent on $1/\varepsilon$, which is all the larger as the particles are closer to each other. If the particles are separated by a distance of the order of their diameter, the resulting force is of the order of the kinetic term found for a single homogeneous fluid (eq. (11.15)). With an appropriate value of ε, the result as predicted by this equation seems to be able to explain that of Armanini and Scotton (1992), who found a dynamic impact of $2.2\rho hU^2$. However a reasonable value of $(1+2/\varepsilon)$ should be larger than 2.2 for a concentrated granular fluid (see eq. (1.12)). As one can see from equation (11.16) the force exerted by the granular front does not depend on the volume arriving at the wall. In fact this result is not valid for a long granular flow. Indeed, sufficiently far from the wall (few L), the longitudinal force is probably mainly supported by the contact of the granular structure with the lateral walls. The same phenomenon occurs in silos: when the stored height becomes larger than a critical value the force exerted at the bottom no longer increases.

Finally, further experimental and theoretical work is needed in the field of granular flow stoppage in order to improve and validate the above approach. Indeed there is still considerable uncertainty surrounding the values of some parameters used in equation (11.16) and the overall validity of the above assumptions concerning physical mechanisms for a wide range of realistic conditions.

11.3 REFERENCES

A, 1973. Dimensionnement des barrages en torrents en béton et béton armé. *Service fédéral des routes et digues*, Berne.

Adachi, K., and Yoshioka, N. 1973. On creeping flow of a viscoplastic fluid past a circular cylinder. *Chem. Eng. Sci.*, 28, 215-226.

Ancey, C.1993. *Lois d'écoulement des matériaux granulaires - Expériences sur des écoulements à surface libre*. DEA Mécanique, INPG, Grenoble.

Ansley, R.W., and Smith, T.N. 1967. Motion of spherical particles in a Bingham plastic. *AIChE J.*, 13, 1193-1196.

Armanini, A., and Scotton, P. 1992. Experimental analysis on the dynamic impact of a debris flow on structures. in *Interpraevent Proc.*, Bern.

Atapattu, D.D., Chhabra, R.P., and Uhlherr, P.H.T. 1990. Wall effect for spheres falling at small Reynolds number in a viscoplastic medium. *J. Non-Newt. Fluid Mech.*, 38, 31-42.

Atapattu, D.D., Chhabra, R.P., and Uhlherr, P.H.T. 1995. Creeping sphere motion in Herschel-Bulkley fluids: flow field and drag. *J. Non-Newt. Fluid Mech.*, 59, 245-265.

B, 1988. Technical standards for the measures against debris flows. *Sabo Division report*, Japan Ministry of Construction.

Bagnold, R.A. 1954. Experiments on a gravity free dispersion of large solid spheres in a Newtonian fluid under shear. *Proc. R. Soc. London*, A 225, 49-63.

Beris, A.N., Tsamopoulos, J.A., Armstrong, R.C., and Brown, R.A. 1985. Creeping motion of a sphere through a Bingham plastic. *J. Fluid Mech.*, 158, 219-244.

Boardman, G., and Whitmore, R.L. 1961. The static measurement of yield stress. *Lab. Practice*, 10, 782-785.

Brookes, G.F., and Whitmore, R.L. 1968. The static drag on bodies in Bingham plastics. *Rheol. Acta*, 7, 188-193.

Chhabra, R.P., and Uhlherr, P.H.T. 1988. Static equilibrium and motion of spheres in viscoplastic liquids. in *Encyclopedia of Fluid Mechanics*, N.P. Cheremisinoff (Ed.), Gulf Publishing Co., Houston, Chapter 21, Vol. 7.

Cho, Y.I., and Hartnett, J.P. 1983. Drag coefficient of a slowly moving sphere in non-Newtonian fluids. *J. Non-Newt. Fluid Mech.*, 12, 243-247.

Coussot, P., and Piau, J.-M. 1995. Effect of an addition of force-free particles on the rheological properties of fine suspensions. *Can. Geotechn. J.*, 32, 263-270.

Davies, T.R.H. 1986. Large Debris Flows: A Macro-Viscous Phenomenon. *Acta Mechanica*, 63, 161-178.

Deymier, C. 1992. *Forces résultant de l'impact de laves torrentielles sur les structures de génie civil*. Internal Report, Cemagref, Grenoble.

du Plessis, M.P., and Ansley, R.W. 1967. Settling parameter in solids piepelining. *J. Pipeline Div.*, ASCE, 93, PL2, 1-17.

Duvault, G., and Lions, J.L. 1976. *Inequalities in mechanics and physics*. Springer, Berlin.

Glowinski, R., Lions, J.L., and Tremoliers, R. 1981. *Numerical analysis of variational inequalities* North-Holland.

Hampton, M.A. 1975. Competence of fine-grained debris flows. *J. Sedimentary Petrology*, 45, 834-844.

Hanks, R.W., and Sen, S. 1983. The influence of yield stresses and fluid rheology on particle drag coefficients. *Proc. 9th Internt. Tech. Conf. Slurry Transp.*, 71-80.

Hill, R. 1950. *The mathematical theory of plasticity*. Oxford University press, Oxford.

Johnson, A.M. 1970. *Physical processes in Geology*. Freeman Cooper and Co, San Francisco.

Otswald, W. 1925. Uber die Geschwindigkeitsfunktion der viskosität disperser systeme. *Kolloid Z.*, 36, 99-117.

Pazwash, H., and Robertson, J.M. 1975. Forces on bodies in Bingham fluids. *J. Hydr. Res.*, 13, 35-55.

Pratx, B. 1994. *Force d'impact des laves torrentielles-Approche expérimentale*. DEA MMGE, Univ. J. Fourier, Grenoble. (in French)

Prager, W. 1954. On slow viscoplastic flow. in *Studies in mathematics and mechanics: R. von Mises presentation volume*, Academic, New York, 208-216.

Takahashi, T. 1991. *Debris flows*. IAHR monograph, Balkema, Rotterdam.

Uhlherr, P.H.T. 1986. A novel method for measuring yield stress in static fluids. *Proc. 4th Nat'l Conf. Rheol.*, Adelaide, 231-235.

Valentik, L. and Whitmore, R.L. 1965. The terminal velocity of spheres in Bingham plastics. Brit. J. Appl. Phys., 16, 1197-1203.

Volarovitch, M.P., and Gutkin, A.M. 1953. Theory of flow of a viscoplastic medium. *Colloid J. (USSR)*, 15, 153-159.

Wan, Z., 1985. Bed material movement in hyperconcentrated flow. *J. Hydr. Eng.*, 111, 987-1002.

Yoshioka, N., and Adachi, K. 1971. On variational principles for a non-Newtonian medium. *J. Chem. Eng. (Japan)*, 4, 217-220.

CHAPTER 12

Spreading, stoppage and deposit

Natural mudflows occurring in mountainous regions are basically transient since they are formed by a finite volume of fluid flowing in the channel of a stream. For the same reason (finite volume supplied) the transient character of many other natural and industrial mudflow types is pronounced. The mathematical treatment of rapid transient flows for which inertia may play a significant role requires specific tools such as numerical models (Laigle and Coussot 1997) or the "method of characteristics" used for example for turbulent water flows (Chow 1959) or viscous flows of Newtonian fluids (Huppert 1982). For natural mudflows overflowing on the alluvial fan inertia may often be neglected. Indeed in that case they are now unconfined and thus can spread in various directions. The flow section increases while the fluid discharge can be assumed to be more or less constant as a first approximation. As a consequence the velocity decreases. Since in addition the initial velocity of channelled mudflows is not high inertia rapidly becomes negligible during overflow on the alluvial fan. This analysis is also supported by the fact that, after the end of the event, mudflows leave large volumes of deposited (static) materials on the surface of the alluvial fan.

Under some conditions slow transient flows may be approached analytically. This makes it possible to predict in particular the remaining fluid layer on a given surface after flow stoppage. Because of fluid yield stress the thickness of the deposits is significant and this constitutes a fundamental characteristic of mudflows. In addition the relationship between slow transient flow characteristics and rheological parameters may also be useful for the interpretation of mudflow deposit forms in rheological terms.

It is worth noting that, for flow stoppage problems, the shear-dependent stress terms do not play any role in the final form of the deposit. Indeed these terms tend towards zero when the fluid stops. Thus the exact form of the shear-dependent terms in the constitutive equation under flow conditions (Herschel-Bulkley, Bingham, Casson, etc. (see Chapters 2 and 7)) is not important, only the yield criterion plays a role.

In the following, we shall always assume that sedimentation is negligible. In the first section we consider the simple case of a "two-dimensional" sheet of fluid flowing over an inclined plane (or line). We start with simple stoppage problems (after uniform or transient flows), then we go on to the general problem of slow spreading. In the second part we consider the case of the unconfined (three-dimensional) spreading of fluid over an inclined plane. We demonstrate that the edge shape of simple deposits can be predicted theoretically. Afterwards the solution for the case of steady spreading from a source is given.

12.1 TWO-DIMENSIONAL FLOW AND STOPPAGE ON AN INCLINED PLANE

12.1.1 *Transient flow and stoppage far from edges*

12.1.1.1 *Stoppage*

Let us consider a uniform fluid sheet lying at rest on an "infinitely" wide and long inclined plane. From the results of 8.2 we know that, for flow to take place, the uniform depth of this sheet must be greater than h_0, which is related to yield stress through equation (8.10). For a shallower fluid depth the fluid remains static. This observation may be used for determining the yield stress.

There are, at first sight, two simple ways of obtaining a situation in which the fluid depth corresponds to the asymptotic one (h_0) limiting rest and flow. The first way involves placing a uniform fluid layer on the initially horizontal plane. Then, one progressively inclines this plane up to the critical slope at which flow is observed (Uhlherr et al. 1984). This technique enters the range of rheological tests under controlled shear stress. Unfortunately it is rather difficult to use in practice (see 4.2.1). The second way involves imposing an approximately steady flow at a given slope and then progressively stopping the upstream supply. Then one has to wait for flow to be completely stopped before measuring the final fluid depth. This method is much easier to achieve in practice since it requires only one test at a given slope. This technique was selected by Coussot and Boyer (1995), who demonstrated from systematic tests that the yield stress determined from accurate rheometrical tests corresponds to that determined from inclined plane tests.

12.1.1.2 *Slow transient flow until stoppage*

Consider the case of a steady uniform flow over an infinitely wide inclined plane whose flow rate progressively decreases until reaching the static situation. In order to establish an analytical formula describing this transient flow we need to assume that inertia effects are negligible. This is especially true for relatively slow flows, which means that the flow depth remains relatively close to the asymptotic depth corresponding to rest state. Then we assume that, at each instant, far from the edges, the fluid depth is uniform, and that, the uniform flow corresponding to the actual fluid depth is reached. Assuming no upstream supply, the conservation of mass integrated over the whole fluid volume gives:

$$\frac{\mathrm{d}(D_0 h)}{\mathrm{d}t} - q = 0 \tag{12.1}$$

where D_0 is the channel length and t the time. The flow rate, q, is given by (8.13) established for negligible surface tension and edge effects. Then we directly obtain the following equation giving the dimensionless fluid depth G as a function of the dimensionless time t^*:

$$\frac{dG}{dt^*} = -\frac{1}{(1+m)(2+m)}(G-1)^{(1+m)}(1+(1+m)G) \quad (12.2)$$

with $t^* = (\frac{\tau_c}{K})^m \frac{\tau_c}{\rho g(\sin i)D_0} t$

The asymptotic depth corresponding to rest state is reached when G drops to 1, which gives us the well-known result $h=h_0$. A similar approach may easily be developed for flow through a channel of complex cross-section using the approximate wall shear stress (see Chapter 8). Equation (12.2) is *a priori* valid only for slowly transient flows. However, with a view to determining the flow time before almost complete stoppage, it is worth noting that the time taken by a possible rapid transient flow phase can easily be neglected, even if inertia effects were not negligible during this phase.

From a theoretical point of view, the time before complete fluid stoppage should be infinite, since, from equation (12.2), the flow depth asymptotically tends towards h_0. Thus the term of apparent complete stoppage should be more appropriate. However it is worth noting that in practice, for natural mud suspensions over an inclined plane, no more motion was observed beyond a given critical time (Coussot and Boyer 1995). This demonstrates that the above theoretical approach is, at least, slightly incomplete from the point of view of either fluid mechanics (edge effects, surface tension) or material behaviour (exact knowledge of constitutive equation for very low shear rates).

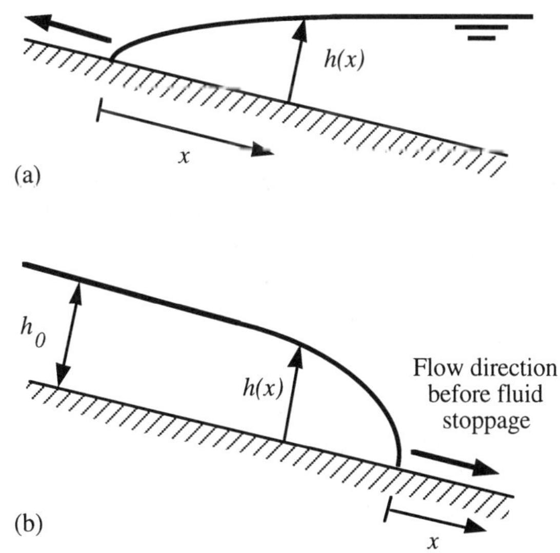

Figure 12.1: Flow front for fluid stoppage on an infinitely wide inclined plane: (a) upward previous flow and (b) downward previous flow.

12.1.2 *Form of the front of a transient flow after stoppage*

Now we are concerned with the geometrical form of the front of a wide (two-dimensional) transient flow on an inclined plane. The solution of this problem under the shallow-water approximation was proposed by Liu and Mei (1989). Here we present a more direct, albeit similar, approach. In the same reference frame as in Chapter 8 we consider an amount of fluid limited by a free surface profile $h(x)$ which does not depend on z. We consider either a previous downward or upward flow with the material front in the flow direction (see Fig. 12.1). We assume that previously the fluid was (everywhere) flowing slowly in the x direction. Thus the shear stress everywhere has overcome τ_c and was maximum at the wall. Now it has just stopped flowing so that the wall shear stress is equal to τ_c at the wall. Assuming a hydrostatic pressure distribution (eq. (8.9)) the momentum equation, applied to a portion of fluid perpendicular to the bottom and whose length is dx, projected along the x-axis, is written:

$$\rho g(\sin i)h dx - \varepsilon\tau_c dx - \rho g\cos i\frac{h^2(x+dx)}{2} + \rho g\cos i\frac{h^2(x)}{2} = 0 \tag{12.3}$$

where ε is equal to 1 and -1 for respectively a downward and upward flow. Equation (12.3) simplifies to:

$$-h\frac{dh}{dx} = (\varepsilon h_0 - h)\tan i \tag{12.4}$$

which, taking the origin (x=0) as the intersection of the free surface profile with channel bottom, may be integrated as:

$$(\tan i)x = h + \varepsilon h_0 \ln(1 - \frac{h}{\varepsilon h_0}) \tag{12.5}$$

It is worth noting that this equation is certainly a very good approximation of the free surface profile as long as the fluid depth varies gradually along the channel. However it might be only a rough approximation when the flow depth tends towards zero since here the assumption of hydrostatic pressure distribution is not sound (as for flow depth approaching critical depth (see Chapter 10)).

For a downward flow, equation (12.5) predicts that h tends towards h_0 when x tends towards $-\infty$. The depth $h_0/2$ is reached for $x = -0.19h_0/\tan i$. Additionally dh/dx tends towards $-\infty$ when h tends towards zero which means that the free surface profile tends to be perpendicular to the channel bottom. This is also true for an upward flow but, now, when x tends towards $+\infty$, h tends towards $+\infty$ as $x\tan i$ which means that the free surface tends to be horizontal. Here the depth h_0 is reached for $x = 0.31h_0/\tan i$.

Equation (12.5) is not valid for i=0, but equation (12.4) may be integrated as:

$$h = \sqrt{\frac{2\tau_c}{\rho g}x} \tag{12.6}$$

In that case it is remarkable that h tends towards $+\infty$ when x tends towards $+\infty$, which is in agreement with the fact that no asymptotic depth exists for a yield stress fluid on a horizontal plane.

12.1.3 *Spreading*

Liu and Mei (1989) proposed a theory for the slow spreading of a thin Bingham fluid layer on a slightly inclined plane on the basis of the lubrication theory, for negligible inertia effects. They assumed the longitudinal lengthscale to be much greater than fluid depth and the longitudinal component of the velocity to be much greater than the vertical one. Here we extend their approach to the case of a Herschel-Bulkley fluid. Under the above assumptions and after eliminating terms depending on z, the motion equations for the two-dimensional flow may be written:

$$0 = \rho g \sin i - \frac{\partial p}{\partial x} + \frac{\partial \tau_{xy}}{\partial y} \qquad (12.7)$$

$$0 = - \frac{\partial p}{\partial y} - \rho g \cos i \qquad (12.8)$$

Considering, as did Liu and Mei (1989), that the external pressure, $P(x,t)$, could be different from the constant atmospheric pressure, (12.8) can be integrated as:

$$p(x,y,t) = \rho g(h-y)\cos i + \mathrm{P}(x,t) \qquad (12.9)$$

Since the shear stress is equal to zero at the free surface, then from (12.7) and (12.9) it is clear that the stress increases linearly with depth:

$$\tau = \tau_{xy} = -M(h-y) \qquad (12.10)$$

where $M = [\frac{\partial P}{\partial x} - \rho g(\tan i - \frac{\partial h}{\partial x})\cos i]$

For a downward flow, M must be negative. Then, introducing the constitutive equation (5.3) into (12.10), this equation may be solved as:

$$u(y) = \frac{1}{m+1}(\frac{-M}{K})^m\, [y_0^{m+1} - (y_0-y)^{m+1}] \quad \text{when } y<y_0 \qquad (12.11a)$$

$$u(y) = \frac{1}{m+1}(\frac{-M}{K})^m\, y_0^{m+1} \quad \text{when } y_0<y<h \qquad (12.11b)$$

where we have taken into account the hypothesis of no slip at the bottom ($u(0)=0$). Here y_0 still corresponds to the critical height above which the fluid remains unsheared and we now have:

$$y_0 = h - \frac{\tau_c}{M} \tag{12.12}$$

The existence of a rigid region results directly from the above rigorous analysis under the initial assumptions (negligible velocity component perpendicular to the plane). It is worth noticing that, from a theoretical point of view, there can be no rigid region within a gradually varied flow (see Chapter 10). However, for a slowly varied flow, that is, here, not too close to the front, this can be a good approximation of reality (Lipscomb and Denn 1984). Here this rigid region varies with time and position in the longitudinal direction (i.e. $y_0 = y_0(x,t)$).

Integrating (12.11) over the fluid depth gives:

$$q = \int_0^h u(y)\mathrm{d}y = \frac{1}{1+m}\left(\frac{A}{K}\right)^m y_0^{(1+m)}\left[h - \frac{1}{2+m}y_0\right] \tag{12.13}$$

and conservation of mass in the entire fluid layer requires:

$$\frac{\partial h}{\partial t} + \frac{\partial q}{\partial x} = 0 \tag{12.14}$$

(12.12)-(12.14) constitute a set of three partial differential equations for the three unknowns $h(x,t)$, $y_0(x,t)$ and $q(x,t)$, which may be solved numerically. In the limit of $\partial P/\partial x=0$ and a Newtonian fluid (with a viscosity μ), this approach leads to the single partial differential equation (Mei 1966, Batchelor 1967):

$$\frac{\partial h}{\partial t} + \frac{g}{\mu}\cos i\,\frac{\partial}{\partial x}\left[\left(\tan i - \frac{\partial h}{\partial x}\right)\frac{h^3}{2}\right] = 0 \tag{12.15}$$

which was found to be in agreement with experimental results (Huppert 1982). A more complete treatment taking into account vertical motions was proposed by Liu and Mei (1990) for Bingham fluids whose behaviour was, in that case, represented by a bi-viscous model (see 8.2.4.1).

For an unpressurized, uniform free surface flow of a limited fluid volume, y_0 is simply related to h through (12.12) and the set of equations (12.1)-(12.2) derives from (12.12)-(12.14). For an unpressurized, stopping fluid layer, τ is equal to τ_c at the bottom and equation (12.4) results from (12.10). Note again that in that case the assumption of a slowly varied flow is inappropriate close to the front of the fluid layer since here the free surface becomes vertical. Nevertheless the experiments of Coussot et al. (1996) showed that this approach gives results that are close to reality.

Solving (12.12)-(12.14) under different conditions, Liu and Mei (1989) demonstrated that, for a constant upstream discharge on an initially uniform fluid layer, a gravity current can advance at a constant speed and with the same profile. For a sudden release of a finite fluid mass on a fluid layer, the flow induced by the new fluid can terminate after the disturbance has travelled a finite distance. Liu and Mei (1989) also proposed a theoretical approach to flow disturbances due to external pressure travelling parallel to the free surface. It was found in particular that a localized pressure gradient pulse leaves behind a permanent footprint.

Other treatments of transient flow properties commonly involve St-Venant equations in a suitable form for yield stress fluids (Laigle and Coussot 1997, O'Brien et al. 1993) with the help of an appropriate local wall shear stress representing viscous dissipation within the fluid layer situated above the considered point, and generally expressed as a function of local fluid depth and mean velocity. In the case of Laigle and Coussot (1997), following Vila's (1987) approach, a special numerical treatment of the equations was developed in order to treat the discontinuity at the flow front correctly. Additionally a realistic wall shear stress expression was used: it was chosen as that corresponding to a uniform flow with the same local fluid depth and flow rate whose use has been proved to be relevant for predicting uniform and gradually varied flows (see Chapters 8 and 10). This approach makes it possible to treat three-dimensional spreading for which inertia is significant.

In the approaches followed by Laigle and Coussot (1997) and Liu and Mei (1989), the most important remaining problem is that of the wall shear stress to be considered close to the front. Indeed in this fluid part there is a recirculation of fluid (for an observer moving with the front velocity) and the free surface slope may be steep. Thus, in contradiction to Liu and Mei's (1989) assumption, there can be no plug within the fluid and there must be a significant vertical velocity component. How far this can induce error in flow prediction remains an open question since comparison of numerical simulations with experiments appeared successful in the case of Laigle and Coussot (1997). Thus there exists at least a range of conditions (not yet well determined) for which, independently of the exact internal velocity profile, the chosen value of the wall stress, which is in fact related to the overall viscous dissipation, remains realistic.

12.2 UNCONFINED SPREADING ON AN INCLINED PLANE: THREE-DIMENSIONAL FLOW

The ideal boundary conditions of Section 12.1 are scarcely encountered in flows of practical interest. Generally, deposits are obtained when, for various reasons, a finite volume of fluid is released over a given surface. Lobes, elongated in the steepest slope direction, are then obtained. Their extent, that is the form of the surface of contact with the bottom, is hard to predict since it depends on flow history before stoppage. Hulme (1974) assumed that, in a steady regime, sufficiently far from boundaries, a tongue of uniform width can form. With a correct three-dimensional treatment it readily appears that this assumption is far from realistic (see 12.2.2). In practice, unconfined free surface flows of a yield stress fluid spread in any direction over an inclined plane. In the following, we first present a mathematical analysis of the form of deposits from a finite volume of fluid over an inclined planar surface. We demonstrate that these forms can be predicted almost exactly, in all directions, independently of the extent of the lobe. In addition, these profiles only depend on the yield criterion of the constitutive equation. Afterwards we focus on the steady unconfined spreading of mud (considered as a Herschel-Bulkley fluid) over an inclined plane.

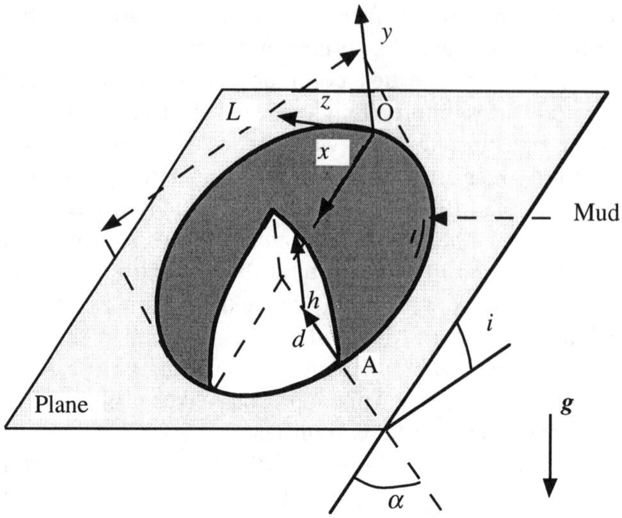

Figure 12.2: Idealized view of the slurry deposit after flow stoppage over an inclined plane.

12.2.1 *Deposit of a finite volume of yield stress fluid over a plane*

12.2.1.1 *General*

The originality of this treatment, proposed by Coussot et al. (1996), is that the deposit is considered as a simple, "asymptotic" flow situation. We consequently assume that fluid stoppage has occurred after a significant flow over the plane, which also means that the volume released was sufficiently large. At this stage, it would be difficult to quantify this condition but this will be clarified below.

Here we consider a finite volume of fluid released on an inclined plane. We use the frame of reference (O,x,y,z) such that the z direction is the horizontal, the x direction the perpendicular to the plane and O the upper fluid point after stoppage on the plane. We shall describe the motion within the framework of the long-wave approximation: the characteristic length of the contact surface with the plane (L) (see Fig. 12.2) is much larger than the characteristic fluid depth (h); the velocity components parallel to the plane (u and w, respectively along the x and z directions) are correspondingly much larger than the component of the velocity (v) along the y direction; the rate of variation of each velocity component is much greater in the y direction than in the perpendicular directions.

From a scale analysis it may be shown that inertia can be neglected in the equations of motion when, for any velocity component V:

$$\frac{V}{\sqrt{gh\sin i}} << 3 \qquad (12.16)$$

Moreover, for all the above assumptions, the equations of motion for a slow flow are:

$$0 = \rho g \sin i - \frac{\partial p}{\partial x} + \frac{\partial \tau_{xy}}{\partial y} \tag{12.17}$$

$$0 = - \rho g \cos i - \frac{\partial p}{\partial y} \tag{12.18}$$

$$0 = - \frac{\partial p}{\partial z} + \frac{\partial \tau_{yz}}{\partial y} \tag{12.19}$$

where ρ is the fluid density, g the gravity and τ_{xy} and τ_{yz} the components of the stress tensor which, from (7.3), are written:

$$\tau_{xy} = [\frac{\tau_c}{\delta} + \frac{1}{2}\mathcal{F}(\frac{\delta}{2})] \frac{\partial u}{\partial y} \tag{12.20}$$

$$\tau_{yz} = [\frac{\tau_c}{\delta} + \frac{1}{2}\mathcal{F}(\frac{\delta}{2})] \frac{\partial w}{\partial y} \tag{12.21}$$

with $\delta = \sqrt{(\frac{\partial u}{\partial y})^2 + (\frac{\partial w}{\partial y})^2}$

Considering the form of δ we can define the angle θ as:

$$\cos\theta = \frac{1}{\delta}\frac{\partial u}{\partial y} \text{ and } \sin\theta = \frac{1}{\delta}\frac{\partial w}{\partial y} \tag{12.22}$$

The boundary conditions are no-slip at the bottom and the existence of a steady free surface. Using atmospheric pressure as a reference, it results from (12.18) that the pressure distribution is of hydrostatic type (eq. (8.9), in which $h-h(x,z)$) here. Then, taking into account (12.20)-(12.21) along with the boundary conditions, (12.17)-(12.19) integrated over the fluid depth yield:

$$\rho g h \sin i - \rho g h \frac{\partial h}{\partial x} \cos i - [\tau_c + \frac{\delta}{2}\mathcal{F}(\frac{\delta}{2})] \cos\theta = 0 \tag{12.23}$$

$$-\rho g h \frac{\partial h}{\partial z} \cos i - [\tau_c + \frac{\delta}{2}\mathcal{F}(\frac{\delta}{2})] \sin\theta = 0 \tag{12.24}$$

where the values of δ and θ given above are taken at the bottom (y=0). From the set of equations (12.23)-(12.24) one can determine the flow characteristics as soon as initial and boundary conditions are completely known. Here we only consider the final stage of flow which, for a yield stress fluid, even neglecting surface tension effects, corresponds to stoppage. Stoppage is consequently a simple, “asymptotic” flow situation. Equations (12.23)-(12.24) are also valid for extremely slow flows. We shall assume that flow over the plane is due to fluid supply from above a fixed point of the plane, the flow direction being vertical. The fluid thus spreads around this particular point. We also assume that the flow rate is sufficient for the flow to be well developed. Here, this means that over the whole contact surface the shear stress exceeded the yield stress at least for a short time during spreading. Then, when in some places the fluid eventually stops, δ locally tends towards zero so that, in the momentum equations (12.23)-(12.24) the term $\delta/2\,\mathcal{F}(\delta/2)$ tends towards zero and disappears. Additionally, at a given point, when the fluid stops, θ reaches a particular value θ_f. Consequently, after complete stoppage, the bottom stress is completely known from (12.20)-(12.21) and the distribution of the final values of θ_f along the fluid-plane contact surface.

12.2.1.2 *Inclined plane ($i \neq 0$)*

Eliminating θ_f, the set of equations (12.23)-(12.24) leads to the following dimensionless equation:

$$H^2\left[\left(1 - \frac{\partial H}{\partial X}\right)^2 + \left(\frac{\partial H}{\partial Z}\right)^2\right] = 1 \qquad (12.25)$$

where the following dimensionless variables have been used:

$$X = \frac{\rho g x(\sin i)^2}{\tau_c \cos i}; \qquad Z = \frac{\rho g z(\sin i)^2}{\tau_c \cos i}; \qquad H = \frac{\rho g h \sin i}{\tau_c}$$

The dimensionless fluid depth (H) is in fact the dimensionless number that we have so far called G.

Boundary and initial conditions are discussed later. It should be noted that H=1 is a solution to equation (12.25). However, since the fluid volume is finite, this solution, which corresponds to an infinite, uniform fluid layer, is not realistic. Since the fluid is spilled vertically above a fixed point on the plane it spreads in each direction from this point and symmetrically relative to the direction of steepest slope. Thus the form of the deposit should be symmetrical relatively to the central, longitudinal plane (z=0), which includes the spilling axis. As a consequence the fluid depth reaches its maximum in this plane, which means that:

$$\frac{\partial H}{\partial Z} = 0 \text{ for } Z=0 \qquad (12.26)$$

Then, using (12.26), (12.25) can be integrated along Z=0, which gives:

$$X = H - \ln\lvert H+1\rvert \quad \text{for } 0 \leq X \leq X_0 \qquad (12.27a)$$

$$X = H + L_0 + \ln\lvert 1-H\rvert \quad \text{for } X_0 \leq X \leq L_0 \qquad (12.27b)$$

with $X_0 = \sqrt{1-\exp(-L_0)} + \ln\lvert 1+\sqrt{1-\exp(-L_0)}\rvert$, where L_0 is the length of the deposit (in the x direction). There the maximum fluid depth is reached for X=X_0 and is equal to $\sqrt{1-\exp(-L_0)}$. Obviously this is also the maximum fluid depth for the whole deposit.

Though H=1 is a possible solution to this equation, a region for which H=1 cannot be linked to the downstream region (defined by eq.(12.27b)) of the deposit for which H necessarily tends towards 0. Consequently the asymptotic fluid depth corresponding to H=1 should theoretically never be reached exactly within such a deposit. In fact one gets closer to it with a larger deposit.

It may appear rather surprising that the asymptotic fluid depth is not exactly reached in the deposit. Indeed, it is well known that a uniform fluid layer can flow only if its thickness overcomes the asymptotic fluid depth (H=1). However this result is only true for an infinite uniform layer. For a finite deposit, there are some regions where the free surface is not parallel to the plane (at least close to the boundaries). In such a region, it is found from equation (12.23) that there can be some (slow) motion if $\rho g h \sin i$ is slightly smaller than τ_c. This remark is also valid for the ultimate part of the deposit in motion just before stoppage. Because the free surface is not parallel to the plane there, slow motion can subsist when the fluid depth is slightly less than the asymptotic depth.

Though we know the value of H along a particular axis (X), equation (12.25) cannot be solved completely because we do not have any lateral boundary conditions (in the Z direction), such as the curve along which H=0. In fact we can obtain an infinity of deposits satisfying (12.25) and (12.27), but with various lateral extents. Simply assuming a symmetrical slow flow is not sufficient to determine the final fluid extent. If the longitudinal extent (L_0) and the total fluid volume are known, the value of the fluid-plane contact surface is limited, but the exact form of the deposit remains indeterminate. In fact, for each particular problem one needs more information concerning the flow history before stoppage. [Considering the simplicity of the assumed experimental procedure here the flow history is in fact the flow rate history.] In our case we at least wish to find some particular relations between the form of the deposit and the rheological characteristics of the material. To this end we shall now consider the fluid depth distribution in particular directions.

Along the curve corresponding to the intersection of the inclined plane with the free surface of the deposit we have H=0. Considering the way in which this deposit is formed, with material being supplied from a fixed point, it is reasonable to assume that the curve H=0 is concave. We shall assume that in each plane perpendicular to the curve H=0 the concavity of the free surface of the deposit remains similar. Then, along the intersecting curve of this plane with the free surface of the deposit, the tangent to the free surface of the deposit is always perpendicular to this plane. This means that, if the equation for the plane perpendicular to the curve H=0 can be written $Z=X\tan\alpha$, where α is an angle ranging from 0 to π (see Fig. 12.2), we have, in this plane:

$$\frac{\partial H}{\partial Z} = \tan\alpha \frac{\partial H}{\partial X} \qquad (12.28)$$

This reasonable asumption makes it possible to solve our problem in an approximate manner. Equation (28) introduced in (25) gives:

$$H^2[1 + 2\cos\alpha\frac{\partial H}{\partial D} + (\frac{\partial H}{\partial D})^2] = 1 \qquad (12.29)$$

where D is the dimensionless distance from the external point (A) of the deposit ($D = \rho g d(\sin i)^2/\tau_c \cos i$ where d is the distance) (see Fig. 12.2). (12.29) can be integrated to give the fluid depth distribution in any direction perpendicular to the periphery of the deposit:

$$D=[1-\cos\alpha[H+\ln\left|\frac{\tan(\frac{\alpha-\arcsin(H\sin\alpha)}{2})}{\tan\frac{\alpha}{2}}\right|]-\cos(\arcsin(H\sin\alpha))] \qquad (12.30)$$

One must keep in mind that, with the above assumptions, expression (12.30) is only approximate. In particular, the validity of equation (12.30) is quite dubious for the central part of the deposit, where fluid depth variations in any direction are rather small. However, since in this region H is very close to 1, the resulting practical uncertainty in the estimated value of H remains negligible. Furthermore, for $\alpha \geq \pi/2$, equation (12.30) cannot be used at a greater distance from A than the critical distance for which H reaches exactly 1.

When α tends towards 0 or π, equation (12.30) leads to:

$$D = (\varepsilon H - \ln|1+\varepsilon H|) \qquad (12.31)$$

where ε is equal to -1 for α=0 and equal to 1 for $\alpha=\pi$. Equations (12.27 a,b), which are derived directly from flow equations, may also be found from (12.31), which demonstrates the consistency of the above approximate approach for the fluid depth distribution in any direction. Equation (12.31) is also the dimensionless form of the expression for fluid depth as a function of distance for two-dimensional flows (eq. (12.5)).

For $\alpha=\pi/2$, equation (12.30) also gives a more precise expression for the free surface profile than the one obtained by Hulme (1974) (see 12.2.1.3):

$$H = \sqrt{2D-D^2} \qquad (12.32)$$

This equation may be used for estimating the yield stress of geophysical fluids from natural, lateral levees resulting from slight overflowing of the fluid onto the banks during its transient motion through a channel. This type of lateral levee is typically encountered with lava and debris flows.

It is worth noting that these results do not depend on $\mathcal{F}$, since they concern a no-shear situation. Equation (12.30) makes it possible to estimate the yield stress of any fluid with a constitutive equation such as equation (7.3), from the form of any such deposit of a finite volume of this material at the surface of a plane. Notice that, due to the dimensionless form of equation (12.30), for a sufficient fluid volume and for a given plane slope and direction α, the depth distributions obtained for different fluids depend only on material parameters (and not on flow history) and are in similarity.

12.2.1.3 *Horizontal plane (i=0)*

Since, in this case, there is no asymptotic fluid depth at rest (sin$i$=0 means that it should be infinite), the solution of equation (12.23)-(12.24) is formally quite different. From (12.23)-(12.24), eliminating θ_f, we find the following dimensionless equation:

$$H^{*2}[(\frac{\partial H^*}{\partial X^*})^2 + (\frac{\partial H^*}{\partial Z^*})^2] = 1 \qquad (12.33)$$

where $X^* = \frac{\rho g x}{\tau_c}$; $Z^* = \frac{\rho g z}{\tau_c}$; $H^* = \frac{\rho g h}{\tau_c}$

Now the deposit is clearly symmetrical by rotation around its central axis. Thus in each cross-section including this central axis, the fluid depth distribution is similar to that obtained on the plane Z^*=0 if we take into account the fact that $(\partial H^*/\partial Z^*) = 0$ through this plane. Thus, if the point corresponding to the intersection of the central axis with the inclined plane is taken as reference (H^*=0, X^*=0), the fluid depth distribution is given by:

$$H^* = \sqrt{2X^*} \qquad (12.34)$$

This also corresponds to the distribution obtained for the two-dimensional case by Liu and Mei (1989) (eq. (12.6)). This means that the deposit has the form of a heap whose height, W, is given by equation (12.34) as a function of the radius (R) of the deposit-plane contact surface. It is worth noticing that here the condition for the validity of the long-wave approximation (W<<2R) implies:

$$W >> \frac{\tau_c}{\rho g} \qquad (12.35)$$

The flow and stoppage of a yield stress fluid on a horizontal plane is precisely the situation encountered with the slump test, which is a comparative test used for fresh concrete in order to estimate its "workability". A slump test basically involves filling up a truncated cone with test material and then allowing it to slump by withdrawing the cone upwards. The loss of height (slump) then gives a measure of the consistency of the material. Since it is recognized that fresh concrete basically behaves as a yield stress fluid (see 6.3.2), some authors (Rajani and Morgenstern 1991, Murata 1984, Christensen 1991, Pashias et al. 1996) have proposed simple analytical developments within the framework of elasto-plasticity to relate the slump to the material yield stress. With the same aim a viscoplastic dynamic numerical model of the slump has been proposed (Tanigawa and Mori 1989). As an example we describe the approach and results of Pashias et al. (1996) in Appendix I.

The present approach (eq. (12.34)) is of a different type since it is an analytical approach to the deposit considered as an ultimate flow situation. In particular, our experimental procedure is intended to provoke flow everywhere within the material whereas the slump test procedure is intended to leave the largest possible volume undeformed. Only the order of magnitude of the result obtained here for the slump (which could be deduced from equation (12.34)) is close to the results obtained by the above authors. Nevertheless the present theory and experimental procedure seem appropriate to determine accurately the yield stress from the final form of the heap. However, the condition for the long-wave approximation to be valid appears rather restrictive in practice. Indeed, for fresh concrete whose yield stress and density are respectively 1000 Pa and 2000 kg/m^3, which is a typically realistic case, (12.34) and (12.35) show that the radius of the base of the deposit should be at least equal to 2.5 m. This is one order of magnitude larger than the corresponding value obtained in slump tests. Clearly, in most cases, in order to use the present theory for the rheological interpretation of slump tests, one should considerably increase the size of the cone.

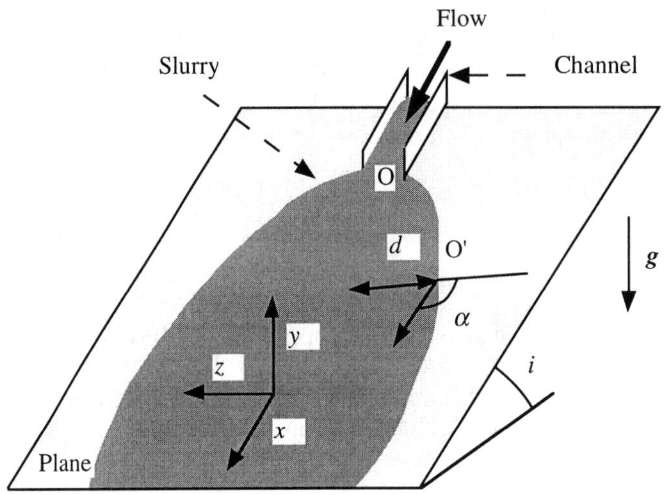

Figure 12.3: Overall view of the unconfined spreading over an inclined plane starting from a source (here a channelled flow).

12.2.2 *Steady spreading from a source*

Hulme's work (1974) has long been the only study of the unconfined flow of yield stress fluids over an inclined plane. His results have been used by volcanologists for analysing field lava deposits (Wadge and Lopes 1991). We emphasize that Hulme (1974) did not consider the three-dimensional form of the constitutive equation and, as a consequence, his analysis is biased. However before describing the simplified three-dimensional approach as proposed by Coussot and Proust (1996), we review Hulme's analysis since it introduces some important elements for understanding the subsequent analysis.

12.2.2.1 *Unconfined spreading with a constant width: Hulme's analysis (1974)*

Volcanic lava flows give rise to deposits of materials which, setting aside thermal changes as a first approximation, might be treated as yield stress fluids at sufficiently low temperatures (Robson 1967, Shaw et al. 1968, Gauthier 1973, Pinkerton and Sparks 1978, McBirney and Murase 1984). This motivated the following pioneering analysis of Hulme (1974). Referring to Figure 12.3 we consider the channelled or confined flow of a yield stress fluid which suddenly spreads over an infinitely wide inclined plane (with a slope i). A constant flow rate (Q) of material is supplied from upstream. Hulme (1974) assumed that, at least far from boundaries, a uniform flow takes place in the form of a tongue of constant width ($2\delta_l$) (see Figs. 12.4-12.5). The velocity, whose components are (u,v,w), and the stress tensor then depends only on y and z. Additionally, the velocity distribution which dissipates the least energy is such that v and w are equal to zero. Thus it is reasonable to suppose (as Hulme (1974) does) that, as in the case of flow through a uniform channel, u is the only non-zero velocity component. From the momentum equation we deduce immediately that, for a Herschel-Bulkley fluid whose constitutive equation is given by (7.3)-(7.4), the pressure distribution is of a hydrostatic type, that is given by equation (8.9), where in the present case atmospheric pressure is used as reference (p_0=0). In order to determine the flow depth profile in a cross-section (Fig. 12.5), Hulme writes the local momentum balance in the z direction:

$$-\rho g(\cos i)h\mathrm{d}h = \tau_2 \mathrm{d}z \qquad (12.36)$$

where τ_2 is the local wall stress amplitude in the z direction.

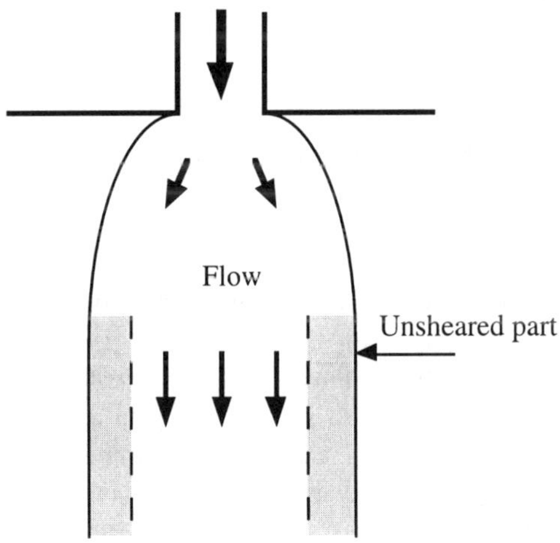

Figure 12.4: View from above of the extent of a previously channelled viscoplastic flow on a wall-free inclined plane, according to Hulme (1974).

Then Hulme assumes that τ_2 can be expressed with the help of the usual constitutive equation under simple shear conditions. He considers that the static situation along the z axis has been reached progressively so that *ultimately τ_2 has reached τ_c everywhere*. Equation (12.36) may then be easily integrated and gives equation

(12.6) where x is now replaced by $(\delta_l\text{-}z)$. Afterwards Hulme writes the local momentum balance in the x direction:

$$\rho g h(\sin i) = \tau_l \tag{12.37}$$

where τ_l is the wall shear stress in the x direction, from which he deduces the size of the lateral static part: this corresponds to the region for which $\rho g h(\sin i)<\tau_c$, limited by the surface where the flow depth reaches the asymptotic fluid depth h_0 given by equation (8.10). Within the framework of the present approximate analysis, the sheared and unsheared parts are assumed to be separated by vertical surfaces. This approach provides relations between the yield stress and the height and width of residual, lateral unsheared deposits. Hulme then assumes that the local flow rate per unit width corresponds to the flow rate per unit width of the uniform free surface flow with the same local flow depth. Thus, by integration over a cross-section he deduces an approximate relation between the total flow rate and the flow width.

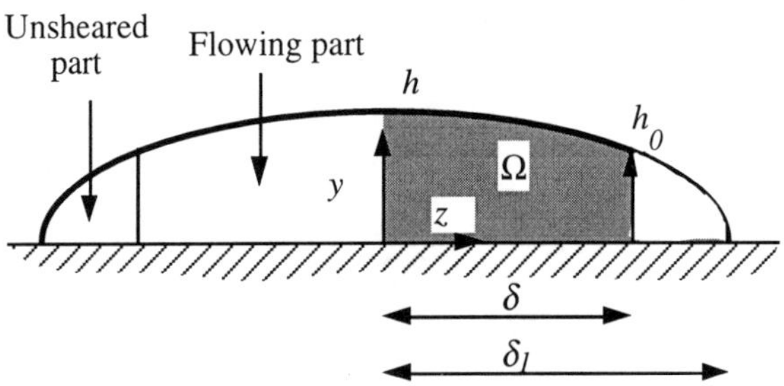

Figure 12.5: Cross-section of fluid spreading on an infinitely wide inclined plane.

12.2.2.2 *Analysis of Hulme's approach*

First let us assume again that flow takes place in the form of a uniform tongue. The above developments remain true up to the point when the wall stress in the z direction (τ_2) is assumed to be equal to the yield stress. This statement proves to be wrong as soon as one considers the constitutive equation in an appropriate three-dimensional form. Here we once again use the general constitutive equation for viscoplastic fluids as given by (7.3). Then, for a velocity field with only one non-zero component in the x direction depending only on y and z, it is apparent from (7.3) that the stress tensor can be written thus:

$$\boldsymbol{T} = \frac{1}{\sqrt{\delta}}(\tau_c + \sqrt{\delta}\,\mathcal{F}(\text{-}\delta))\begin{pmatrix} 0 & \dfrac{\partial u}{\partial y} & \dfrac{\partial u}{\partial z} \\ \dfrac{\partial u}{\partial y} & 0 & 0 \\ \dfrac{\partial u}{\partial z} & 0 & 0 \end{pmatrix} \tag{12.38}$$

where now $\delta=(\frac{\partial u}{\partial z})^2+(\frac{\partial u}{\partial y})^2$

From (12.38) it appears that the stress vector perpendicular to the plane is **T.y**, the one non-zero component of which is directed along the x axis. Thus there is no wall stress component directed along the z axis.

Figure 12.6: Mudflow extent a certain time after leaving the channel. (The squares drawn on the plane have 30 cm edges). From Proust's (1995) tests.

Therefore, when applying the momentum balance to a lateral volume Ω (Cf Fig. 12.5) or a smaller portion of fluid, the only external forces in the z direction are due to the fluid pressure on the wetted sides of the volume. Thus, according to (8.9), if the free surface is curved, there is a net positive force directed in the z direction so that the fluid part under consideration should tend to spread laterally. This contradicts our initial hypothesis: the free surface cannot be curved if the flow width remains constant. For unsheared regions, equation (7.3a) applies and the wall shear stress may reach the yield stress in a particular direction. This problem has been analysed in 12.2.1 and it appears that, within these static fluid parts, the fluid depth should be equal to or smaller than the asymptotic fluid depth (h_0) everywhere.

The general consequence is that, as long as the fluid flows uniformly in the x direction, the free surface of sheared regions must be parallel to the bottom while that of unsheared regions may be curved. The transition between the central sheared part and the lateral unsheared parts can only be achieved via a curved free surface, which, as a consequence of the above results, must be static. Since the fluid depth of unsheared regions cannot exceed h_0, the only possible flow depth of the central part is h_0, which corresponds to complete flow stoppage. This means that the tongue of uniform width assumed by Hulme cannot exist, except under static conditions.

It is worth noticing that, though we reasoned here with the help of equation (7.3), the above statement would be obtained with any (simple) yield stress fluid even if another three-dimensional form of the constitutive equation were used, as long as this model includes a "volume-based" yield criterion. Indeed this criterion is only a mathematical representation of the fact that, as soon as the fluid flows in any direction, the yield criterion is overcome so that there is no particular reason for the shear stress in other directions to correspond to yield stress.

It is clear from the above analysis that, as long as the fluid flows, the free surface is curved in the z direction so that the fluid should tend to spread laterally. Thus the width of the fluid sheet increases with distance from the channel exit and with time. Locally this lateral spreading will stop when the resistance to flow in the z direction balances the pressure gradient (see next section). Since the fluid reaches a point further from the exit at a later time, the lateral spreading stabilizes downstream later (see Fig. 12.6). In the following, we shall only consider the plane length along which the steady state situation (constant width at a given distance) has been reached at a given time. This situation apparently exists in practice (see Fig. 12.6). Now we intend to determine the fluid depth and velocity as a function of distance from the channel exit.

12.2.2.3 *Theoretical analysis of the steady regime*

12.2.2.3.1 General

Here we use the long-wave approximation for the steady three-dimensional flow of a Herschel-Bulkley fluid down a wide inclined plane. This approach has already been discussed and successfully used by Ng and Mei (1994) and Liu and Mei (1990, 1994), in the case of the two-dimensional spreading of a sheet of mud (considered as a Bingham fluid) over an inclined plane. In the present case, it is assumed that the characteristic depth h is much less than the lateral extent $2\delta_l$, which is much less than the longitudinal scale (in the x direction). Correspondingly, for the velocity components, we have $u >> w >> v$. So from a scale analysis it may be shown that inertia can be neglected in the equations of motion when:

$$\frac{u}{\sqrt{gh(\sin i)}} << 10 \qquad (12.39)$$

Even for rapid flows, this condition can become true not far downstream of the source (or channel exit) because lateral spreading induces a considerable decrease in velocity. Under these hypotheses, with the help of (7.3)-(7.4), the equations of motion in steady state are again given by (12.17)-(12.19) where now, for a downward flow, that is flow in the $+x$ direction ($\partial u/\partial y > 0$):

$$\tau_{xy} = [\tau_c + K(\frac{\partial u}{\partial y})^n] \qquad (12.40)$$

$$\tau_{yz} = \left[\frac{\frac{\partial w}{\partial y}}{\frac{\partial u}{\partial y}}\right] \left[\tau_c + K\left(\frac{\partial u}{\partial y}\right)^n\right] \quad (12.41)$$

Additionally the mass conservation is again given by (7.5) and the boundary conditions are no-slip at the bottom, the existence of a steady free surface and steady volume discharge (Q) supplied at the source (x=0). Furthermore, by symmetry, there can be no flow through the central plane (z=0) and the fluid depth is maximum in this plane. These conditions are written as follows:

$$\boldsymbol{u}(x,0,z)=0 \quad (12.42)$$

$$(-p\boldsymbol{I} + \boldsymbol{T}).\boldsymbol{n} = 0 \text{ and } \boldsymbol{u}.\boldsymbol{n} = 0 \text{ for } y=h(x,z) \quad (12.43)$$

$$\int_{x=0} u\mathrm{d}y\mathrm{d}z = Q \quad (12.44)$$

$$\mathrm{w}(\mathrm{x,y,0}) = 0 \quad (12.45)$$

$$\frac{\partial h}{\partial z} = 0 \text{ for } z=0 \quad (12.46)$$

where $\boldsymbol{I}$ is the unit tensor and $\boldsymbol{n}$ the outer normal. It appears from (12.18) that the pressure distribution is of hydrostatic type. Then, as long as the plane is not too close to the horizontal, we have $(\partial h/\partial x) << \tan i$, so that $(\partial p/\partial \mathrm{x})$ may be neglected in equation (12.17). Then (12.17) integrated over the fluid depth taking into account (12.43) leads to:

$$0 = \rho g h \sin i - \tau_l \quad (12.47)$$

where τ_l is the component of the stress tensor defined by equation (12.14) taken at y=0. From (12.40) and (12.47) it is apparent that there is no flow if $\tau_l < \tau_c$, which means that the fluid does not move in the regions where:

$$h < h_0 \quad (12.48)$$

Considering that the fluid is initially supplied along the central longitudinal axis, the fluid depth should decrease from this axis in the z direction. The static regions should then take the form of two symmetrical lateral bands where the fluid depth falls below the asymptotic depth. In a steady regime these bands remain fixed. Thus we shall consider them separately. It is worth noticing that, within the framework of the present approach, it is sufficient to consider that these rigid regions are separated from sheared regions by a surface perpendicular to the plane, though this separating surface can in fact be slightly curved or inclined in the y direction.

12.2.2.3.2 Sheared part

Here we shall integrate the equations of motion over the fluid depth and determine the solution in the particular case n=1/3.

12.2.2.3.2.1 General dimensionless equations

Considering the sheared region separately we now have a new boundary condition:

$$\boldsymbol{u}=0 \text{ for } h=h_0 \tag{12.49}$$

Following Ng and Mei (1994) and Liu and Mei (1994), we shall assume that, locally, the velocity profile in the main direction of flow (i.e. x) is that of the steady uniform flow with the same local fluid depth. This leads to a similar result to that obtained (see Chapter 10) by considering that the bottom shear stress is that of the corresponding uniform flow with the same local fluid depth. Thus this assumption is consistent with equation (12.47). Hence, for a fluid whose constitutive equation is given by (7.3)-(7.4), the velocity distribution is given by equation (8.11), corresponding to the uniform case where h is the local fluid depth. Once again, as long as the flow depth variations with x or z are slow enough, the corresponding uniform profile including a rigid part should be a good approximation of reality.

Taking into account (12.41) and (12.43), equation (12.19) integrated from h to y gives:

$$(y\text{-}h)\frac{\partial p}{\partial z}\frac{\partial u}{\partial y} = \frac{\partial w}{\partial y}[\tau_c + K(\frac{\partial u}{\partial y})^n] \tag{12.50}$$

Using (12.49), equation (12.50) integrated over the fluid depth gives exactly:

$$\text{-}u^*\frac{\partial p}{\partial z} = w^*\rho g\sin i \tag{12.51}$$

where u^* and w^* are the two components of the mean velocity over the fluid depth:

$$u^* = \frac{1}{h}\int_0^h u\mathrm{d}y \qquad \text{and} \qquad w^* = \frac{1}{h}\int_0^h w\mathrm{d}y \tag{12.52}$$

From (12.52) and the velocity distribution, we can obtain the expression for u^*, which is in fact that of Q obtained from (8.13). Taking into account (12.43) and (12.52), equation (7.5) integrated over the fluid depth gives:

$$\frac{\partial(hu^*)}{\partial x} + \frac{\partial(hw^*)}{\partial z} = 0 \tag{12.53}$$

hu^* and hw^* are the two components of the volume discharge per unit width over the fluid depth. We now introduce the following dimensionless variables:

$$U = U(X,Z) = \frac{hu^*}{h_0^2}(\frac{K}{\tau_c})^m\ ; \qquad W = W(X,Z) = \frac{hw^*}{h_0^2}(\frac{K}{\tau_c})^m\ ; \qquad Q_0 = \frac{Q\tan i}{h_0^3}(\frac{K}{\tau_c})^m$$

where X, Z and H have been defined above. The normalized set of equations governing the flow, which results from (12.44-12.46), (12.49), (12.51) and (12.53), is then as follows:

$$\frac{\partial U}{\partial X} + \frac{\partial W}{\partial Z} = 0 \qquad (12.54)$$

$$U = \frac{1}{(m+1)(m+2)}(H\text{-}1)^{m+1}((m+1)H+1) \qquad (12.55)$$

$$-U\frac{\partial H}{\partial Z} = W \qquad (12.56)$$

with the boundary conditions:

$$\int_{X=0} U\mathrm{d}Z = Q_0 \qquad (12.57)$$

$$W(X,0) = 0 \qquad (12.58)$$

$$\frac{\partial H}{\partial Z} = 0 \text{ for } Z=0 \qquad (12.59)$$

$$U(X,L(X))=W(X,L(X))=0 \qquad (12.60)$$

where $L(X)$ is the dimensionless half-width of the sheared region at a distance X from the source ($L=\delta\ \tan i/h_0$ where δ is the half-width of the sheared region). When useful, $Z/L(X)$ will be referred to as ζ.

12.2.2.3.2.2 Solution for n=1/3

In the following we only consider in more detail the case $n=1/m=1/3$, which appears to correspond to many practical situations. A similar procedure can be followed and would give the solution for any other particular n value. In each case the basic operation consists in inverting equation (12.55). In particular, for n=1/3, it is possible to obtain a simpler set of equations by using, instead of (12.55), the expression:

$$H = 1 + 1.19U^{2/9} \qquad (12.61)$$

which approximates H to within 10% for (H-1) ranging from 0.1 to 10. This range includes many practical situations of yield stress fluid flows. Indeed a flow for which (H-1)<0.1 would generally be extremely slow (when K is of the same order as τ_c) so that in practice the fluid could, as a first approximation, be considered as static. When (H-1)>10, yield stress effects become negligible and the present theory may be unnecessary or even irrelevant if inertia becomes significant. From (12.54), (12.56) and (12.61) we obtain:

$$\frac{\partial U}{\partial X} = 0.264\ \frac{\partial}{\partial Z}(U^{2/9}\frac{\partial U}{\partial Z}) \tag{12.62}$$

The simplest approach for solving this equation consists in assuming that the velocity distribution in a given cross-section (at a given distance X from the source) is similar to that obtained at another distance, that is:

$$U(X,Z) = \phi(X)\psi(\frac{Z}{L(X)}) \tag{12.63}$$

where ϕ, L and ψ are three positive functions. Introducing (12.59) in (12.58), we obtain:

$$\phi^{-11/9}L^2\frac{d\phi}{dX} - [\phi^{-2/9}L\frac{dL}{dX}]\ [\zeta\frac{1}{\psi}\frac{d\psi}{d\zeta}] = 0.264\ \psi^{-7/9}\frac{d^2\psi}{d\zeta^2} + 0.059\ (\frac{d\psi}{d\zeta})^2\psi^{-16/9} \tag{12.64}$$

For this equation to be satisfied there are two possibilities. First $[\zeta/\psi)\ (d\psi/d\zeta)]$ may be constant. This does not correspond to a solution of the present problem since in that case ψ should be proportional to a power of ζ but then $\psi(1)$ could not be equal to 0 though required by (12.59) and (12.61). Therefore the remaining solution is that $[\phi^{-11/9}L^2\ (d\phi/dX)]$ and $[\phi^{-2/9}L\ (dL/dX)]$ be separately constant. This means that:

$$L(X) = k(X+A)^a \tag{12.65}$$

$$\phi(X) = k'(X+A)^{-9/2+9a} \tag{12.66}$$

where k, k', A and a are parameters which can be determined from boundary conditions (12.57-12.60). If the dimensionless half-width of the flow at the source is l_0, we have:

$$l_0 = k(A)^a \tag{12.67}$$

From (12.64) integrated from 0 to L, then from 0 to X using (12.57-12.60), we find:

$$[L(X)\phi(X)]\ [\int_0^1 (\psi(\zeta)d\zeta] = \frac{Q_0}{2} \tag{12.68}$$

which implies that a must be equal to 9/20. Then $\psi(\zeta)$ may be integrated from (12.64) in which (12.65-12.66) have been introduced and (12.58-12.60) taken into account. In particular, from (12.68), one finds that:

$$k=2.17Q_0^{1/10} \tag{12.69}$$

Finally, the exact solution based on the approximate equation (12.61) is:

$$U(X,Z) = 0.59\ Q_0^{9/10}(X+A)^{-9/20}\left(1-\left(\frac{Z}{L}\right)^2\right)^{9/2} \tag{12.70}$$

$$W(X,Z) = 0.27\ Q_0^{9/10}(X+A)^{-29/20}\ Z\left(1 - \left(\frac{Z}{L}\right)^2\right)^{9/2} \tag{12.71}$$

$$H(X,Z) = 1 + 1.06\ Q_0^{2/10}(X+A)^{-1/10}\left(1 - \left(\frac{Z}{L}\right)^2\right) \tag{12.72}$$

$$L(X) = l_0\left(\frac{X}{A}+1\right)^{9/20} \tag{12.73}$$

with $A=0.18\ \left(\frac{l_0^{20/9}}{Q_0^{2/9}}\right)$

It is now possible to check whether the long-wave approximation is valid for this solution: W is much smaller than U if:

$$x \gg \frac{9}{20}\frac{h_0}{\tan i} \quad \text{or} \quad \left(\frac{l_0^{20/9}}{Q_0^{2/9}}\right) \gg 2.5 \tag{12.74}$$

This means that, in some cases, not far from the source, the above solution may not be valid. In other cases equations (12.70-12.73) should give us the velocity and fluid depth distributions for the unconfined, steady, spreading of a yield stress fluid (with $n=1/3$) over an inclined plane for $0.1<(H-1)<10$ and if conditions (12.39) and (12.74) are respected. The results concerning the fluid depth distribution were confirmed by Coussot and Proust's (1996) experiments. The theoretical lateral extent nevertheless appeared to be an overestimate of reality by almost 30% but this is probably due to the fact that in these tests the characteristic longitudinal length was not much smaller than the characteristic width. As a consequence, in particular, the boundaries were not far enough from the part of flow under study. It remains that the experimental rate of increase of the lateral extent with distance from the source was in good agreement with the theoretical prediction.

12.2.2.3.3 Unsheared part

In a steady state the lateral unsheared regions are fixed, so that everything suggests that they could also be the lateral parts of a completely deposited mass. We have seen that a surface perpendicular to the bottom roughly separates sheared and unsheared regions where $h=h_0$. In order to predict the fluid depth distribution of these lateral unsheared regions, we can use the three-dimensional analysis of 12.2.1 concerning mud deposits. The fluid depth distribution is then given as a function of the angle α and the distance from the deposit edge (Cf eq. (12.30)).

12.3 REFERENCES

Batchelor, G.K. 1967. *Introduction to Fluid Dynamics*, Cambridge University Press, Cambridge.

Christensen, G. 1991. *Modeling the flow of fresh concrete: the slump test.* Ph.D. thesis, Princeton University, Princeton.

Chow, V.T. 1959. *Open-channel hydraulics.* Mc Graw-Hill, Civil Engineering Series, New-York.

Coussot, P., and Boyer, S. 1995. Determination of yield stress fluid behaviour from inclined plane test. *Rheol. Acta*, 34, 534-543.

Coussot, P., and Proust, S. 1996. Slow, unconfined spreading of a mudflow. *J. Geophys. Res.*, 101, 25,217-25,229.

Coussot, P., Proust, S., and Ancey, C. 1996. Rheological interpretation of deposits of yield stress fluids. *J. Non-Newt. Fluid Mech.*, 66, 55-70.

Gauthier, F. 1973. Field and laboratory studies of the rheology of Mount Etna lava. *Phil. Trans. R. Soc. Lond.*, A274, 83-98.

Hulme, G. 1974. The interpretation of lava flow morphology. *Geophys. J. Roy. Astrophys. Soc.*, 39, 361-383.

Huppert, H.E. 1982. Flow and instability of a viscous current down a slope. *Nature*, 300, 427-429.

Laigle, D., and Coussot, P. 1997. Numerical modelling of mudflows. in press in *J. Hydraul. Eng.*

Lipscomb, G.G., and Denn, M.M. 1984. Flow of Bingham fluids in complex geometries. *J. Non-Newt. Fluid Mech.*, 14, 337-346.

Liu, F.K., and Mei, C.C. 1990. Approximate equations for the slow spreading of a thin sheet of Bingham plastic fluid. *Phys. Fluids*, A 2, 30-36.

Liu, K.F., and Mei, C.C. 1989. Slow spreading of a sheet of Bingham fluid on an inclined plane. *J. Fluid Mech.*, 207, 505-529.

Liu, K.F., and Mei, C.C. 1994. Roll waves on a layer of a muddy fluid flowing down a gentle slope - A Bingham model. *Phys. Fluids*, 6, 2577-2590.

McBirney, A.R., and Murase, T. 1984. Rheological properties of magmas. *Ann. Rev. Earth Planet. Sci.*, 12, 337-357.

Mei, C.C. 1966. Nonlinear gravity waves in a thin sheet of viscous fluid. *J. Maths. Phys.*, 45., 266-288 .

Murata, J. 1984. Flow and deformation of fresh concrete. *Matér. Constr (Paris)*, 17, 117-129.

Ng, C., and Mei, C.C. 1994. Roll waves on a shallow layer of mud modelled as a power-law fluid. *J. Fluid Mech.*, 263, 151-183.

O'Brien, J.S., Julien, P.Y., and Fullerton, W.T. 1993. Two-dimensional water flood and mudflow simulation. *J. Hydraul. Eng.*, 119, 244-261.

Pashias, N., Boger, D.V., Summers, J., and Glenister, D.J. 1996. A fifty cent rheometer for yield stress measurement. *J. Rheol.*, 40, 1179-1189.

Pinkerton, H., and Sparks, R.S.J. 1978. Field measurements of the rheology of lava. *Nature*, 276, 383-384.

Proust, S. 1995. *Physical modelling of stoppage conditions for debris flows.* DEA Mécanique et Ingénierie, IMF-ENGEES, Strasbourg. (in French)

Rajani, B., Morgenstern, N. 1991. On the yield stress of geotechnical materials from the slump test. *Can. geotech. J.* 28, 457-462.

Robson, G.R. 1967. Thickness of Etnean lavas. *Nature*, 216, 251-252.

Shaw, H.R., Wright, T.L., Peck, D.L., and Okamura, R. 1968. The viscosity of basaltic magma: an analysis of field measurements in Makaopuhi lava lake, Hawaii. *Am. J. Sci.*, 266, 225-264.

Tanigawa, Y., Mori, H. 1989. Analytical study on deformation of fresh concrete. *J. Eng. Mech. Div. (ASCE)*, 115, 493-508.

Uhlherr, P.H.T, Park, K.H., Tiu, C., and Andrews, J.R.G. 1984. Yield stress fluid behaviour on an inclined plane. in *Advances in Rheology*, B. Mena, A. Garcia-Rejon, and C. Rangel-Nagaile (Ed.), Mexico City, 2, 183-190.

Vila, J.P. 1987. Schémas numériques en hydraulique des écoulements avec discontinuités. *IAHR Congress on Topics in Hydraulic Modelling*, J.A. Cunge and A. Ackers (Ed.), Lausanne, Switzerland, 120-125.

Wadge, G., and Lopes, R.M.C. 1991. The lobes of lava flows on Earth and Olympus Mons, Mars. *Bull. Volcanol.*, 54, 10-24.

Appendix 1: MOTION EQUATIONS IN CYLINDRICAL COORDINATES

The coordinates of a point M, i.e. the components of **OM** in the cylindrical basis $(\boldsymbol{e}_r,\boldsymbol{e}_\theta,\boldsymbol{e}_z)$, are (r,θ,z). A vector $\boldsymbol{u}$ and a tensor $\boldsymbol{T}$ may be written in matrix form as:

$$\boldsymbol{u} = \begin{pmatrix} u_r \\ u_\theta \\ u_z \end{pmatrix} \text{ and } \boldsymbol{T} = \begin{pmatrix} \tau_{rr} & \tau_{r\theta} & \tau_{rz} \\ \tau_{r\theta} & \tau_{\theta\theta} & \tau_{\theta z} \\ \tau_{rz} & \tau_{\theta z} & \tau_{zz} \end{pmatrix}$$

Consequently we have:

$$\nabla . \boldsymbol{u} = \text{div } \boldsymbol{u} = \frac{1}{\mathrm{r}}\frac{\partial(ru_r)}{\partial r} + \frac{1}{r}\frac{\partial u_\theta}{\partial \theta} + \frac{\partial u_z}{\partial z} \quad \text{(A.1.1)}$$

$$\nabla \boldsymbol{u} = \textbf{grad } \boldsymbol{u} = \begin{pmatrix} \frac{\partial u_r}{\partial r} & \frac{1}{r}(\frac{\partial u_r}{\partial \theta} - u_\theta) & \frac{\partial u_r}{\partial z} \\ \frac{\partial u_\theta}{\partial r} & \frac{1}{r}(\frac{\partial u_\theta}{\partial \theta} + u_r) & \frac{\partial u_\theta}{\partial z} \\ \frac{\partial u_z}{\partial r} & \frac{1}{r}\frac{\partial u_z}{\partial \theta} & \frac{\partial u_z}{\partial z} \end{pmatrix} \quad \text{(A.1.2)}$$

$$\nabla . \boldsymbol{T} = \textbf{div } \boldsymbol{T} = \begin{pmatrix} \frac{\partial \tau_{rr}}{\partial r} + \frac{1}{r}\frac{\partial \tau_{r\theta}}{\partial \theta} + \frac{\partial \tau_{rz}}{\partial z} + \frac{\tau_{rr} - \tau_{\theta\theta}}{r} \\ \frac{\partial \tau_{r\theta}}{\partial r} + \frac{1}{r}\frac{\partial \tau_{\theta\theta}}{\partial \theta} + \frac{\partial \tau_{\theta z}}{\partial z} + \frac{2\tau_{r\theta}}{r} \\ \frac{\partial \tau_{rz}}{\partial r} + \frac{1}{r}\frac{\partial \tau_{\theta z}}{\partial \theta} + \frac{\partial \tau_{zz}}{\partial z} + \frac{\tau_{rz}}{r} \end{pmatrix} \quad \text{(A.1.3)}$$

so that the momentum equation writes:

$$\rho(\frac{\partial u_r}{\partial t} + u_r\frac{\partial u_r}{\partial r} + \frac{u_\theta}{r}\frac{\partial u_r}{\partial \theta} + u_z\frac{\partial u_r}{\partial z} - \frac{u_\theta^2}{r}) = b_r + \frac{1}{r}\frac{\partial(r\tau_{rr})}{\partial r} + \frac{1}{r}\frac{\partial \tau_{r\theta}}{\partial \theta} + \frac{\partial \tau_{rz}}{\partial z} - \frac{\tau_{\theta\theta}}{r} - \frac{\partial p}{\partial r}$$

(A.1.4)

$$\rho(\frac{\partial u_\theta}{\partial t} + u_r\frac{\partial u_\theta}{\partial r} + \frac{u_\theta}{r}\frac{\partial u_\theta}{\partial \theta} + u_z\frac{\partial u_\theta}{\partial z} + \frac{u_r u_\theta}{r}) = b_\theta + \frac{1}{r^2}\frac{\partial(r^2\tau_{r\theta})}{\partial r} + \frac{1}{r}\frac{\partial \tau_{\theta\theta}}{\partial \theta} + \frac{\partial \tau_{\theta z}}{\partial z} - \frac{1}{r}\frac{\partial p}{\partial \theta}$$

(A.1.5)

$$\rho(\frac{\partial u_z}{\partial t} + u_r\frac{\partial u_z}{\partial r} + \frac{u_\theta}{r}\frac{\partial u_z}{\partial r} + u_z\frac{\partial u_z}{\partial z}) = b_z + \frac{1}{r}\frac{\partial(r\tau_{rz})}{\partial r} + \frac{1}{r}\frac{\partial \tau_{\theta z}}{\partial \theta} + \frac{\partial \tau_{zz}}{\partial z} - \frac{\partial p}{\partial z}$$

(A.1.6)

Appendix 2: SHEAR RHEOMETRY CALCULATIONS

A.2.1 *General considerations*

The following computations concern viscometric flows in some simple geometries. They aim at determining the experimental shear rate-shear stress curve from specific measurements. We assume that the material under consideration is incompressible, homogeneous and continuously deformed during the tests and that flows are laminar. Moreover we assume that a unique relation between steady state shear stress (τ) and shear rate (γ) exists for this material. In practice a set of data of torque (C) (or pressure difference) vs rotation velocity (Ω) (or flow rate) is obtained experimentally. For the following theoretical developments we assume that a general mathematical relationship $C(\Omega)$ has been deduced from these results. Obviously the experimental curve never covers the whole range of Ω which thus limits the theoretical interpretation in terms of shear rate-shear stress curve. Apart from this practical limitation the aim of these calculations will be to determine $\tau(\gamma)$ or, which is equivalent, the shear rate function, $\gamma=\zeta(\tau)$, as a function of $C(\Omega)$. More details about the following calculations may be found in Astarita et al. (1964) for incline plane rheometer and Coleman et al. (1966), Tanner (1988), or Piau (1979) for other geometries. The results *a priori* concern steady flows but when the shear rate is homogeneous within the suspension they can be used to predict instantaneous properties of thixotropic fluids. This is no longer true for rheometrical flows with heterogeneous shear. Indeed the appropriate computations involve derivatives, of an experimental curve corresponding to different flow intensities for which the material state could be heterogeneous within the fluid, without any consideration of internal material state (see 4.2.2). As a consequence the following computations must be considered as rough approximations as long as they are used to predict instantaneous properties of thixotropic fluids.

A.2.2 *Coaxial cylinder rheometer*

This rheometer is composed of two coaxial cylinders of different radii. Here we consider the case when the inner cylinder can rotate at a given rotation velocity (Ω) (Cf Fig. A.2.1) while the outer cylinder is fixed. The torque applied to the inner cylinder is C. In the following all variables are described in a cylindrical frame of reference ($\boldsymbol{e}_r, \boldsymbol{e}_\theta, \boldsymbol{e}_z$) which rotates with the inner cylinder. We assume that the only non-zero velocity component is that along $\boldsymbol{e}_\theta$, u_θ, which only depends on r. This is a reasonable velocity distribution from which flow equations may generally be solved. It follows that:

$$D = \frac{1}{2}\begin{pmatrix} 0 & \gamma & 0 \\ \gamma & 0 & 0 \\ 0 & 0 & 0 \end{pmatrix} \text{with } \gamma = r\frac{\mathrm{d}\omega}{\mathrm{d}r} = r\frac{\mathrm{d}(u_\theta/r)}{\mathrm{d}r} \tag{A.2.1}$$

and that:

$$\tau = \tau_{r\theta}(r) \tag{A.2.2}$$

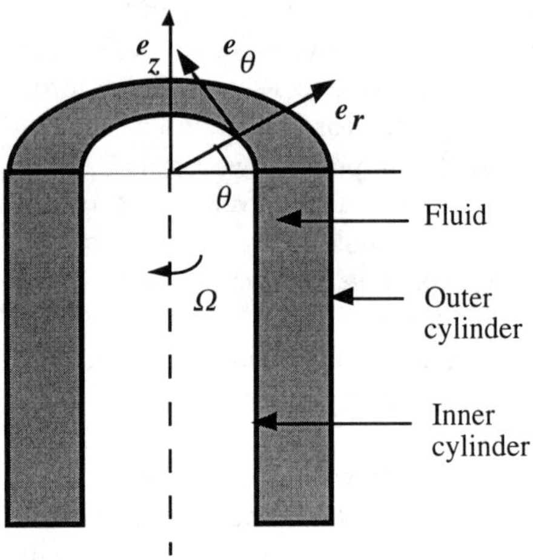

Figure A.2.1: Coaxial cylinder rheometer.

In steady state the momentum equations may be written:

$$\frac{\partial(\tau_{rr})}{\partial r} = \frac{N_1}{r} - \rho\frac{u_\theta^2}{r} \tag{A.2.3}$$

$$\frac{\partial(r^2\tau_{r\theta})}{\partial r} = 0 \tag{A.2.4}$$

$$\tau_{zz} = \rho g z + p(r) \tag{A.2.5}$$

From (A.2.4) it results that the torque by unit length at a distance r is given as:

$$C = 2\pi r^2 \tau_{r\theta} \tag{A.2.6}$$

and is constant. We also have:

$$\Omega = \int_{r_i}^{r_e} \frac{d\omega}{dr} dr = \int_{r_i}^{r_e} \frac{1}{r} \zeta(\frac{C}{2\pi r^2}) dr \tag{A.2.7}$$

If $(\frac{r_e}{r_i} - 1) << 1$, that is, for a small gap, we have:

$$\Omega \approx \zeta(\frac{C}{2\pi r_i^2})(\frac{r_e - r_i}{r_i}) \tag{A.2.8}$$

from which one may directly deduce the value of ζ for each data pair (C,Ω).

If $(\frac{r_e}{r_i} - 1) >> 0.1$ the above method cannot be used. We consider J defined as:

$$J = \frac{C}{2\pi r^2} \tag{A.2.9}$$

Then:

$$dJ = -\frac{2}{r}\frac{C}{2\pi r^2} dr = -\frac{2}{r} J dr \tag{A.2.10}$$

and (A.2.7) becomes:

$$\Omega = -\frac{1}{2} \int_{J_i}^{J_e} \frac{\zeta(J)}{J} dJ \quad (\text{with } J_i = \frac{C}{2\pi r_i^2} \text{ and } J_e = \frac{C}{2\pi r_e^2}) \tag{A.2.11}$$

whose derivative relatively to C gives:

$$\frac{\partial\Omega}{\partial C} = \frac{1}{2} [\frac{\zeta(J_i)}{J_i} (\frac{\partial J_i}{\partial C}) - \frac{\zeta(J_e)}{J_e} (\frac{\partial J_e}{\partial C})] = \frac{1}{2} (\frac{\zeta(J_i)}{C} - \frac{\zeta(J_e)}{C}) \tag{A.2.12}$$

Let us take:

$$Y(J_i) = 2C\frac{\partial\Omega}{\partial C}(J_i, J_e) = 2C\frac{\partial\Omega}{\partial C}(J_i, \beta J_i) \tag{A.2.13}$$

with $\beta = (\frac{r_i}{r_e})^2$, then:

$$Y(J_i) = \zeta(J_i) - \zeta(\beta J_i) \tag{A.2.14}$$

and:

$$\zeta(J_i) = \sum_{n=0}^{\infty} Y(\beta^n J_i) \qquad \text{(A.2.15)}$$

The series in equation (A.2.15) converges all the faster as β is larger. For mud suspensions it converges rapidly because these materials are shear-thinning yield stress fluids: for $\beta^n J_i$ smaller than a given value the corresponding Y value is equal to zero. Each value of ζ for a given J_i will be deduced from a set of data pairs (C,Ω) from which Y is found. A similar analysis may be carried out for normal stress differences from equation (A.2.3) neglecting the inertia term. If normal stress differences become significant they can affect the flow field assumed above (in particular this may give rise to the Weissenberg effect).

A.2.3 *Cone and plate rheometer*

This rheometer is composed of a troncated cone and a disk with common axis. The cone top must (geometrically) fall at the disk center. The material is set up between these tools so as to be sheared by their relative rotation (Cf Fig. A.2.2). For small angle and when inertia is neglected the shear rate is given as:

$$\gamma = \frac{\Omega}{\alpha} \qquad \text{(A.2.16)}$$

and is independent of r and θ. Then the shear stress is constant within the gap and, by integration, one directly obtains the shear stress expression for a given shear rate as a function of the total applied torque (C):

$$\tau(\gamma) = \frac{3C}{2\pi(R^3)} \qquad \text{(A.2.17)}$$

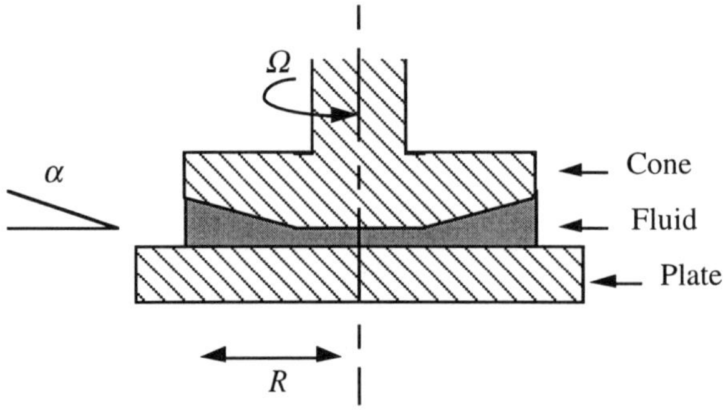

Figure A.2.2: Cone and plate rheometer.

A.2.4 *Parallel plate rheometer*

This rheometer is composed of two coaxial disks (Cf Fig. A.2.3). Their relative rotation shears the material which has been squeezed between them. Considering the symmetry of the problem the shear stress appears to depends only on the distance (r) from the axis. Then the shear rate also depends only on r and is given as:

$$\gamma = \gamma(r) = \frac{\Omega r}{h} \tag{A.2.18}$$

The torque exerted by the upper plate can be written:

$$C = \int_0^R 2\pi r \tau(r) r \mathrm{d}r = \frac{2\pi}{(\Omega/h)^3} \int_0^{(R\Omega/h)} \gamma^2 \tau(\gamma) \mathrm{d}\gamma \tag{A.2.19}$$

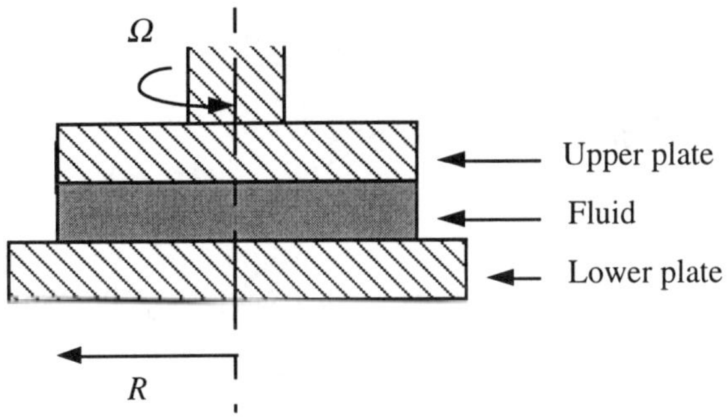

Figure A.2.3: Parallel plate rheometer.

Then, by differentiating (A.2.19) relatively to $\gamma(R)=\gamma_R$, one obtains the expression for the shear stress at the periphery:

$$\tau(\gamma_R) = \frac{3C}{2\pi R^3} + \frac{\gamma_R}{2\pi R^3} \frac{\mathrm{d}C}{\mathrm{d}\gamma_R} \tag{A.2.20}$$

A set of data pairs (τ,γ_R) can thus be deduced from a set of data pairs (C,Ω). In practice, for shear-thinning yield stress mud suspensions (with n around 1/3), the second term of the right hand-side of (A.2.20) is small compared to the first one. For a Newtonian fluid this term is not negligible and one finally obtains:

$$\tau(\gamma R) = \frac{2C}{\pi R^3} \tag{A.2.21}$$

A.2.5 *Capillary rheometer*

This rheometer is simply composed of a cylindrical duct through which the material is injected with the help of driving pressure difference between the extremities of this duct (Cf Fig. A.2.4). The momentum equation can be solved completely assuming that the velocity has only one non-zero component, u, directed along the tube axis and which only depends on the distance (r) from this axis. In this case the momentum equation predicts that the pressure distribution is such that $p_g=p+\rho gz$, is constant over a cross-section. Additionally we obtain:

$$\frac{\Delta p_g}{L} = A \text{ and } \tau(r) = -\frac{r}{2}\frac{\Delta p_g}{L} = J \tag{A.2.22}$$

where L is the tube length, A a constant and Δp_g the pressure difference between the extremities of the tube. Additionally the flow rate may be estimated through:

$$Q = 2\pi \int_0^R ur\mathrm{d}r = 2\pi \left(\left[u\frac{r^2}{2}\right]_{0,R} - \int_0^R \frac{\mathrm{d}u}{\mathrm{d}r}\frac{r^2}{2}\mathrm{d}r\right) = -2\pi \left(\int_0^R \zeta(J)\frac{r^2}{2}\mathrm{d}r\right) \tag{A.2.23}$$

Using only the variable J, the flow rate can be expressed as:

$$Q = \frac{8\pi}{\left(\frac{\Delta p_g}{L}\right)^3} \int_0^{\frac{\Delta p_g}{L}\frac{R}{2}} \zeta(J)J^2\mathrm{d}J \tag{A.2.24}$$

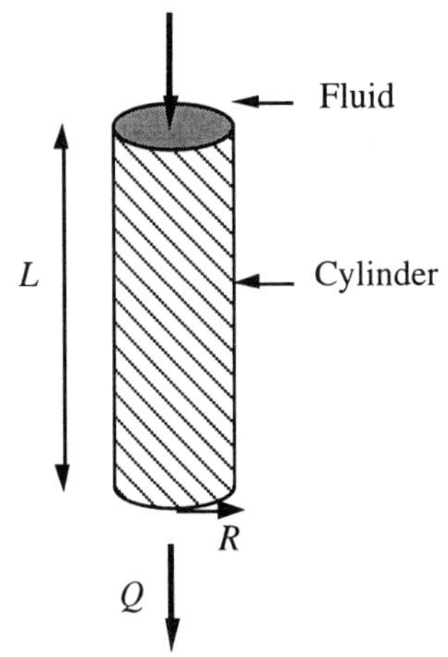

Figure A.2.4: Capillary rheometer.

Differentiating equation (A.2.24) relatively to $\Delta p_g/L$ finally gives the expression for the shear rate along internal wall:

$$\gamma_R = \zeta\left(-\frac{R}{2}\frac{\Delta p_g}{L}\right) = \frac{1}{\pi R^3(\Delta p_g/L)^2}\frac{\mathrm{d}(Q(\Delta p_g/L)^3)}{\mathrm{d}(\Delta p_g/L)} \tag{A.2.25}$$

Equations (A.2.22) and (A.2.25) applied to a set of data pairs $(Q,(\Delta p_g/L))$ give a set of data pairs $(\gamma_R, \tau(\gamma_R))$.

A.2.6 *Inclined plane rheometer*

The inclined plane rheometer is not a conventional rheometrical technique but this appears as a useful technique for mud suspensions. The geometry consists in an inclined rectangular channel and a system recirculating the material (Coussot and Boyer 1995, Astarita et al. 1964) (Cf Fig. A.2.5). The uniform depth corresponding to a given flow rate or free surface velocity is recorded. With a set of such data the following theoretical analysis may be carried out (Coleman et al. 1966, Astarita et al. 1964). On may also use this system to directly determine fluid yield stress (Uhlherr et al. 1991) (see Chapter 12). An appropriate experimental procedure for the inclined plane test was described in Coussot and Boyer (1995) and is summarized in Figure A.2.5.

For a steady uniform flow on an infinitely wide inclined plane, neglecting edge and surface tension effects, the discharge by unit width may be written as follows:

$$q = \int_0^h u(y)\mathrm{d}y = \int_0^h (h-y)\frac{\mathrm{d}u}{\mathrm{d}y}\,\mathrm{d}y \tag{A.2.26}$$

where y is the current position perpendiculary above the channel bottom (see Fig. A.2.5). We now use the shear stress τ at the position y as a new variable in (A.2.26). From momentum equations we deduce (see Chapter 8):

$$\tau = \tau_p\frac{h-y}{h} \tag{A.2.27}$$

equation (A.2.26) becomes:

$$q = \int_0^{\tau_p} \frac{h^2}{\tau_p{}^2}\,\tau\zeta(\tau)\mathrm{d}\tau \tag{A.2.28}$$

Differentiating (A.2.28) and using the fact that τ_p is equal to $\rho g h \sin i$, we obtain:

$$\left(\frac{\mathrm{d}q}{\mathrm{d}h}\right)_i = h\zeta(\tau_p) \tag{A.2.29}$$

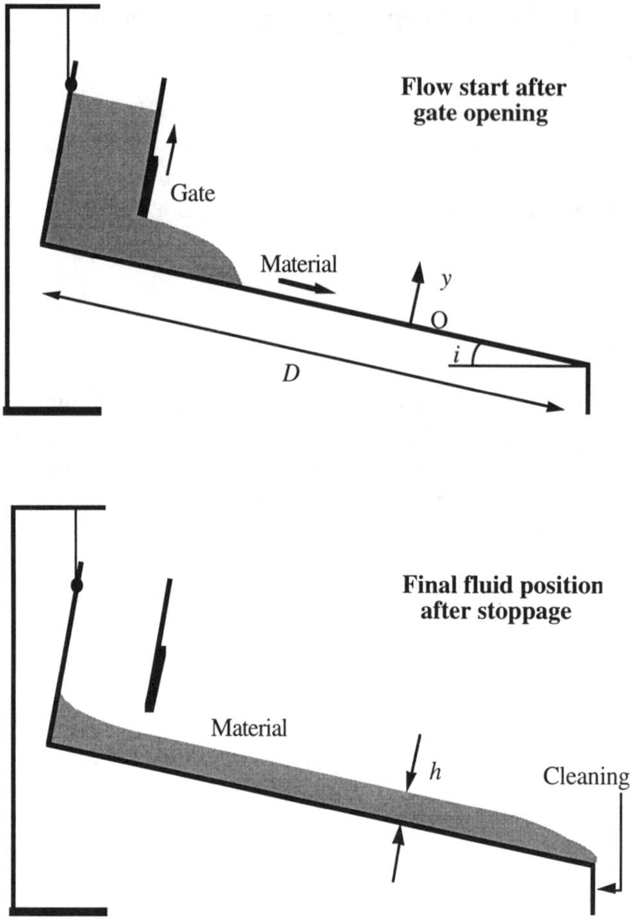

Figure A.2.5: Inclined plane test for the direct determination of the yield stress.

A similar analysis can be carried out using the free surface velocity (V_s) instead of q (Astarita et al. 1964). We have:

$$V_s = \int_0^h \frac{\mathrm{d}u}{\mathrm{d}y}\,\mathrm{d}y = \int_0^{\tau_p} \frac{h}{\tau_p}\,\zeta(\tau)\mathrm{d}\tau \tag{A.2.30}$$

from which we deduce:

$$\left(\frac{\mathrm{d}V_s}{\mathrm{d}h}\right)_i = \zeta(\tau_p) \tag{A.2.31}$$

Using results of channel flow tests, it is possible to determine, either from equation (A.2.29) or equation (A.2.31), the function ζ, that is the flow curve (in practice only a set of data pairs). Astarita et al. (1964) proposed a more complete analysis taking into account possible wall slip. For fine mud suspensions flows, this effect appeared negligible (Coussot 1994). In general this assumption can be validated by the technique of Astarita et al. (1964) or by a direct comparison of these results with rheometrical data using another technique.

A.2.7 REFERENCES

Astarita, G., Marrucci, and Palumbo, G. 1964. Non-Newtonian gravity flow along inclined plane surfaces. *Ind. Eng. Chem. Fund.*, 3, 333-339.

Coleman, B.D., Markowitz, H., and Noll, W. 1966. *Viscometric Flows of Non-Newtonian Fluids*. Springer Tracts in Natural Philosophy, Volume 5, Springer Verlag, Berlin.

Coussot, P., and Boyer, S. 1995. Determination of yield stress fluid behaviour from inclined plane test. *Rheol. Acta*, 34, 534-543.

Piau, J.M. 1979. Fluides non-newtoniens. *Techniques de l'Ingénieur*; A710, 1-16, A711, 1-24.

Tanner, R.I. 1988. *Engineering Rheology*. Clarendon Press, Oxford.

Uhlherr, P.H.T., Fang, T., and Zhou, Z. 1991. Yield stress from static equilibrium on an inclined plane. *China-Japan Int. Conf. Rheol.*, Beijing, China, 316-318.

Appendix 3: SLUMP TEST INTERPRETATION - Pashias et al. (1996) approach for determining fluid yield stress from slump test with a cylindrical geometry)

When a yield stress fluid is released over a horizontal plane by evenly lifting the cylinder (of radius R) containing the material, there remains an undeformed, cylindrical volume of material lying over a deformed part (see Fig. A.3.1). Within the initial (undeformed) material the maximum shear stress in the plane (x,z) may be found from elasticity theory. It is equal to:

$$\tau_0(y) = \frac{1}{2}\rho g y \qquad \text{(A.3.1)}$$

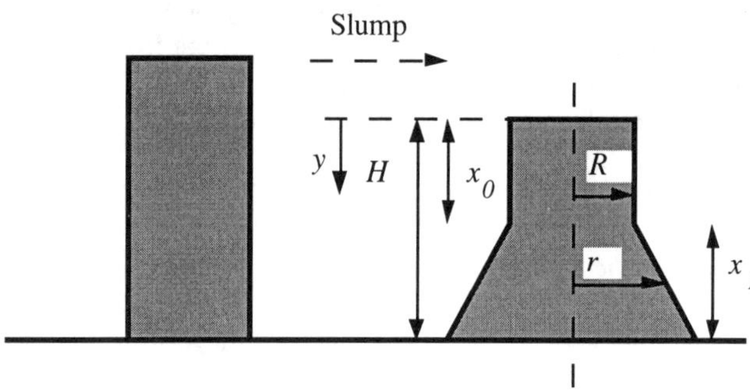

Figure A.3.1: Initial and final state of the fluid sample during the cylindrical slump test.

The horizontal layers of the fluid are assumed to be simply compressed during the slump. Thus the material volume above each layer remains constant and the thickness of a layer situated at y after the slump writes:

$$\mathrm{d}y = \frac{R^2}{r(y)^2}\,\mathrm{d}y_0 \qquad \text{(A.3.2)}$$

where $\mathrm{d}y_0$ is the initial thickness of a layer and $r(y)$ the radius of the final layer. It is now assumed that the shearing force along a given layer is proportional to the weight of material above this layer. Additionally the shear stress along each horizontal plane is supposed to be constant. Then we have:

$$\tau_0(y)R^2 = \tau(y)r(y)^2 \qquad \text{(A.3.3)}$$

in which it must be taken into account that $\tau(y)$ reaches τ_c everywhere in the deformed region when flow stops.

x_1 is found from:

$$x_1 = \int_{x_0}^{H} \mathrm{d}y \tag{A.3.4}$$

in which one takes into account (A.3.1-A.3.3) and where H is the final material height. Then we obtain:

$$x_1 = \frac{2\tau_c}{\rho g} \ln\frac{H}{x_0} \tag{A.3.5}$$

Moreover x_0 corresponds to the depth for which τ_c is reached exactly:

$$x_0 = \frac{2\tau_c}{\rho g} \tag{A.3.6}$$

Appendix 4: VANE RHEOMETRY

A vane, in general consisting of four thin blades arranged at equal angles around a small cylindrical shaft, is immersed in the fluid contained in a cup. Then it is rotated around its axis and the torque-time response is recorded. For a fixed rotation rate the torque-time curve for yield stress fluids exhibit a peak after a linear elastic deformation region (Nguyen and Boger 1983). One assumes that the torque (T) at this peak corresponds to the minimum force necessary for flow to take place, which is associated to the fluid yield stress. For the calculations the simplest approach consists in assuming that the fluid yields, or starts flowing, in the region where at the same time the shear stress is large and the fluid not retained by the blades. The corresponding yielding region is then a cylindrical surface defined by the vane dimensions. This assumption has been verified by both visual observation and numerical simulation (see Nguyen and Boger (1992)). However we emphasize that the peak and decrease of the torque-time curve remains to be explained. It can be due either to slight changes in boundary conditions during flow (this is common for parallel plate or cone-plate rheometry since the free surface form may slightly change during flow (see Chapter 4)) or to true, particular, rheological characteristics of the fluid.

Under above assumptions the torque balance gives:

$$T = (2\pi R^2 h)\tau_c + 4\pi \int_0^R \tau(r) r^2 \mathrm{d}r \tag{A.4.1}$$

where R and h are radius and length of the vane, respectively. τ is the shear stress at the end surface. As a first approximate one assumes that τ is constant and equal to the yield stress. Then equation (A.4.1) reduces to:

$$T = 2\pi R^3 \left(\frac{h}{R} + \frac{2}{3}\right)\tau_c \tag{A.4.2}$$

A simple, more realistic profile for τ may be assumed which yields slightly different expressions. For example, when (Nguyen and Boger 1983):

$$\tau = \left(\frac{r}{R}\right)^p \tau_c \tag{A.4.3}$$

we obtain:

$$T = 2\pi R^3 \left(\frac{h}{R} + \frac{2}{3+p}\right)\tau_c \tag{A.4.4}$$

Though still inexact, (A.4.4) is probably closer to reality and thus makes it possible to estimate the error associated with the use of equation (A.4.2). For vanes with $h/R>4$ and $p>1$, this error is smaller than 3.7% (Nguyen and Boger 1983).

Index